P9-CKU-258

Alien Species in North America and Hawaii

IMPACTS ON NATURAL ECOSYSTEMS

GEORGE W. COX

ISLAND PRESS
Washington, D.C. • Covelo, California

COVER PHOTOS:
Upper left: Mountain goat; see Chapter 15 for discussion. (Photo by George W. Cox)
Lower left: Miconia; see Chapter 13 for discussion. (Photo by George W. Cox)
Lower right: European green crab; see Chapter 5 for discussion. (Photo by Edwin D. Grosholz)

Library of Congress Cataloging-in-Publication Data
Cox, George W., 1935–
Alien species in North America and Hawaii : impacts on natural ecosystems / George W. Cox.
p. cm.
Includes bibliographical references (p.) and index.
ISBN 1–55963–679–3 (cloth)
ISBN 1–55963–680–7 (paper)
1. Nonindigenous pests—United States. 2. Biological invasions—United States. 3. Biological diversity conservation—United States. I. Title.
SB990.5U6 C68 1999 99–16652
577'.18—dc21 CIP

Printed on recycled, acid-free paper

Manufactured in the United States of America
10 9 8 7 6 5 4 3 2 1

To William DeMott Stull,
mentor and lifelong friend

Contents

Preface

This book has grown out of a long-standing interest in biological conservation and has been encouraged by the growth of a curriculum in conservation biology at San Diego State University. In the region around San Diego, California, one can clearly observe the extent to which invasive exotics have modified terrestrial, freshwater, and marine ecosystems. In grasslands on the coastal marine terraces and in interior valleys, exotic annual grasses and forbs have so fully displaced native grassland species that the nature of the original vegetation is uncertain. In the Colorado Desert, exotic shrubs are displacing native shrubs and desiccating streams. Along the Santa Margarita River, brown-headed cowbirds, newcomers to California, threaten native riparian bird species. Freshwater impoundments are filled with exotics, some introduced deliberately, others inadvertently. Coastal marine waters teem with exotic plants and animals, with more species appearing annually.

In the course of fieldwork in other parts of North America, I have had the opportunity to observe exotics and their impacts in many other situations: cheatgrass and knapweeds on the Columbia Plateau, Australian melaleuca and Brazilian pepper in South Florida, feral hogs and banana poka on Kauai, masses of dead alewives on the beaches of Lake Michigan, and exotic game fish in the Colorado River. Where I now live in New Mexico, I am reminded daily of the seriousness of exotic invasions by the Russian olives, saltcedars, and Siberian elms invading washes near my home.

Many concerned environmentalists are aware of the threats posed by the continuing invasion of exotics. The public at large, however, remains largely unaware of the seriousness of the threat. The opinions of many people are determined by personal perceptions of the hunting and fishing value or landscaping attractiveness of species that are aggressive invaders of natural ecosystems. Concern for the welfare of individuals of exotic species frequently overshadows concern for the survival of North America's native species and distinctive ecosystems. An education gap exists, and if we are to preserve the ecological uniqueness of the continent, this gap must be filled.

In this book, I examine the issues of exotics and their impacts in all of North American and Hawaiian native ecosystems, land, freshwater, and marine. I consider not only those protected as wilderness or natural biotic preserves, but also wild lands and waters exploited with light-to-moderate intensity through grazing, timber harvest, hunting, and fishing—the ecosystems on which the survival of the bulk of native biodiversity ultimately depends. To cover this vast topic, I have focused on regions and ecosystem types that are most affected by exotics, on several groups of exotics that occur throughout the continent, and on topics of theoretical interest and policy importance.

The large number of exotic species that have invaded North America, however, precludes the discussion of each one. In each chapter, I have included the species that are most important to the topic in question. To reduce the clutter of detail in the text, in most chapters I have placed the scientific names of these exotics, together with their native regions, in a table. For native species that come into discussion, however, I have given the scientific name at first mention.

I especially thank the following individuals for reviewing one or more chapters and making valuable suggestions for improvement: Ellen Bauder, Robert J. Behnke, Carl Bock, Jane Bock, Doug Bolger, Richard Brewer, Boyd Collier, Robert H. Cowie, Jeffrey A. Crooks, Deborah Dexter, Robert Doren, Lisa Ellis, David J. Jude, Fred Kraus, Andrew M. Liebhold, Thomas E. Lodge, Svata Louda, Earl D. McCoy, Peter B. Moyle, Marcel Rejmánek, Anthony Ricciardi, Terrel D. Rich, Robert E. Ricklefs, Don C. Schmitz, L. K. Thomas Jr., Raul Valdez, and Richard H. Yahner. I also thank the following individuals for helping me locate important literature materials: Jay Diefendorfer, Francis G. Howarth, Thomas R. Hrabik, David R. Klein, Barbara Kus, Richard N. Mack, David Pimentel, M. Eric Reeves, Raul Valdez, Charles van Riper III, and Susan L. Williams. Barbara Dean, Executive Editor for Island Press, gave many valuable suggestions for organization and presentation of the material.

George W. Cox
Santa Fe, New Mexico

Part I

Introduction

Chapter 1

The Threat of Exotics: Biotic Pollution

> It is not just nuclear bombs that threaten us. . . . An ecological explosion means the enormous increase in numbers of some kind of living organism—it may be an infectious virus like influenza, or a bacterium like bubonic plague, or a fungus like that of potato disease, a green plant like the prickly pear, or an animal like the grey squirrel.
>
> —Charles Elton, *The Ecology of Invasions by Animals and Plants**

In 1962, in her book *Silent Spring,* Rachel Carson warned that pollution by chemical pesticides might still the spring voices of songbirds, as well as decimate wildlife of many other sorts. Today, the threats of chemical pollution to biodiversity are still with us. But a new threat to the integrity of the forests, prairies, lakes, streams, and coastal oceans of North America and Hawaii has emerged. When we look at these ecosystems, we often are forced to ask, "Where have all our native plants and animals gone?" Too often, the answer is, "They've been displaced by species alien to North America and Hawaii." Are we facing a new crisis—one that threatens our native plants and animals, our unique ecosystems, and even the productivity of our lands and waters?

A worldwide ecological explosion is now in progress. Thousands of species of microbes, plants, and animals are being introduced, deliberately

*Charles Elton (1958). *The Ecology of Invasions by Animals and Plants*. Methuen & Co., London, England.

and inadvertently, to new land areas, seas, and fresh waters. The rate of introduction is several orders of magnitude greater than the rate that prevailed only a few hundred years ago. In many regions, these new colonists are running wild, disrupting the dynamics of ecosystems, pushing native species toward extinction, and causing billions of dollars of direct damage to human enterprises. This book examines the seriousness of the threat of these invaders to North America, and what can be done to reduce this threat.

The Invaders

Who are these invaders? Most often termed *exotic species,* they are also synonymously termed *alien, nonindigenous,* or *nonnative* species. Exotic species are forms whose occurrence in a region is the result of human activities that have aided their dispersal across geographical barriers or have created favorable conditions for their establishment. Some exotics not only become established in new regions, but also become serious invaders of natural and seminatural lands and waters; these are termed *invasive exotics.* Next to total ecosystem transformation, invasive exotics now pose the most serious threat to native ecosystems and their member species in North America and in many other parts of the world.

The line between exotic species and natural invaders is not always sharp. Any species' range will shift as it tracks, by dispersal, the geographic distribution of environmental conditions that favor it. As Pleistocene continental glaciers retreated, tree species in eastern North America gradually dispersed northward, in response to natural climatic change. With the deforestation that accompanied settlement of eastern North America by European colonists, many species from the prairie region of the Midwest spread eastward, clearly as a response to human activity. But both natural and human-induced changes in environmental conditions often occur together, and since organisms respond to both, it is not always easy to tell if a change in geographical range by a species is due to one or the other or both. For example, does the change in range of a species due to climatic warming, which may be partly natural and partly human-induced, make it an exotic? Regardless of how such cases are evaluated, no doubt exists that human activities alone are enabling scores of species to invade new, and often distant, locations. Such species are clearly nonindigenous.

Exotic species include many species that have been associated with humans, both beneficially and detrimentally, for hundreds to thousands of years. Many can be regarded as having a mutualistic (mutually beneficial) relationship with humans. Human mutualists include, for example, food and fiber plants, livestock animals, useful microbes, horticultural plants, and pets. Species kept for display in zoos, aquariums, and botanical gardens are rapidly entering a mutualistic relationship with humans, as well. These classes of exotics have contributed enormously to human welfare and well-being, even though some of these species, when introduced to new areas with benefits in mind, have created serious problems. Goats, pigs, dogs, and cats introduced to oceanic islands, for instance, have led to massive damage to natural vegetation and native biotas.

The benefits that our mutualists yield derive, of course, partly from the fact that humans grow, nourish, and protect them very carefully. But they also result from a fact at first serendipitous, then later deliberate. Species moved to new regions often escape many predators, parasites, and diseases that plague them in their area of origin. Nowadays, a deliberate strategy of agriculture and forestry is to introduce potentially productive species to new regions, where they are free to realize maximum productivity. The planting of pines in Australia, and eucalypts in areas outside Australia, is an example of this strategy.

Another category of exotic species comprises an unwelcome group of long-standing human associates: diseases, parasites, and their vectors. Human populations are, in effect, biotic hosts for such agents to exploit. The huge human population presents a massive evolutionary target for these species. International travel and commerce mean that disease agents and their vectors can be introduced suddenly to almost any point on earth. The influenza and AIDS viruses have repeatedly demonstrated this capacity. Some of the greatest public health dangers we now face are vector-borne diseases, such as malaria, or infectious diseases, such as ebola, that can suddenly invade new world regions. The Asian tiger mosquito is a recent example of an exotic disease vector that has invaded North America (Rai 1991). (See Table 1.1 for scientific names and native regions of exotics mentioned in this chapter.) First identified in Houston, Texas, in 1985, it was probably introduced to Texas in shipments of used tires from Japan. This mosquito has now spread to at least twenty-five states and occurs as far north as Iowa and Illinois (Craig 1993). It is known to breed in tiny pools of water in urban areas, such as those that accumulate in discarded tires, beverage cans, and other containers. Billions of such potential breeding sites exist in North America. This mosquito is a vector for several exotic viral diseases of animals and humans, including human dengue fever, yellow fever, and various forms of encephalitis.

Still other exotics consist of nonindigenous weeds, pests, and diseases of crops, forest plantations, and domestic or captive animals. The latter plants and animals are themselves mostly exotics, often having escaped

Table 1.1. Scientific names and native regions of exotic species.

PLANTS	
Australian melaleuca, *Melaleuca quinquenervia*	Australia
Brazilian pepper, *Schinus terebinthifolius*	Brazil
purple loosestrife, *Lythrum salicaria*	Eurasia
Russian olive, *Eleagnus angustifolia*	Asia
salt cedars, *Tamarix* spp.	Eurasia
Scotch broom, *Cytisus scoparius*	Europe
PLANT DISEASES	
chestnut blight, *Cryphonectria parasitica*	Asia
Dutch elm disease, *Ophiostoma ulmi*	Asia
MARINE INVERTEBRATES	
European green crab, *Carcinus maenas*	Europe
zebra mussel, *Dreissena polymorpha*	Asia
TERRESTRIAL INVERTEBRATES	
giant African snail, *Achatina fulica*	Africa
rosy wolfsnail, *Euglandina rosea*	Florida, USA
INSECTS	
Asian tiger mosquito, *Aedes albopictus*	Asia
FISH	
sea lamprey, *Petromyzon marinus*	Atlantic Ocean

these nonindigenous pests when they were introduced to North America. Exotic pests of this type have simply succeeded in following their original hosts to North America. In addition, some other exotic pests and diseases transported to North America have found suitable new hosts that lack resistance to them or new habitats suitable to their growth. Most weeds and arthropod pests of plant and animal agriculture are exotics. Their vigor is often a reflection of the fact that they, too, have been freed from their own predators, parasites, and diseases, enabling them to exploit croplands, pastures, or domesticated plants and animals with little restraint.

In addition to all of these kinds of exotics, however, is the pervasive and accelerating invasion of nonindigenous plants, animals, and microbes into the ecosystems of our wild lands and waters—the ecosystems we prize because of the native species that compose them. Increasingly, these invasions are threatening the ecological and evolutionary integrity of parklands, wilderness areas, public forests, rangelands, lakes, streams, and coastal marine areas. Chestnut blight and Dutch elm disease have virtually eliminated or greatly reduced in abundance two of the most important trees of the eastern deciduous forest. Purple loosestrife has taken over vast areas of freshwater marshes. The Everglades are becoming an exotic plant showcase, featuring Australian melaleucas and Brazilian pepper trees. Asian salt cedars and Russian olives are taking over the riparian woodlands of the western states and provinces. Scotch broom is turning West Coast prairies into shrublands. In the Great Lakes, sea lampreys attack salmonid fish at the top of the food chain and zebra mussels the phytoplankton at the base of the food chain. European green crabs and other exotic invertebrates dominate the marine community in San Francisco Bay.

Unfortunately, efforts to combat problem exotics of all types are often themselves the cause of introductions of additional exotics. The strategy of classical biological control is the introduction of biotic agents in deliberate attempts to reduce populations of detrimental species to levels at which their impacts are negligible. Most such introductions are done only after extensive screening to assure that the introduced biological control species themselves will not create worse problems. Sometimes, however, screening and impact analysis are inadequate. For example, a predatory snail, the rosy wolfsnail, introduced to many Pacific islands to control the giant African snail, has proved to be the most serious threat to hundreds of varieties of native land snails (Hadfield et al. 1993). Furthermore, all species are capable of evolution, and when an exotic species is suddenly placed into a new evolutionary setting, the ultimate consequences of adaptive evolution cannot be stated with certainty (see Chapter 19).

Impacts of Invaders

In 1958, British ecologist Charles Elton characterized the role of humans in assisting the invasion of new areas by animals and plants as "one of the great historical convulsions of the world's fauna and flora." Humans, in our migrations to all corners of the earth, have never traveled alone. Our companions, known and unknown, have discovered that we are the most effective agent of biogeographic spread in the history of life on earth. Humans have excelled as agents of introduction of species to new world regions. In their migrations, humans have profoundly changed the landscapes that they have colonized, creating environments favorable to the invasion of many other species by their own dispersal. Over the short span of history, human migrations have resulted in extinction of many species (Steadman 1995). Over a longer time span, perhaps, the biotic introductions promoted by humans may stimulate the origin of new species and may ultimately offset the initial spasm of extinctions. These direct and indirect roles make the geographical history of humans much more than an accounting of where and when *Homo sapiens* migrated.

Conservation scientists now recognize that invasive exotics are not just additions to the regional biota but are one of the greatest threats to native species and ecosystems. On a global scale, the threat posed to native species by exotics is ominous. Worldwide, introduced species are identified as the major cause of 25 percent of extinctions of fish, 42 percent of those of reptiles, 22 percent of birds, and 20 percent of mammals (Reid and Miller 1989). Intercontinental invasions of nonindigenous species are thus leading toward gradual homogenization of the earth's biota. Good invaders are colonizing suitable areas on all continents, while local endemic species are declining toward extinction. A homogenized global biota would also be a substantially impoverished biota. The fragmentation of the earth's land into continents and islands has permitted the evolutionary diversification of groups of organisms to a greater degree than would be the case if all land were a single mass. Considering mammals, for example, the mathematical relation of number of species to land area predicts a total of only about 2,000 species, whereas about 4,200 species exist (Vitousek et al. 1994). A global homogenization of the mammalian fauna thus might lead to loss of more than half the world's mammal species.

The direct and indirect economic impacts of exotics run far into billions of dollars annually. Agricultural weeds, crop and rangeland pests, and diseases of crops, livestock, and wildlife are largely exotics. Exotic fungi and invertebrates damage stored foods and clothing. Exotic plants and animals are pests of homes and commercial buildings, landscaped areas, timberlands, and aquacultural enterprises. Many human disease agents, including the AIDS virus, as well as disease vectors, are exotic species. Exotic aquatic plants and animals clog waterways, foul industrial cooling systems, and damage boats and docking facilities. Exotic large animals, including free-ranging and feral livestock and other domestic animals, degrade rangelands, encourage erosion, and pollute surface and subsurface waters. Exotic plants and animals create control costs in parks, wildlife refuges, and nature preserves. All of these categories of exotics cause direct economic damage and control expense.

Indirect economic damage by exotics—damage not assessed in normal market actions—likewise involves many categories of impact. Exotics may alter basic environmental processes, such as the productivity of lands and waters, the frequency of fire, and the streamflow patterns, in ways that are not easily evaluated in dollar terms. They may reduce recreational or aesthetic values of natural landscapes. They may also destroy existence values that many people attach to unique ecosystems, such as the Everglades; to native species, such as the cutthroat trout; or to spectacular natural phenomena, such as the concentration of bald eagles that once wintered on the Flathead River in Montana. Although ecological economists are developing methods to measure such costs, their assessment remains incomplete and inadequate.

History of Concern About Exotics

The seriousness of the threat of invasive exotics was drawn to the attention of the scientific community in 1958 by Charles Elton's book *The Ecology of Invasions by Animals and Plants.* Although weeds, pests, and diseases, many of them exotics, have long plagued agriculture, horticulture, forestry, and aquaculture, Elton brought attention to the fact that many invade natural ecosystems, where their impacts can be serious, but less evident, in direct economic terms.

General concern about invasive exotics did not emerge, however, until the 1980s, when comparative studies of ecosystem structure and function revealed the degree of invasion and impact of exotics in many ecosystems. A host of conferences and symposia during the 1980s resulted in books that documented the scope of the problem (e.g., Mooney and Drake 1986; Groves and Burdon 1986; Kornberg and Williamson 1987) and documented its seriousness in particular areas (e.g., Stone and Scott 1985). The International Council of Scientific Unions, through its Scientific Committee on Problems of the Environment (SCOPE), also initiated a study of the ecology of biological invasions. This study culminated in a 1984 conference at Asilomar, California, and the publication of a volume that examined biological invasions in detail (Drake et al. 1989).

The attention focused on the problems of exotics by these efforts stimulated a flood of research in the late 1980s and the 1990s. The publication of information on exotics has accelerated greatly in the 1990s. In 1993, the U.S. Office of Technology Assessment (OTA) prepared a summary of the impacts of harmful nonindigenous species in the United States. Cronk and Fuller (1995) have summarized information on invasive plants in various world regions. Williamson (1996) has presented a concise summary of information on theoretical aspects of biological invasions. Many additional symposia have examined aspects of exotic invasions (e.g., McKnight 1993; Pysek et al. 1995; Brock et al. 1997). Detailed analyses of problems created by exotics have been conducted for Florida (Simberloff et al. 1997) and Hawaii (Stone et al. 1993). A new SCOPE study of exotics was organized in 1997. This effort, focusing on exotics and global change, is based on the conclusion that global change is likely to increase the incidence and impact of exotics. The study will update the science of invasion ecology, which is now beginning to move into an experimental phase. It will also address ways in which society needs to respond to the threat of exotic invasions, specifically through technologies designed to prevent invasions and conduct rapid assessment of the probable impact of invaders. It will also bring together experts from other fields in order to address aspects such as risk analysis, economic assessment, and legal and institutional needs. The first meeting of this SCOPE panel was held in January 1998.

These efforts are part of a maturing science of invasion ecology, in which systematic and experimental approaches to basic questions of exotic invasions are being emphasized (Vermeij 1996). These approaches address three phases of the invasion process: dispersal, establishment, and integration. Studies focused on the dispersal phase examine how characteristics of the donor region, those of the dispersing organisms, those of natural and human agents that influence dispersal, and those of the recipient region affect the movement of species from one region to another. Studies of establishment relate to how biotic and abiotic conditions in the recipient region affect the survival, reproduction, and spread of exotic species. Studies of integration relate to how exotic species interact with the communities and ecosystems they invade, both initially and as the interacting species evolve.

Organization of This Book

The overall objective of this book is to evaluate exotic invasions of North America in the light of the emerging science of invasion ecology. The examination of exotics and their impacts continues in Chapter 2 with an overview of North American invaders and their threats. Chapter 2 describes the kinds and numbers of invader species, the regions most heavily invaded, the general origins of invaders, and the major sorts of threats that they pose to North American ecosystems. Chapter 3 considers in detail how exotic species are dispersed to new regions and how the physical and biotic features of regions such as North America influence their establishment and spread.

In Part II of the book, Chapters 4 through 13 emphasize the fact that patterns of exotic invasion are not uniform throughout North America but differ markedly for different regions and for different terrestrial, freshwater, and marine ecosystems. To fully understand the significance of exotic invasions, we must examine these specific patterns. A series of ten chapters focuses on the most seriously invaded regions or regional ecosystems of North America and Hawaii and describes each system's principal invasive exotics and their impacts. Each chapter also considers the effectiveness of current prevention and control efforts and discusses future options to deal with the challenges of invasive exotics.

Several major groups of exotics, mostly animals, show patterns of invasiveness that cut across regions and ecosystem types. Part III, Chapters 14 through 16, considers these groups of exotics. This section begins with a chapter on exotic game and fish species and continues with chapters on native North American species that have invaded areas outside their original ranges and on animals that are domesticates, commensals, or close associates of humans. These groups of exotics pose some of the most difficult issues of management policy, because many are beneficial in some situations and detrimental in others.

Invasions of exotic species raise several theoretical issues that must also be addressed if we are to gain an understanding of the long-term nature of the impacts of exotic species. Part IV, Chapters 17 through 19, focuses on these issues. Chapter 17 examines specific questions of invasiveness of species and invasibility of communities—why certain species are able to invade communities in new regions and others are not, and why communities differ in their degree of resistance to invasion. Chapter 18 examines the extent to which invasive species not only affect the species composition of invaded communities but also change the way that the total ecosystem functions. This chapter addresses how exotics modify basic ecosystem properties such as fertility, productivity, and trophic structure. Finally, because invasive exotics can act as powerful agents of natural selection, Chapter 19 discusses evolutionary processes involving exotics and the species with which they interact.

The enormous economic and ecological impact of exotics requires that we consider aspects of management and public policy. Part IV, Chapters 20 and 21, focuses on these topics. Chapter 20 looks in detail at the direct and indirect economic impacts of exotics and evaluates current efforts to control the most destructive species. Since this technology is young, and derived largely from traditional agricultural and forestry pest control technology, the discussion centers on how control of exotics can be improved by an integrated management strategy that relies heavily on biological control.

Chapter 21 examines public policy relating to exotics. At present, laws and policies relating to exotics, at all levels from local to international, are fragmented, uncoordinated, and inadequate. Governmental regulation of both international and intranational movements of exotic species has been permissive. Efforts to eliminate or control exotics of proven detrimental impact have usually been delayed until the problem has grown so large that containment or eradication is impossible. Coordination of regulation at state or provincial, national, and international levels has been almost nonexistent. The need to formulate effective management policy and practice is urgent. Chapter 21 examines public policy toward exotics and suggests ways in which policy can be improved.

The North American Battleground

As pointed out in Chapter 3, North America is the largest battleground in the global war against invasive exotics. It has the greatest number of exotic species and the most diverse patterns of invasion of lands, fresh waters, and coastal marine waters. As world leaders in environmental science, technology, and policy, the United States, Canada, and Mexico must devise a winning strategy or risk the extreme degradation of some of the world's most pristine natural ecosystems.

Chapter 2

North American Invaders: The Invited and the Uninvited

> We suggest that biological invasions by notorious species like the zebra mussel, and its less famous counterparts, have become so widespread as to represent a significant component of global environmental change.
>
> —Peter Vitousek et al., "Introduced Species: A Significant Component of Human-Caused Global Change."*

In 1988, numerous specimens of a new mollusk, the zebra mussel (*Dreissena polymorpha*), were collected in western Lake Erie and Lake St. Clair. Native to southern Russia, the zebra mussel was apparently introduced to the Great Lakes by ballast water discharges from an ocean cargo vessel. The abundance of the species in 1988 indicated an earlier introduction, perhaps as early as 1985. Within just two more years, however, zebra mussels had been recorded in all the Great Lakes and in the St. Lawrence River. By 1991, they had invaded the Hudson, Mohawk, Susquehanna, Ohio, Illinois, and Mississippi Rivers. A second very similar mussel, now known as the quagga mussel (*Dreissena bugensis*), appeared in western Lake Erie in 1989. By 1991, quaggas had spread into Lake Ontario and the St. Lawrence River. Zebra and quagga mussels have shown the most explosive population growth and range expansion of any exotics brought to the New World. It appears likely that by A.D. 2000 zebra

*Peter Vitousek, Carla D'Antonio, Lloyd Loope, Marcel Rejmánek, and Randy Westbrooks (1997). "Introduced Species: A Significant Component of Human-Caused Global Change." *New Zealand Journal of Ecology* 21:1–16.

mussels will have colonized fresh and brackish waters—even isolated lakes and ponds—throughout much of the United States and southern Canada.

Zebra and quagga mussels reach enormous densities, up to 627,000 per square yard or more, completely covering submerged hard substrates and stable sandy or muddy sediments. Prior to invasion of lakes by these mussels, large attached invertebrates of this sort were essentially absent from North American freshwaters. Dense growths of mussels have created a costly fouling problem, especially by clogging water intakes for city water systems, power plant cooling systems, and boat engines. Zebra and quagga mussels have also altered the food-chain dynamics of the lower Great Lakes, consuming a large fraction of the primary production from the open-water system and directing it into benthic (bottom-dwelling) organisms. The mussels themselves have become food for certain fish that can crush their shells. From a conservation viewpoint, zebra mussels pose a serious threat to North American freshwater clams (family Unionidae). In western Lake Erie, where native unionid clams occurred at a density of several individuals per square meter, all were essentially extirpated by 1991. All living unionids and their dead shells were fouled by zebra mussels. Some individual unionid shells were covered by 15,000 zebra mussels, a quantity equal to about five times the weight of the live unionid. The invasion of these mussels has created enormous economic costs, transformed lake ecology, and threatened a unique fauna of native animals.

Exotics and North American Wilderness

Zebra and quagga mussels are just two of the many exotics that have invaded North America in the past few decades. As the Lake Erie example shows, some of these exotics can profoundly alter the ecosystems they invade. They can displace the native biota, disrupt the function of unique ecosystems, and threaten even the most American of environmental characteristics—wilderness.

Wilderness as a sustaining counterpoint to civilization is a uniquely North American concept. Wilderness has both social and ecological aspects and values. National, state, and provincial parks, wildlife refuges, and wilderness areas were pioneered in North America to preserve the ecosystems, and their rich plant and animal life, that dominated the landscape European colonists encountered and later shaped into emerging societies. These reflect the social values of wilderness. To environmental scientists, wilderness ecosystems are of special importance because they provide the best baseline for understanding ecosystem function, and especially the impacts of urban-industrial civilization on the dynamics of ecosystems. This represents the major ecological value of wilderness.

The pre-European landscape itself, of course, was not free from the impacts of humans. Hunting, farming, and the use of fire by Native Americans and by the Polynesian colonists of Hawaii had created a landscape quite different than that which would have existed in their absence. We can only speculate about the nature of this landscape when it was free of all human influence—a landscape with, perhaps, a rich megafauna of now-extinct animals in mainland North America and a distinctive avifauna of flightless birds in Hawaii. So, we accept the landscape with which pre-Columbian Native Americans and Hawaiians interacted as our benchmark of wilderness.

What is "benchmark" or "natural" in this context can be, however, very uncertain. Ecologists now realize that ecosystems have not been static, nor have their floras and faunas been unchanging (Sprugel 1991). Nevertheless, just prior to European colonization, ecological interactions among plants, animals, and humans created and maintained ecosystems that were diverse in their characteristics and high in biological productivity. The massive extinctions of large animals that occurred when retreat of continental glaciers allowed human hunters to enter the main portion of North America were far in the past. Eleven thousand or more years of ecological adjustment and evolutionary adaptation had led to a new landscape that tended toward a steady-state condition. Periodic disasters—major forest fires, hurricanes, and floods—certainly disrupted this quasi equilibrium. Climatic cycles still created a background of change in controlling factors of the environment. But directional change was slow, and variation around prevailing ecological states was the norm. This is as close to "natural" as pre-European conditions came.

Furthermore, we remain uncomfortably ignorant of the exact conditions that prevailed in much of North America and Hawaii at the time of European settlement. Paleoecological studies give us a hint of the nature of the pre-European landscape, but most of the details are guesswork at best. Likewise, we are largely ignorant of how typical the conditions at a given location and time were influenced by varying climate, natural disasters such as fires and hurricanes, and the gradually changing patterns of Native American life. Recent studies of the paleoecology of fire in the Yellowstone National Park region, for example, have shown how poorly we understand the behavior of forest ecosystems of that region over periods of hundreds to thousands of years.

The Exotic Invasion and Its Impacts

With European discovery and colonization, however, a major ecological revolution began. This revolution proceeded in phases as waves of human activity swept across the continent. Wildlife was decimated, forests were cut, lands were cleared and plowed, minerals were mined, and waters were dammed, diverted, and polluted. Yet, after all these assaults, major remnants of pre-European ecosystems have remained until today. These are the areas in which parks, wilderness areas, and wildlife refuges were established by federal, state, and provincial governments to preserve. In addition, several private organizations, such as the Save the Redwoods League and the Nature Conservancy, were founded to acquire and preserve pristine examples of natural ecosystems.

Ominously, the wave of exotics that is now sweeping across North America is threatening our most protected ecosystems as well as the productive ecosystems that we depend on for food, fiber, and other harvests. Although exotic invasions began with first settlement (see Chapter 4), the pace has increased greatly since World War II. Exotic plants, animals, and microorganisms of all types are invading North American ecosystems of all types—urban areas, agricultural ecosystems, and even our wild lands and waters. Like the era of change following the arrival of Paleo-Indian hunters (Martin 1973), an era of profound ecological change is accompanying this invasion. Many of our parks, nature preserves, forests, and waters are beginning to lose the integrity of composition and function that we have committed ourselves to preserve. The ecological health of our productive farms, forests, and waters is at risk.

Evidence now shows that the ecological and economic costs of this exotic invasion are great. The productivity, stability, and resilience of North American and Hawaiian native ecosystems are being degraded as exotics multiply, impact native species, and alter ecological processes. In addition, our farmlands, timberlands, rangelands, and productive waters are suffering enormous damage from exotic weeds, pests, and diseases. The human population itself is being confronted with an increasing number of exotic diseases and disease vectors. Exotics are imposing an enormous burden on North American economies. The direct economic damage and control expense related to exotics in the United States have recently (Pimentel et al. 1999) been estimated at about $122 billion annually. (Economic impacts of exotics and their implications are examined in detail in Chapter 20.)

The ecological costs of exotics are great, as well. Nonindigenous species have already contributed to the extinction of many native species, and many conservation scientists regard them as the most rapidly growing threat to biodiversity at large. Exotic species have been identified as a major or contributing cause of declines of 49 percent of all species in the United States identified as imperiled by federal agencies and The Nature Conservancy (Wilcove et al. 1998). For 100 (15.8 percent) of the species federally listed as threatened or endangered, exotic predators or herbivores are considered the primary threat; for 45 (7.1 percent) species, competitors are the primary threat; and for 5 (0.8 percent), exotic diseases are the main threat (Simberloff 1996).

Exotic species are much more serious threats to some groups of organisms than to others. For terrestrial vertebrates, for example, exotics are the main cause for listing of 10 percent and a contributing cause for listing of 26 percent of species listed as threatened or endangered in the United States (OTA 1993). For fish, exotics are the most important cause of listing for 6 percent and a contributing cause of listing for 51 percent of species. For plants, exotics are the primary cause of listing for 6 percent and a contributing cause of listing for only 16 percent of species. Exotics have also been identified as major obstacles to recovery of many listed species, including freshwater fish and freshwater mussels (Richter et al. 1997).

The evolutionary impacts of exotics cannot be overlooked. Invasive exotics are changing the course of evolution for virtually all members of North American ecosystems. As soon as exotics enter their new environment, natural selection begins to adapt them to their new biotic and abiotic surroundings. In addition, the exotics themselves impose new patterns of selection on the native species with which they interact. As a result of these evolutionary influences, native species are changed forever. Even if exotic species could be eliminated, the assemblage of native species has acquired a "ghost of past exotic influence," which precludes its return to the exact state that prevailed prior to exotic invasion.

Interestingly, over hundreds or thousands of generations, natural selection may also be working toward an "evolutionary control" of detrimental exotics. Just as the huge human population represents a potential host for any disease that can crack our defenses, successful exotics are targets for biotic exploitation or defensive adjustment by other species. In time, evolutionary adjustments of members of the biotic community at large, a process known as counteradaptation (Ricklefs and Cox 1972), may transform the exotic into a "peaceful" member of the community. How quickly counteradaptation can occur and how fully it can mitigate the impacts of the most disruptive exotics are evolutionary unknowns but are topics of great interest to evolutionary ecologists (see Chapter 19 for more detail).

Magnitude of the Exotic Invasion

North America has become the most important world battleground in the war on exotics. More exotics have invaded North America than any other continent. An estimated 6,600 species of terrestrial, freshwater, and marine organisms have been introduced into the United States since European discovery (Table 2.1). This estimate is probably conservative. These exotics include all groups of plants, animals, and microorganisms, but more than 5,600 of the species are terrestrial plants and arthropods. Exotics now make up somewhere between 2 and 8 percent, or roughly 5 percent, of the U.S. continental biota. For Canada, more than 970 exotic terrestrial and freshwater species have been recorded. At least 400 exotic plant and animal species have been introduced into estuaries and coastal marine waters around North America (Ruiz et al. 1997). The invasions are continuing. Just between 1980 and 1993, more than 205 new exotics were found or identified in the United States.

We can only estimate the number of exotic species now present in North America. The true number of exotics in North America or other heavily invaded world regions is almost impossible to determine. The long history of human influences on habitat conditions and dispersal processes, beginning before biotic inventories of native species existed, means that the status of many species is uncertain. This is especially true

Table 2.1. Estimated numbers of intercontinental exotic species in the United States and Canada.

	United States		*Canada*	
Taxon	*Number*	*Percent of Biota*	*Number*	*Percent of Biota*
Plants[1]	2,100	10.8	940	22.3
Plant pathogens	239[6]	—	n/a	—
Marine species	400[8]	—	175[7]	—
Mollusks				
Nonmarine	91[6]	4	n/a	—
Marine[4]	34	—	15	—
Arthropods[5]	>3,500	3.5	n/a	—
Freshwater fish	70[2]	8	10[3]	—
Terrestrial vertebrates	142[6]	6	20[6]	—
TOTAL	>6,576			

[1]Vitousek et al. (1997); [2]Courtenay et al. (1984); [3]Crossman (1984); [4]Carlton (1992b); [5]Author's estimate, based on Frank and McCoy (1995b); [6]OTA (1993); [7]Author's estimate, based on various sources; [8]Ruiz et al. (1997).

in North American coastal waters. Many species of uncertain origin, termed "cryptogenic" species, occur in certain ecosystems. Many of these must certainly be exotics that reached North America before the first biological surveys were made (Carlton 1996a). In San Francisco Bay, where about 234 nonindigenous species have been recognized, a reasonable guess is that about 125 more species fall in the cryptogenic category (Cohen and Carlton 1998). In Chesapeake Bay, with a longer history of intense human impact, cryptogenic species may be as much as three times as numerous as species known to have been introduced.

Numbers of species aside, exotics have invaded all regions and all natural ecosystems in North America and Hawaii, although the extent of invasion differs greatly. The terrestrial ecosystems of Hawaii, Florida, and California have perhaps suffered the heaviest invasions. Some ecosystem types, such as the California valley grassland, have been so altered by exotics that the original composition and structure are unknown. Others, such as southwestern piñon-juniper woodlands, have only been lightly invaded. Coastal waters, and particularly estuaries such as Chesapeake Bay and the San Francisco Bay and Delta ecosystem, may now have more exotic than native species. Freshwater lakes and rivers have also been invaded heavily, with the North American Great Lakes being one of the prime examples of invasions by exotics at all levels in the food chain (Mills et al. 1993a).

Hawaii is the most invaded region of the United States, with at least 4,598 exotic species established in the wild (Eldredge and Miller 1998) (see Chapter 13). About 4,275 species of exotic flowering plants are in cultivation in Hawaii, of which about 908 have become invasive into wildlands (Loope and Mueller-Dombois 1989; Eldredge and Miller 1998). Exotic insects in Hawaii number about 2,598 species (Eldredge and Miller 1998). There are probably about thirty exotic land mollusks that are established in this state (Howarth and Medeiros 1989). More land birds, thirty-eight species in all, have been introduced successfully to Hawaii than to any other region on earth. In addition, eighteen exotic mammals, seventeen exotic reptiles and amphibians, and nineteen exotic freshwater fish are now established in Hawaii.

In continental North America, the extent of invasion varies greatly from region to region, with Florida and the Gulf Coast being the most heavily invaded region (see Chapter 9). An estimated 925 species of exotic plants are established in the wild in Florida, and perhaps 25,000 are in cultivation, but not established in the wild (Frank and McCoy 1995a). About 988 exotic insects, nearly 8 percent of the total insect fauna, are known from Florida, as well (Frank and McCoy 1995b). California and other

Pacific Coast areas are the next most heavily invaded region. At least 674 species of exotic plants, about 11 percent of the flora, have become naturalized in California (Mooney et al. 1986).

Introductions to coastal marine waters have been numerous, especially along the Pacific Coast of North America. At least 400 species of marine plants and animals are proven exotics. Although the overall impact of these exotics on the native biota is still uncertain, some groups of exotics may have very serious ecological and human health impacts. Some biologists suspect, for example, that the increased incidence of toxic red and brown tides in coastal waters may be due to the transoceanic transport of dinoflagellates and other toxic algae in ship ballast water (Morton 1997).

Numerous introductions have occurred to inland waters, with over 140 exotic plants and animals having been recorded in the Great Lakes, for example (see Chapter 6). For freshwater fish, on average, about 8 percent of the fauna in different regions of the United States consists of exotics from other continents (OTA 1993). Another 17 percent of species in various locations in the United States consist of North American species introduced beyond their original ranges. In most other groups of organisms, however, the percentage of exotics is small. Only sixteen species of freshwater mollusks, or about 1.6 percent of the North American fauna, are exotic (Morton 1997). These include four bivalves and a dozen snails. In recent years, however, a number of other exotic invertebrates have been introduced into the Great Lakes in ballast water of cargo ships.

Invasiveness also varies greatly among taxonomic groups (Pysek 1998). Worldwide, four families stand out in terms of number and proportion of plant species that appear as exotics: grasses, composites, legumes, and crucifers. In continental North America, several plant and animal groups are highly represented among exotics. Among grasses, for example, about 11.2 percent of the 1,398 species listed by Hitchcock (1950) for the United States are exotic. For terrestrial vertebrates, about 6 percent of North American species are exotic. The percentage is even higher for freshwater fish, as noted earlier. For terrestrial mollusks, 37 species, or about 6.8 percent, of the North American fauna consist of exotics (Morton 1997).

Sources of the Invaders

The sources of exotics differ for various groups of plants and animals. For exotic flowering plants and ferns of eastern North America, Europe has been the principal source. Of 1,051 exotic species in eastern North America, 76.4 percent are from Europe or Eurasia at large. An additional 11.5 percent are from eastern Asia. Thus, 87.9 percent are from Eurasia

(Foy et al. 1983). For California, the situation is much the same. Of 600 species whose origins are from known locations outside North America, 79.7 percent are from Eurasia (Rejmánek et al. 1991). The second greatest source area is Central and South America, which contributes 10.7 percent of California's exotics. For Florida and Hawaii, however, plant exotics come largely from tropical and subtropical regions. Many of the plant exotics in Florida come from Central and South America, and many of those in Hawaii come from Asian regions.

For terrestrial animals, most exotics in continental North America again are from Eurasia. Exotic terrestrial vertebrates come almost entirely from Eurasia, except in southern Florida, where several invaders come from the West Indies and Central America. For insects, 66.2 percent of the 1,683 species that had invaded the United States, as of the early 1980s, came from Europe and the Middle East, with an additional 13.8 percent from eastern Asia (Sailer 1983). About 14.3 percent came from Central and South America. In Hawaii, exotics come from many geographical areas, both temperate and tropical.

Aquatic animals are of more varied origins. Of the forty-two species of freshwater fish that have been introduced successfully into continental North America (Courtenay et al. 1984), the largest number, eighteen, are from Central and South America. Many of these are aquarium fish that have established populations in the wild in the southern states and California. The remaining species come about equally from Europe, Asia, and Africa. Two species of saltwater fish from the Pacific coast of Mexico have also been introduced into the highly saline Salton Sea in California.

Exotic marine organisms differ in origin on the Atlantic and Pacific Coasts of North America (Carlton 1992a). For mollusks, most species that have invaded the Atlantic Coast come from the northeast Atlantic. Those that have invaded the Pacific Coast come almost equally from the western Pacific and the northwestern Atlantic.

North America is not the only important battleground in the exotics wars. In terms of the ratio of exotics to native species, Australia and New Zealand have exotic invasion problems even worse than those of North America. Australia has about 222 species of noxious weeds and about 1,500 other species of terrestrial and freshwater exotics. But exotic plants make up about 21.0 to 43.6 percent of the flora of southern Australia. In New Zealand, the situation is worse yet. The 1,570 exotic plants there constitute 46.7 percent of the total flora (Heywood 1989). Because of the evolutionary uniqueness of plant and animal life in Australia and New Zealand, conservation biologists in these countries are alarmed about the impacts of exotics, and they have developed extensive research programs on exotic ecology and explored new strategies of control. North Americans have much to learn from efforts in these two countries. Recently, as an example of national concern about exotics, an Australian legislator proposed the eradication of the domestic cat, including both feral and domestic animals.

Problem Exotics and the "Tens Rule"

Sheer numbers do not give an accurate accounting of the seriousness of exotic threats. Only a fraction of established exotics become serious invasives. In fact, a general relationship appears to exist between total numbers of exotics and those that create serious problems. This relationship, a statistical generalization, is the tens rule (Williamson and Fitter 1996). The tens rule involves three 10 percent relationships. The rule states that of the species brought into a region by natural or human-assisted dispersal, about 10 percent will appear in the wild. Of the species that appear in the wild, about 10 percent will establish self-reproducing populations there. And, of these exotics that establish in the wild, about 10 percent will prove to be significant problem species. In other words, of every 1,000 species that reach a new region, about 100 will survive or grow at least temporarily, 10 will become firmly established, and 1 will prove to be a seriously invasive exotic.

The tens rule is, of course, a very rough generalization, subject to considerable variation from place to place and from one group of organisms to another. The best examples of the tens rule pertain to the last of the three relationships, the fraction of established species that become serious invasives. In Hawaii Volcanos National Park, for example, about 475 exotic plants have become established (Williamson 1996). Of these, about 53, or 11.2 percent, are judged to be problem species. Some obvious violations of the tens rule can also be noted. For the exotic birds released into the wild in Hawaii, for example, 38, or 54 percent, of 70 species have become established. On the other hand, of those established, 3 or 4 (8 to 10 percent) might be considered agricultural pests or threats to certain native bird species. This is more or less in agreement with the tens rule.

Likewise, of the 6,600 or so exotics established in North American terrestrial, freshwater, and coastal marine environments, only a fraction cause serious harm. According to the tens rule, about 660 would be expected to be seriously harmful. According to the U.S. Office of Technology Assessment study (OTA 1993), the fraction of high-impact terrestrial and freshwater exotics varies from about 3.6 percent, for fish, to about 17.6 percent, for terrestrial vertebrates. This study identified 214 high-impact terrestrial and freshwater animals. Considering the additional large number of high-impact plants and plant pathogens that are known, 600–700 exotics of seriously harmful nature appear reasonable.

The increasing numbers of nonindigenous species, their continuing spread, and their increase in local abundance, in combination, mean that the ecological impacts of nonindigenous species are growing exponentially. This growth also contains a "time bomb" element—many species exhibit a lag, or "incubation" period, before they begin to spread or increase in abundance. Many established exotics survive at very low abundance or in very restricted locations for many years. Then, perhaps because of some evolutionary adjustment, the development of a population capable of high dispersal potential, or the occurrence of human activity that brings about environmental change that favors the species, the period of innocuous survival ends with a sudden explosive spread (Crooks and Soulé 1999).

Exotics, Biodiversity, and Ecosystems

Many issues arise as a result of this accelerating invasion of nonindigenous species. The mass invasion of exotics poses a serious threat to survival of many North American native species. The threat of extinction is especially great for island areas and freshwater lakes, where evolution has adapted species to specialized local conditions. Although the number of extinctions of native North American species is still small, exotics are extirpating natives in numerous local areas (Hobbs and Mooney 1998). Even if they survive, natives are gradually becoming relict populations confined to small refuges.

Other issues relate to the underlying processes of ecosystem fertility, productivity, and stability. Exotics are exerting detrimental impacts on the structure and dynamics of our natural ecosystems. Exotics can modify the way that nutrients cycle, energy is fixed and passed along food chains, fire and other disturbances occur, and the landscape itself evolves. These effects can be amplified by stresses such as overgrazing and other destructive land use practices. In some cases, ecosystems can be transformed to states that are stable and dominated by exotics, from which recovery is impractical. In the Florida Everglades, for example, exotic trees are transforming open sawgrass marshes into dense forests (see Chapter 9).

Questions about invading species and their impacts on biodiversity and ecosystem function are especially relevant now because of concern about global change (Beerling 1995). Will climatic change increase the rate of invasion of world regions by exotic species? One study that has begun to examine this question is the Global Change and Terrestrial Ecosystems project of the International Geosphere-Biosphere Program. This project has concluded that the alien species component of continental biotas is likely to increase and that regional biotas are likely to become more homogeneous. The implications of such change for global biodiversity and the health of the biosphere are enormous.

The Ecology of Invasive Species

To address the issues raised in the last section, we must learn much more about the ecology of invasive species. Where do most invaders come from? What are the dispersal routes for invaders? How are they transported from source areas to new recipient areas? What factors make regions and ecosystems vulnerable to exotic invasion? What determines why certain species establish successfully and others do not? What determines differences in ecological impact of established exotics? How can invasions be prevented? How can impacts of established exotics be reduced?

These questions form the focus of the rest of this book. This chapter has considered the numbers and origins of exotics for North America at large. Chapter 3 will continue to address basic questions about exotics by examining the mechanisms of dispersal of exotics to North America and other new regions. It will also consider the patterns of spread once they have become established in new regions. This final portion of the introduction will set the stage for the examination of the patterns of invasion of different regions and ecosystems in North America.

Chapter 3

A Brief History of Invasions: Human History—An Exotic's Perspective

Take heed the fate of flightless rails,
the kiwi of the north,
in nameless graves are moldering those,
who numberless went forth.

These tiny birds, by all beloved
despite peculiar habits,
owe their demise to rodents, the
productive race of rabbits.

—Lewis Walker and Loring Hudson, *Midway in Verse—or Worse**

Laysan Island, a coral atoll slightly over 1,000 acres in area, lies nearly 800 miles west of Kauai, one of the main islands of Hawaii. Twenty-five species of plants originally occurred on the island, five of them endemic subspecies. Three land birds, including the flightless Laysan rail (*Porzana palmeri*) and the Laysan teal (*Anas laysanensis*), were endemic to this tiny island. The island was also a major seabird nesting area, and had major guano deposits. In the late 1800s, guano miners introduced European rabbits (not really rodents!) to the island to provide variety in their diet. The plague of rabbits that resulted largely devegetated the island, eliminating

* Lewis Walker and Loring Hudson (1945). *Midway in Verse—or Worse*. Hester Colorgraphic Studios, San Diego, CA.

all but four of the native plants and ultimately driving two of the land birds to extinction (Carlquist 1980). The Laysan teal (*Anas laysanensis*) narrowly escaped; the six surviving individuals in 1911 have enabled the population to recover. In 1891, the Laysan rail was translocated to Midway Island, nearly 400 miles farther west. Midway had no rabbits, and the rail flourished.

Lewis Walker and Loring Hudson were U.S. Navy personnel stationed on Midway Island during World War II, and on many days they had considerable time on their hands. So they recorded in verse the role of exotic animals—rabbits and rats—in the extinction of the Laysan rail. Black rats were accidentally introduced to Midway from navy ships during World War II. The rail, first driven to the edge of extinction by introduced rabbits on Laysan, unfortunately fell prey to the black rat. The last rails on Midway were seen in 1944. The rabbits of Laysan were eradicated in 1923, but the extinctions they precipitated are forever.

Laysan and Midway Islands are among the most remote places on earth. Yet exotics reached these islands, with disastrous effect. This example demonstrates that humans are now the prime agent of long-distance dispersal. It also shows the efficiency with which species such as the black rat are able to take advantage of human assistance to cross seemingly insurmountable barriers. This chapter examines natural and human-assisted dispersal processes that introduce species to new continents, islands, and ocean areas (Table 3.1). It also considers their patterns of dispersal once they have become established in new areas.

Dispersal as a Natural Process

Dispersal is an essential ecological process for all species, but mechanisms of dispersal vary greatly in nature and effectiveness. Their effectiveness spans a spectrum from those of species whose propagules are concentrated close to the parent organism to those of species whose propagules are discharged into air or water currents that transport them widely. Some species rely on abiotic processes, such as wind and flowing water; others, on animals that carry the propagules to specific types of habitats.

Natural dispersal mechanisms have achieved remarkable success in enabling organisms to cross formidable barriers, given a long span of geological time. The Galápagos Islands, west of South America, for example, owe their terrestrial biota to dispersal over at least 800 miles of ocean. Wind-carried spores and seeds account for some of the plant colonizations, and winds must have aided many insects and other arthropods in reaching the islands. A number of land birds have reached the islands. In

Table 3.1. Scientific names and native regions of exotic species that illustrate patterns of intercontinental dispersal and invasion.

Species	Native region
PLANTS	
African rue, *Peganum harmala*	North Africa
autumn olive, *Elaeagnus umbellata*	Asia
baby's breath, *Gypsophila paniculata*	Eurasia
gorse, *Ulex europaeus*	Europe
Japanese honeysuckle, *Lonicera japonica*	Asia
multiflora rose, *Rosa multiflora*	Asia
kudzu, *Pueraria lobata*	Asia
puncture vine, *Tribulus terrestris*	Africa
Russian olive, *Elaeagnus angustifolia*	Asia
Scotch broom, *Cytisus scoparius*	Asia
water spinach, *Ipomoea aquatica*	South America
witchweed, *Striga asiatica*	Asia
MARINE INVERTEBRATES	
European green crab, *Carcinus maenas*	Europe
Japanese oyster, *Crassostrea gigas*	Asia
FISH	
bighead carp, *Aristichthys nobilis*	Asia
TERRESTRIAL INVERTEBRATES	
golden nematode, *Globodera rostochiensis*	Asia
rosy wolfsnail, *Euglandina rosea*	Florida, USA
INSECTS	
Asian tiger mosquito, *Aedes albopictus*	Asia
flowerhead weevil, *Rhinocyllus conicus*	Eurasia
Mediterranean fruit fly, *Ceratitis capitata*	Europe
AMPHIBIANS	
African clawed frog, *Xenopus laevis*	Africa
REPTILES	
brown tree snake, *Boiga irregularis*	South Pacific Islands
MAMMALS	
black rat, *Rattus rattus*	Europe
cattle, *Bos taurus*	Eurasia
Channel Island fox, *Urocyon littoralis*	Mainland California
dog (dingo), *Canis familiaris*	Eurasia
domestic cat, *Felis cattus*	Eurasia
European rabbit, *Oryctolagus cuniculus*	Europe
goat, *Capra hircus*	Eurasia
nutria, *Myocastor coypus*	South America
pig, *Sus scrofa*	Eurasia
Polynesian rat, *Rattus exulans*	Pacific Islands
reindeer, *Rangifer tarandus*	Eurasia
small Indian mongoose, *Herpestes auropunctatus*	India
sheep, *Ovis aries*	Eurasia

the case of finches, only a single species is likely to have colonized the islands, giving rise by speciation to thirteen daughter species. Many plants and flightless animals, including at least two small mammals, made the water crossing on rafts of vegetation. The colonization of the Hawaiian Islands involves similar processes. In both cases, the time over which colonizations occurred is measured in tens of millions of years, longer than the age of the oldest existing islands. In both cases, generations of islands originating over volcanic "hot spots" in the ocean floor have risen and been destroyed by erosion, those that we see now being just the current generation. Similar patterns can be seen in the biotas of many other oceanic islands.

Remarkable cases of long-distance dispersal have also been noted between continents, as between temperate North and South America. A number of desert and Mediterranean-climate plants show hemispheric disjunct distributions, occurring in Argentina and Chile in South America and in the southwestern United States. One species, creosote bush (*Larrea tridentata*), is of South American origin but is the dominant shrub over vast areas of the North American deserts. The mechanisms of such dispersal are not completely known, although transport by migrating birds is one possibility. Mud adhering to the legs of birds, for example, can contain a surprising number of seeds (Foy et al. 1983).

Thus, no geographical barrier on earth has proven insuperable to dispersal processes, so that colonization by previously "exotic" species is a natural phenomenon. But, like extinction, the colonization of world regions by exotics is now a process that has been elevated to an extraordinary level, with unnatural results. Doubtless, even natural colonizations had serious impacts on the invaded ecosystems. But in most such cases, evolutionary adjustments enabled the invaded biotas to absorb the newcomers without massive extinction. The current rate of exotic invasions, however, is far in excess of the rate of evolutionary integration of the new species into colonized ecosystems.

Aboriginal Humans as Agents of Dispersal

Humans, as already noted, are now the prime agent of long-distance dispersal. This role is not entirely recent. As humans colonized the islands of Melanesia, Micronesia, and Polynesia between 30,000 and 1000 B.P., they carried with them forty to fifty species of plants and animals (Loope and Mueller-Dombois 1989). These included Polynesian rats, dogs, and pigs, bringing these terrestrial predators to islands where they were previously unknown (Steadman 1995). The effect of these introductions, together with predation by the humans themselves, was the extinction of perhaps 2,000 species of birds, many of which were flightless species endemic to specific oceanic islands or archipelagos. In the case of the Hawaiian Islands, at least sixty endemic species became extinct during the Polynesian era. Around Australia, islands visited by Aboriginal humans, and to which the dingo was introduced permanently or temporarily, much more frequently lack kangaroos than islands not visited by humans (Abbott 1980). Dingos likely contributed to the extirpation of many of these kangaroo populations.

Off the coast of California, the island fox may be another example of human-assisted colonization. The fox occurs on six of the eight Channel Islands, including San Clemente and Santa Catalina, which have never been connected to continental California. These foxes are not derivatives of the mainland gray fox (*Urocyon cinereoargenteus*) but appear to be surviving populations of a small fox that once lived on the nearby mainland, perhaps a relative of several species of small foxes now confined to southern Mexico and Guatemala. How it reached the islands is an intriguing question. The foxes might well have been introduced to San Clemente and Santa Catalina by Indians, perhaps as tame animals carried by canoe from island to island with traveling groups of hunters and fishermen.

European Exploration and Dispersal of Exotics

The age of European exploration and colonization opened an era of introduction of plants and animals to newly settled parts of the world, particularly the Americas, Australia, South Africa, and eastern Asia. The ships themselves carried mice and rats. With the livestock transported to these new lands came the seeds of weeds tangled in their pelage and associated with feed and bedding. Seed grain carried contaminant seed of grainfield weeds, and probably many insects. The spread of many of these exotics through newly colonized regions was rapid. For example, by A.D. 1600 the species that dominated the weed flora of central Mexico were probably the same as those that predominated in the Iberian Peninsula (Crosby 1986). Similar invasions of exotics took place in eastern North America, South America, and Australia.

In the marine environment, the wooden hulls of ships of early exploration, whaling, and commerce provided ideal habitats for boring and fouling marine invertebrates (Carlton 1989). Shipworms (actually bivalve mollusks) and gribbles (boring isopod crustaceans) lived in the wood of the hulls. Sessile organisms such as marine algae, sponges, hydrozoan coelenterates, sea anemones, bryozoans, mollusks, barnacles, tube-dwelling polychaete worms, and tunicates formed dense growths on submerged portions of the hulls. In these marine jungles small larvae and adults of a host of other marine invertebrates and fish sheltered themselves. Many of these species disembarked in ports and coastal waters of foreign continents. Before biologists had surveyed the distribution of marine life, they thus had achieved cosmopolitan or pantropical distributions as a result of ship traffic.

In addition, early wooden-hulled cargo ships often carried dry ballast, which was used when a light or partial cargo did not give the weight to make the ship trim in the water, that is, to prevent winds and waves from capsizing it. This material consisted of soil, sand, and rock from near-shore areas and contained propagules of coastal plants and animals of many sorts (Carlton 1989). Where this ballast was discarded, many of these species grew and became established. In a soil ballast dump in Oregon, for example, Nelson (1917) found ninety-three plant species that were not known elsewhere in the state, along with thirty-two native plants and eighty-eight exotics that occurred in other locations in Oregon.

The seafarers that explored the world's oceans and later hunted whales and seals also took more deliberate actions to introduce exotics. To assure themselves of a meat supply on future visits, or in the case of shipwreck, they often introduced goats to remote islands (Fosberg 1972). Cattle, sheep, and European rabbits were also released on some islands in the southern oceans, with no thought to ecological consequences (Chapuis et al. 1994). When permanent settlements or stations were established on such remote islands, other animals, such as reindeer, were often introduced. Around permanent settlements, domestic animals such as cats and dogs often became feral.

The voyages of discovery and colonization also took back to Europe useful and interesting plants and animals from the areas visited. Exotic crop plants and domestic animals from the Americas were introduced to Europe, doubtlessly accompanied by some of the species that have become exotic pests there. Nevertheless, the flow of exotics was strongly from Europe to newly discovered temperate regions (Crosby 1986).

Dispersal of Exotics in the Modern Era

As settlement followed exploration, commerce in plants and animals increased. The colonies in North America and elsewhere imported crop plants, forages, medicinal plants, horticultural plants, livestock animals, pets, game animals, and insects with real or hoped-for economic value. Many people in European colonies throughout the world assumed that

European species were inherently desirable, and some of those interested formed acclimatization societies to promote the introduction of the familiar plants and animals of Europe. In the late 1800s in New Zealand, for example, introductions of freshwater fish by acclimatization societies resulted in exotics now constituting twenty of the forty-six species known for the country. As sea trade expanded, plants and animals were also translocated between other continental regions. Temperate species from eastern Asia were moved to Europe, North America, and temperate regions of the Southern Hemisphere. Tropical species were introduced widely throughout the tropics and subtropics.

World Wars I and II introduced military transport, by air as well as water, as an effective means for human-assisted dispersal of a variety of species. Puncture vine, native to the Sahara Desert, may have been introduced to North America on the tires of military vehicles and aircraft returning from Europe after World War I (Foy et al. 1983). Black rats inadvertently introduced to Midway Island by navy ships during World War II killed the last individuals of the Laysan rail. The brown tree snake, native to New Guinea and neighboring areas, was introduced accidentally to the island of Guam in the late 1940s or early 1950s, probably in military shipments of fruit. An arboreal, nocturnal predator on eggs, young, and adult birds, this snake has nearly eliminated native forest bird species, driving six species to extinction and reducing the remaining four species to fewer than a hundred individuals on the main island (Savidge 1987). In Texas and New Mexico, a desert shrub, African rue, was apparently introduced inadvertently at a World War II airfield. Witchweed and the golden nematode are other pests that are believed to have entered North America on returning military equipment (OTA 1993).

Following World War II, international trade and travel grew enormously and also became faster. These changes greatly increased the potential for dispersal of exotic species. As a result, many pathways of introduction, ranging from unintentional to deliberate, now exist. Imported plant materials such as seeds, horticultural and forestry stock, grain, fruit and vegetable produce, bulk cotton and other commodities, veneer logs, and processed timber can carry a variety of living contaminants. These include weed seeds, plant pathogens, larval or adult arthropods and other invertebrates, and even some kinds of vertebrates. Animal materials include livestock, aquacultural stock, pets, and commodities such as wool and hides, all of which can also carry arthropods and disease agents. Fish and shellfish imported for aquaculture have been a major route of introduction of a variety of diseases, parasites, and predators of fish and shellfish. Even inorganic commodities, such as sand and gravel, coal, and metal ores, can be contaminated with pathogens, seeds, and arthropods. Crating and packing materials may also contain pests and pathogens. Containerized cargo of any sort can shelter almost any exotic from microorganisms to snakes and rodents. Cargos that contain even small volumes of water can transport other exotics. As noted in Chapter 1, the Asian tiger mosquito apparently reached the United States as larvae in water trapped in used tires shipped to Houston, Texas, for recycling (Craven et al. 1988).

Vehicles themselves are often agents of transport of exotics. Commercial and recreational movement of vehicles across international boundaries has increased enormously. Cars, trucks, trains, planes, and ships can be contaminated with soil, plant seeds, and arthropods. Military vehicles, often moved under emergency conditions and with little or no inspection, are a potentially serious carrier of exotics. Such vehicles are often heavily contaminated with mud and plant debris after their use abroad.

In the past few decades, by far the most serious means of inadvertent introduction of aquatic organisms is in the ballast water of cargo ships. When modern cargo ships are traveling empty or lightly loaded, water is taken into tanks or holds to make the ship ride lower in the water and give it stability. Fresh, brackish, or ocean water may be used as ballast. Being essentially unfiltered water, a great variety of aquatic organisms are normally present. In or near the destination port, the ballast water is discharged, and any surviving organisms are introduced to the new environment. About 35,000 to 40,000 cargo ships now ply the world oceans, and water ballasting has been practiced for about a hundred years (Pierce et al. 1997), so that this is probably the most significant route of dispersal of aquatic organisms. Many species of freshwater, estuarine, and marine animals, and some plants, have been introduced to North America by this route (see Chapters 5 and 6). Considerable evidence now exists that dinoflagellates and other algae that produce paralytic toxins are being transported around the world in ballast water of marine cargo ships (Hallegraeff and Bolch 1991). This transport may be a major contributor to the increased incidence of toxic algal blooms.

Deliberate Introductions of Exotics

Some introductions, although intentional, lead to unanticipated invasion of natural ecosystems. Many pernicious weeds were actually imported for horticultural or agricultural use and have simply escaped into the wild. In the western United States, some twenty-eight species of weeds are escapes from horticulture and nine are escapes from agriculture. In the eastern United States, kudzu vine and Japanese honeysuckle are two notorious escapes from horticulture (see Chapter 8). Other examples of horticultural escapes with serious consequences are gorse and Scotch broom (see Chapter 12). In eastern North America, baby's breath is a horticultural plant that is now becoming a serious weed in dunelands of the north-central states. Western agricultural escapes include several perennial grasses introduced for pasture improvement (see Chapter 10). Similarly, animals imported for commercial fur production, such as the nutria, or as pets, such as various parrots, have escaped or been released. The African clawed frog, once used widely in human pregnancy tests, is an example of a species that has accidentally escaped and become established in the wild in some places.

Most states and provinces now have laws preventing private releases of exotic plants and animals. Illegal importations and introductions still occur, however. These are especially frequent in aquatic environments. Aquarium plants and fish, dumped into natural waters, have been the source of many introductions. In Mississippi, for example, the bighead carp has become established in the wild either by escape from aquaculture or by illegal introduction. Illegal stocking of exotic trout has hampered recovery efforts for the Gila trout (*Oncorhynchus gilae*) in New Mexico, according to news reports in 1998. Illegal importation and planting of designated noxious weeds, such as water spinach in Florida, also occurs frequently.

Truly deliberate and legal introductions of exotics have also been the source of many problem exotics. In the early 1900s, plants such as multiflora rose, autumn olive, and Russian olive were actively promoted as food and cover species for wildlife (see Chapters 4 and 11); now they are among North America's most out-of-control exotics. Until quite recently, many state fish and game departments encouraged the introduction of exotic mammals, birds, and fish in the name of improved hunting and fishing (see Chapter 14). Private landowners also import many such species, with the intent of keeping them in confinement. But many ultimately escape. In Texas, for example, more than 160,000 animals of 163 species now legally exist under confined, but often free-range, conditions (Teer 1991) (see Chapter 14). Exotic aquacultural species, such as the Japanese oyster on the Pacific coast of North America (Hanna 1966), are another type of deliberate introduction.

Deliberate introductions for biological control have led to the establishment of over 237 exotic insects and several plant pathogens in the United States. In the past two decades, introductions for biological control have been made only after extensive testing to assure that the biological agent will not have negative effects on nontarget species. In the 1800s and early 1900s, however, some introductions for biological control were done very casually, as in the case of the small Indian mongoose and the rosy wolfsnail in Hawaii (see Chapter 13). Inadequate screening and faulty evaluation led to the introduction of the flowerhead weevil for biological control of Eurasian thistles (Louda et al. 1997). This weevil, introduced in 1972, has begun to utilize native North American thistles of the genus *Cirsium*, some of which are endangered species (see Chapter 20).

Pattern of Spread of Established Exotics

How do invading species behave once established? In particular, what patterns and rates of spread do they show? These questions form the focus of another branch of invasion ecology. Pattern and rate of spread depend on life history features of the organism in question and are affected by the abiotic and biotic conditions encountered by the species in its new environment. Once established in a new region, some species may spread largely by natural processes, whereas the spread of others can be aided by human activities. In North America, the spread of many exotic animals has been largely by natural means. Many insects disperse actively or passively through the air. The spread of many freshwater plants and animals, on the other hand, is aided by human activities (see Chapters 6 and 7). Many weedy plants have spread along the routes of barge canals, highways, and railroads, suggesting that human transport of their seeds has been important.

A number of ecologists have examined the rates of spread of exotics and have attempted to derive mathematical models of spread (Hastings 1996). The simplest of these models relates the rate of spread to the per capita population growth rate and a coefficient of diffusion that describes the intrinsic ability of the organism to disperse. This formulation predicts that the spread in distance from an establishment point will be constant, at least after an initial "gearing up" phase of somewhat slower spread. On an area basis, this would correspond to a linear increase in the square root of the area occupied.

For many exotics, however, the pattern of spread is not uniform in all directions but is constrained by the distribution of suitable habitat. This is clearly the case for organisms inhabiting streams and coastal aquatic or terrestrial habitats but may also be the case for other terrestrial animals. Carey (1996), for example, suggested that the Mediterranean fruit fly, or medfly, is spreading through coastal southern California along a set of branching pathways channeled by topography and host availability. Along the Pacific Coast, the green crab is spreading northward much faster than southward from its point of introduction in San Francisco Bay, presumably as a result of ocean current patterns (Grosholz 1997).

The Problem Is Us

Humans have thus become the principal agent of dispersal of exotic species to new continents, islands, and waters. In many cases, they are also the main agent of intraregion spread. The mobility of the human population, and of the vehicles it employs, is enormous. International travel and tourism are expanding rapidly. The volume of commerce in plants, animals, and commodities that can harbor living organisms is also growing. Indeed, worldwide free trade is considered an ultimate goal by many political economists.

The potential for exotic organisms both to reach new areas and to spread through them is now at its greatest level. The overwhelming importance of humans as agents of dispersal of exotics thus raises serious issues about increased freedom of travel and trade. How can we manage these activities to reduce the spread of harmful exotics? (Chapter 21 returns to this question and considers some ways to reduce the risk of introduction of exotics to North America.)

As shown in this chapter, the mechanisms of dispersal of exotic species are also very diverse. The mechanisms responsible for introducing exotics to different regions and ecosystem types in North America differ greatly. Part II examines regional patterns of exotic invasion in North America and Hawaii and identifies the principal ways in which humans have encouraged the invasion and spread of exotics.

Part II

Regional Perspectives

Chapter 4

The Eastern Seaboard: Exotics Discover America

> When English pathfinders topped the Appalachian and proceeded into Kentucky in the last decades of the eighteenth century, they found white clover and bluegrass waiting for them. The plants either had crept over the mountains clinging to the coats of trader's horses and mules or, more likely, had entered with the French in the late seventeenth and eighteenth century.
>
> —Alfred Crosby, *Ecological Imperialism**

In 1814, in a flora of the northeastern United States, botanists John Torrey and Asa Gray noted that purple loosestrife (Table 4.1) was growing in freshwater marshes near several major port cities from Philadelphia to Boston. Native to Europe, it had probably reached the east coast of North America in soil used as ship ballast, or as seeds embedded in imported wool, probably in the late 1700s. Later, it was probably imported deliberately as an ornamental plant, because of its attractive spikes of magenta flowers, and for medicinal reasons, based on its reputation as an effective astringent and insect repellent. By the 1830s, it had become well established in New England wetlands. In the mid-1800s, its popularity as a horticultural plant led to its establishment in Michigan, Washington State, and near Vancouver, British Columbia. The construction of canals linking major rivers to the Great Lakes also contributed to the spread of loosestrife. Canal boats often carried soil and sand as ballast, and loosestrife

*Alfred Crosby (1986). *Ecological Imperialism: The Biological Expansion of Europe 900–1900*. Cambridge University Press, Cambridge, England.

Table 4.1. Scientific names and native regions of early exotics that were introduced to the eastern seaboard.

PLANTS	
Australian pine, *Casuarina equisetifolia*	Australia
bittersweet nightshade, *Solanum dulcamara*	Eurasia
Brazilian pepper, *Schinus terebinthifolius*	South America
Canada bluegrass, *Poa compressa*	Europe
Canada thistle, *Cirsium arvense*	Europe
cockleburs, *Xanthium* spp.	Europe
common mullein, *Verbascum thapsus*	Europe
common plantain, *Plantago major*	Europe
common reed, *Phragmites communis*	Europe
common yarrow, *Achillea millefolium*	Europe
crabgrass, *Digitaria sanguinalis*	Europe
dandelion, *Taraxacum officinale*	Europe
English buckthorn, *Rhamnus cathartica*	Europe
gorse, *Ulex europeaus*	Europe
Japanese honeysuckle, *Lonicera japonica*	Asia
Japanese knotweed, *Fallopia japonica*	Asia
Johnsongrass, *Sorghum halapense*	Africa
Kentucky bluegrass, *Poa pratensis*	Europe
kudzu, *Pueraria lobata*	Asia
melaleuca, *Melaleuca quinquenervis*	Australia
multiflora rose, *Rosa multiflora*	Asia
Norway maple, *Acer platanoides*	Europe
purple loosestrife, *Lythrum salicaria*	Eurasia
Scotch broom, *Cytisus scoparius*	Europe
stinging nettle, *Urtica dioica*	Europe
tamarisk, *Tamarix* spp.	Eurasia
Tatarian honeysuckle, *Lonicera tatarica*	Asia
water hyacinth, *Eichhornia crassipes*	South America
water starwort, *Callitriche stagnalis*	Europe
white clover, *Trifolium repens*	Europe
MARINE INVERTEBRATES	
European periwinkle, *Littorina littorea*	Europe
INSECTS	
European honeybee, *Apis mellifera*	Europe
MAMMALS	
feral hog, *Sus scrofa*	Eurasia
FISH	
common carp, *Cyprinus carpio*	Eurasia

seeds in this ballast helped the species to spread inland to the Great Lakes region. Cargo vessels plying the Great Lakes also aided dispersal of the plant. By 1900, in eastern North America, it had spread north to Newfoundland, west to Michigan, and south to North Carolina and Tennessee.

Still, purple loosestrife was only a marsh plant of minor significance. Most botanists would have considered it an attractive addition to the native wetland vegetation. In the 1930s, more than 130 years after it arrived in North America, purple loosestrife began its explosive invasion of wetlands all across North America. In the Northeast, the small initial colonies expanded into vast monocultures of loosestrife. Between the 1940s and the 1980s, the growth of irrigated farming and the construction of interstate highways, which both disturbed and created wetlands, led to its colonization of the interior West. By 1985, it was essentially continent-wide in distribution (Thompson et al. 1987) and was spreading at a rate of about 480,000 acres per year. Growing in dense, tall, monospecific stands, purple loosestrife choked out native wetland vegetation over thousands of acres, giving it the ecological nickname of "purple plague."

Exotics Reach the New World

Introductions of exotics into the New World began with Columbus's second voyage, in 1493, on which he brought seeds and cuttings of a variety of crop plants, probably a few weeds, and livestock animals to the island of Española, in the West Indies. Later Spanish colonizing expeditions brought still other crop plants, weeds, and domestic animals to the West Indies, Mexico, and Central America.

From Mexico, weeds and domestic animals spread north as Spanish settlement expanded. Some were probably introduced into the Rio Grande Valley of present-day New Mexico by Juan Oñate's colonization party in 1598, however, no records exist of the weed introductions associated with settlement of this interior region. We do know that many weeds reached Baja California and Alta California with the establishment of the Spanish missions in the 1600s and the 1700s, their presence recorded in the adobe bricks of early buildings (see Chapter 12).

In the 1600s, establishment of the English colonies brought rapid change to the eastern seaboard, as forests were cleared and farms established. To this disturbed landscape came a host of farm and garden weeds, some so early that their acquired common names imply that they are North American natives (Foy et al. 1983). Some, such as common dandelion and common plantain, were brought as garden plants and pot herbs.

Others, such as common mullein and common yarrow, were medicinal plants. Still others, including Kentucky bluegrass and white clover, were imported as pasture plants. Many early plant colonists, including Canada thistle, also came as contaminants in seed and livestock feed and bedding. The first imported livestock animals also brought with them weeds such as cockleburs, their seeds tangled in animal pelage.

The majority of these early immigrants have not invaded native ecosystems, although many are pernicious weeds of farmland, roadsides, and lawns. Some of them have become members of early successional communities of abandoned farmland and other severely disturbed sites. A few, such as dandelion, Kentucky bluegrass, white clover, and common yarrow, are now fully naturalized members of moist grasslands and meadows throughout North America. As Crosby (1986) noted, Kentucky bluegrass and white clover, their seeds apparently carried to western Kentucky by the pack animals of early explorers and traders, were already established when settlers from the East Coast reached this region.

During the 1700s, new groups of European settlers came to North America, bringing their own crops, horticultural plants, and weeds. The growing population led to increased commerce between the North American colonies and Europe. The result was that many weeds came in the rubble, soil, or sand used as ballast in the sailing vessels that carried cargo between Europe and North America (see Chapter 3). In 1815, for example, seventeen partially or fully ballasted ships sailed from Poole, England, to North America. On arriving in North American ports, the ballast was discarded into a dump area near the wharf. Dry ballast continued to be a major route of introductions through about 1880. In addition to purple loosestrife, the common reed and stinging nettle probably reached North America in this manner.

In the 1800s, several new modes of invasion and spread of weeds appeared. The commercial seed trade emerged, and a great variety of agricultural, ornamental, and medicinal plants were advertised in catalogs and distributed by mail or sold by local seed houses. The first seed catalog appeared in 1771, and by the early 1800s several catalogs listing hundreds of species had appeared. After about 1865, mail-order commerce increased greatly, because of inexpensive postal rates and efficient delivery systems. Many of the cataloged species had already become naturalized, so that this commerce served to assist their geographical spread.

For many of North America's worst exotics, the seed trade was probably the primary mode of introduction (Mack 1991). Crabgrass and Johnsongrass were imported as grain and forage crop plants, respectively. Both failed to become useful crop species and are now pernicious weeds.

Other highly invasive species sold by early seed companies include Japanese honeysuckle, kudzu, and bittersweet nightshade, three weedy vines now major pests in eastern forests. Two of Florida's most invasive trees, Brazilian pepper and Australian pine, were sold as ornamentals during the 1800s. Multiflora rose was imported as an ornamental shrub before 1811. Its widespread planting over a hundred years later led to its becoming seriously invasive (see Chapter 8). Scotch broom, gorse, English buckthorn, and tamarisk are other woody taxa that were imported to the East very early and have become extremely invasive in different parts of North America in the mid- and late 1900s. Several of the most troublesome aquatic weeds, including water hyacinth, were also distributed commercially. Overall, during the 1800s, at least 139 plant species that have become naturalized in the United States were distributed by one or more commercial seed houses.

Beginning in the late 1800s, organized federal efforts encouraged the importation of "useful" plants. The U.S. Department of Agriculture, established in 1862, had as one of its charges the naturalization of foreign plants. In 1887, the Office of Plant Introduction was created within the agency, specifically to establish gardens to acclimate and propagate imported species. This agency brought about 200,000 plant species and varieties into the United States (Sailer 1983). One of the most notorious importations was the melaleuca tree that has become a scourge of Florida wetlands (see Chapter 9).

Insects were slower to initiate colonization of North America, but they followed many of the same routes (Sailer 1983). Not surprisingly, one of the earliest introductions was the honeybee, which was brought to Virginia perhaps as early as 1622 and to Massachusetts by 1640. The arrival of exotic insects was relatively slow. By the American Revolution, no more than 6 species of harmful exotic insects were recognized in the new United States. In the last quarter of the 1700s, the number of exotic insects on the East Coast was probably no more than 28. Between 1800 and 1860, about 70 more species invaded. Between 1860 and 1910, however, some 475 new exotics were recorded.

Before 1820, over 90 percent of the insect immigrants were beetles, attesting to the importance of dry ballast—soil, sand, and rock—as the medium in which exotic insects were transported (Sailer 1983). The origin of many of these beetles can be traced to ports in southwest England, and their arrival points to Nova Scotia, Prince Edward Island, and the St. Lawrence River area. By 1840–1860, with increased transoceanic commerce in living plants, lepidoptera became the most numerous invaders, arriving as eggs or larvae on their plant hosts. During this era, little concern was given to the possibility that imported plants might carry unwanted insects. By 1900, with faster transport and more diverse plant cargos, about 40 percent of new exotics were homopterans—aphids, whiteflies, scales, and the like. Many of these insects entered on plants brought in by the U.S. Department of Agriculture and its Office of Plant Introduction. By 1920, about 194 species of exotic homopterans had reached North America, most of them proving to be serious agricultural pests.

Exotic marine invertebrates began to reach the east coast of North America at least as early as the mid-1700s. Many of these arrived as fouling organisms on the hulls of ships or in commercial shipments of oysters. Some of these species arrived so early, and have been introduced to so many other locations, that their identity as exotics is obscured. Since about 1880, water ballasting has been practiced by ocean cargo ships, and this has been the vehicle for transoceanic introductions of exotic marine organisms (see Chapters 5 and 6). For Chesapeake Bay, the result has been the introduction of at least 116 species (Ruiz et al. 1997).

In the early 1800s, the European periwinkle and the European green crab were introduced to the North American Atlantic coast, probably in oyster shipments. Periwinkles now range from Nova Scotia to Virginia, and European green crabs from Nova Scotia to New Jersey.

The exotics introduced to the eastern seaboard in colonial times, like the early human colonists of North America, were just the first wave of an invasion that has grown steadily in magnitude. Several of these early colonists, however, have become serious invaders of natural ecosystems and are now the subject of control efforts. Their influence, and prospects for their control, can be seen by examination of case histories of seven species: Kentucky bluegrass, purple loosestrife, Canada thistle, European honeybee, European periwinkle, common carp, and domestic swine.

Case Histories of Colonial Exotics

Kentucky Bluegrass

Native to a widespread area of Eurasia, Kentucky bluegrass was introduced to North America in the 1600s, probably before 1685, when William Penn planted some in his Philadelphia yard (Crosby 1986). A perennial, Kentucky bluegrass spreads mostly by rhizomes and is tolerant of repeated grazing or cutting. It also produces large quantities of seed apomictically. These characteristics made it popular as a lawn and pasture grass, and it has spread across North America, occurring now in every state and Canadian province and at all elevations below alpine tundra. Its close relative, Canada bluegrass, is also widespread in similar habitats. Both are cool-season grasses and are most abundant in the cool, moist regions of southern Canada and the northern United States.

Kentucky bluegrass, the familiar turf grass of lawns and pastures, has become a common member of tall- and mixed-grass prairies and montane meadows (Sather 1995a). It colonizes the spaces between bunchgrasses and competes directly with native cool-season grasses. Because it is tolerant of grazing, and because livestock tend to prefer both native cool-season and warm-season grasses, Kentucky bluegrass tends to increase at the expense of native plants in grazed areas. Continuous grazing of tall-grass prairie eventually converts the community into bluegrass sod, a new stable state in which Kentucky bluegrass remains dominant even if grazing is discontinued.

Kentucky bluegrass has become so firmly established in cool, moist grasslands across North America that its eradication is essentially impossible. In many locations, however, its abundance can be reduced greatly, and native species promoted, by the use of fire and seasonal grazing. Where a strong component of warm-season grasses exists in the flora, repeated spring burning or intensive spring grazing, combined with occasional burning, tends to reduce bluegrass and encourage warm-season grasses that grow and reproduce primarily in summer. In Kansas and Nebraska, for example, burning in spring for three consecutive years usually converts bluegrass-dominated grassland to grassland dominated by native warm-season grasses. Farther north, the effect of spring burning is weaker, and replacement of bluegrass by native grasses takes longer.

Purple Loosestrife

As noted at the chapter outset, purple loosestrife, a tall, slender perennial with attractive magenta flowers, has become a serious invader of areas of wet or periodically flooded soil. Mature plants reach a height of 6.5 feet and proliferate thirty to fifty shoots from a single rootstock, so that a single plant may become 5 feet in diameter. Mature plants may set 1,000 seed capsules per shoot and produce 2.7 million or so seeds annually. The seeds fall to the ground or water beneath plants and appear to be dispersed by the water movement and by any agent—birds, muskrats, livestock, humans, vehicles—that can carry seed-containing mud or plant debris from place to place. Germinating seeds or seedlings that become located on wet mud quickly put down roots, and the plant becomes firmly established within three to four weeks. Purple loosestrife is thus well adapted to freshwater marshes, stream and pond margins, and damp floodplains. Areas with cattails (*Typha* spp.) and other large, emergent wetland plants are typical sites for invasion of purple loosestrife.

Its translocation to North America freed it from a number of herbivore pests in its native areas, enabling it to become an aggressive invader of wet meadows and the borders of streams and lakes, especially where wetland vegetation had been disturbed. Purple loosestrife is a competitor of cattail species, invading marshes and gradually overwhelming the cattails. In New England marshes, rare or threatened species of bullrush and spikerush have also been displaced by purple loosestrife in some localities.

The replacement of cattails by purple loosestrife has the potential to severely impact marsh wildlife in many places, although many suggested effects are still largely speculative (Anderson 1995; Hager and McCoy 1998). Muskrats and marsh wrens, cattail marsh specialists, might be excluded completely by replacement of cattails by purple loosestrife. Nesting habitat of other marsh species, including the black tern and various species of waterfowl, might be destroyed by loosestrife invasion. Populations of mink, a muskrat predator, could also be reduced where loosestrife becomes dominant. The invasion of shallow, open marshes by purple loosestrife might also displace the bog turtle (*Clemmys muhlenbergii*), a declining reptile in the eastern United States.

Efforts to control purple loosestrife have met with only limited success. Most herbicides, although usually killing the present-year's shoots, do not kill the perennial root system or destroy the seed bank. Thus, resprouting and germination quickly reestablish the stand. Killing or weakening loosestrife plants by increasing the water level has been effective in a few locations, but this option is not always available. Plowing, disking, and seeding in competitor species can sometimes be effective if the wetland can be drained. Mowing or clipping has also been used, but these procedures create fragmented stems that may produce roots and regrow as vegetative propagules (Brown and Wickstrom 1997).

The best hope for success in controlling purple loosestrife is biological control. Its insect associates make it a likely candidate for biological control (Hight and Drea 1991). Some fourteen species of insects that feed exclusively on purple loosestrife in Europe have been identified. After extensive testing in Europe to determine if any threat existed to North American plants, five of these—two leaf beetles (*Galerucella calamariensis* and *G. pusilla*), two seed weevils (*Nanophyes marmoratus* and *N. brevis*), and one root weevil (*Hylobius transversovittatus*)—have been released in the United States and Canada (Mullin 1998). As of 1997, releases had been made in thirty states and six Canadian provinces. Preliminary indications are that these agents may cause major reductions in stand biomass of purple loosestrife. The fact that purple loosestrife is not an aggressive weed in Europe, where it is associated with a host of specific herbivorous insects, suggests that the potential for effective control by importing such insects to North America is great.

Canada Thistle

Canada thistle, a true thistle with purple or white flower heads, is one of the most troublesome cropland weeds in North America. Like Kentucky bluegrass, it is native to Eurasia, not Canada, but was an early immigrant to eastern North America, arriving in the late 1700s (Moore 1975). It quickly spread throughout southern Canada, the northern United States, and the Rocky Mountains south to New Mexico. Since it is a perennial and tends to have separate male and female plants (Donald 1994), one would think it poorly qualified to be a noxious weed. Propagating actively by rhizomes, however, in a year or two one plant can give rise to hundreds of offspring plants that form dense patches of shoots up to 6 feet or more in height. The female plants also produce large numbers of seeds; male plants, fewer seeds with poorer germination. Genetic variability within populations of Canada thistle is high, and several varieties with distinct morphology are recognized. Evolutionary resistance to herbicides also tends to develop quickly.

Canada thistle is still a serious weed of cropland across southern Canada and the northern United States, causing production losses counted in millions of dollars annually (Sather 1995b). It also invades native prairie, reducing native plant diversity. Disturbance of various sorts, including ungulate grazing, deposition of pocket gopher heaps, and soil erosion or deposition, favors establishment of Canada thistle by seed.

Controlling Canada thistle is difficult. Herbicides are not very effective because of the thistle's deep root and rhizome system. Mowing of infested areas several times during the summer for several years can eliminate the thistle by preventing its reproduction and exhausting its root reserves. Several potential biological control insects have been introduced to North American areas infested with Canada thistle (Louda and Masters 1993). So far, none appears to have exerted effective control.

European Honeybee

The honeybee is an example of the success that is frequently achieved by transporting an agricultural species to a new continent. Native to western Eurasia and Africa, the honeybee was domesticated in the Middle East several thousand years ago. A number of varieties exist in domestication. In North America, freed of various parasites and predators, domesticated honeybees became highly productive and were carried to new areas as settlement proceeded. They soon became feral and have spread through most native ecosystems of North America.

Until recently, the honeybee has been considered a thoroughly beneficial introduction. Several conservation biologists (e.g., Buchmann and Nabhan 1996; Kearns and Inouye 1997) have now suggested that honeybees have displaced native pollinators and have probably contributed to changed composition of the vegetation. Others have reviewed information on these suggested impacts (Huryn 1997) and found little solid evidence to support serious impacts of any sort, especially in North America.

The suggested disruptive effects of honeybees include competition with native pollinators, inadequate pollination of native plants, physical damage to flowers of native plants, interspecific transfers of pollen leading to hybridization, and encouragement of exotic plants by their effective pollination. Honeybees are unable to pollinate some New World plants that require the anthers to be vibrated, or "buzzed," with a particular frequency in order for pollen to be released. Pollination of these species is affected only by certain native bees and flies.

Several studies have shown that exploitation of nectar and pollen resources by honeybees partially displaces or reduces populations of native bees such as bumble and carpenter bees. In Arizona, for example, foraging by honeybees was concentrated on the most productive patches of the shindagger agave (*Agave schottii*), whereas bumble and carpenter bees were largely restricted to less productive patches (Schaffer et al. 1979). By manipulating nectar availability and honeybee abundance, the competitive basis of this segregation was confirmed (Schaffer et al. 1983). On Santa Cruz Island, off the coast of southern California, removal of feral honeybee colonies led to increased visitation of manzanita shrubs (*Arctostaphylos* sp.) by native bees (Wenner and Thorp 1994). Nevertheless, complete displacement of native pollinators from large regions or from any plant species on which they forage has not been documented.

On the other hand, honeybees in many instances pollinate exotic plant species, many of which come from Eurasia, as do honeybees themselves. Among the exotics pollinated effectively by honeybees are purple loosestrife and yellow star thistle (see Chapter 10). Whether honeybees are more effective pollinators than native insects would be in their absence is uncertain, however, and other characteristics of these pernicious weeds may be more significant to their success than pollination.

Feral populations of honeybees occur throughout temperate North America. Because of the economic importance of honeybees, no efforts have been made to control these populations, although some conservation biologists have suggested that they be removed from natural preserves. North American honeybees are now being attacked by an exotic parasite, the varroa mite, which is native to Asia. Colonies of feral bees throughout much of eastern North America have been severely reduced in numbers. Secondary impacts related to inadequate pollination of fruiting plants used by birds and mammals have been reported (e.g., Nickens 1996).

European Periwinkle

The European periwinkle is a common intertidal snail of European intertidal habitats ranging from marshes and eelgrass beds to rocky shores. The periwinkle appears to have been introduced near Pictou, on the northern coast of Nova Scotia, in the 1840s (Bertness 1984). From there it has spread south along the Atlantic coast of the United States, reaching Cape Cod in 1870 and New Jersey in 1890. It now occurs south to Delaware, Maryland, and Virginia. From Nova Scotia south to New York, it is the most abundant littoral herbivorous snail, especially on protected shores. On protected rocky shores, it commonly reaches densities of 400–800 animals per square yard throughout the intertidal zone.

The periwinkle has had a profound effect on intertidal ecology in its zone of high abundance. Bertness (1984) conducted experimental removals of periwinkles on a bare, rocky intertidal area in Rhode Island. After removals, he found that sediment accumulated rapidly, and invertebrates typical of soft sediments colonized the site. Eventually, with continued sediment accumulation, marsh grass (*Spartina alterniflora*) also colonized the site. Bertness concluded that a major effect of these snails has apparently been to strip away shallow sediments overlying hard intertidal substrates. By feeding on the shoots and rhizomes of marsh grass and bulldozing the shallow sediments, periwinkles have thus apparently eliminated extensive areas of shallow sediments and fringing salt marsh along the New England coast.

At least one native species has been largely displaced by the periwinkle. In North America, the mud snail (*Ilyanassa obsoleta*) occupied a range of habitats similar to those of the periwinkle from Nova Scotia to Florida. As the periwinkle has spread southward, the mud snail has been displaced from habitats with hard substrates and is now confined largely to mud- and sand flats. The mud snails actively avoid periwinkles and migrate out of areas where periwinkles reach densities of about twenty individuals per square meter. Mud snails prefer to lay their eggs on hard substrates, where they are usually grazed by periwinkles. Thus, displacement seems to result from active avoidance and egg predation. The displacement of mud snails may have significant secondary effects. Mud snails plow through soft sediments, disturbing them and reducing their suitability for polychaete worms. Aggregations of mud snails that result from their movement out of areas occupied by periwinkles may also lead to destruction of colonies of tube-dwelling amphipods. Thus, the invading periwinkle may have modifying effects on the intertidal community at large.

As for most exotic aquatic invertebrates, no strategy for control of this species has been proposed. The European periwinkle has likely become an abundant and permanent member of the intertidal fauna of the Atlantic Coast.

Common Carp

Common carp, a large, coarse, omnivorous fish belonging to the minnow family, were first introduced to North America in 1831, when they were privately imported and released into the Hudson River (Courtenay et al. 1984). For several decades, they were propagated only on a small scale in the eastern United States. In the 1870s, however, the techniques of large-scale propagation of fish in hatcheries, imported from Europe, took hold, and federal, state, and private hatcheries appeared quickly (Moyle 1986). Common carp became one of the first popular food and sport fish. Railroad cars designed to hold up to 20,000 carp were used to ship this novel species all over the United States. In Mexico, carp were first imported to Valley of Mexico in 1872, using stock brought from Haiti (Courtenay and Kohler 1986). In 1880, carp were introduced to Ontario, Canada, from the United States (Crossman 1984). By 1890, carp had become established in streams and lakes throughout North America (Moyle 1986).

Common carp are still regarded as a sport fish, but their popularity has declined greatly. Carp are now considered an undesirable species in many situations, especially in the midwestern United States. In many ponds and lakes their feeding and spawning activities have been linked to destruction of rooted aquatic vegetation and increased turbidity (Taylor et al. 1984; Moyle et al. 1986). In turn, the altered environmental conditions cause increased egg and larval mortality of various native fish, reducing their abundance. The loss of rooted aquatic vegetation and increased turbidity are especially detrimental to fish that hunt visually and prey on invertebrates.

Carp removals have been used successfully to restore native fish faunas. To remove carp, however, all fish must be killed by application of rotenone or other piscicides, and the desired fish restocked. Removing carp from waters with rare or endangered native species is thus very difficult.

Feral Hog

Domestic hogs were first introduced to North America by Hernando de Soto, who released swine in Florida in 1539 (Wood and Barrett 1979). They soon became a staple of Florida Indians, who hunted and sold them to European settlers. Swine were also brought to North America by English colonists in the 1600s, and they soon became an abundant, free-ranging animal in the south Atlantic region and in Florida.

Free-ranging hogs were a common exotic in eastern forests well into the twentieth century, and feral populations, in many cases genetically mixed with European wild boar introduced later, still occur in many places (Wood and Lynn 1977). Hogs have had a severe impact on forest ecology in many places (see Chapters 14 and 16). Hogs cause severe damage to many tree seedlings, particularly those of longleaf pine (*Pinus palustris*). They also feed heavily on acorns, beechnuts, and other mast. Consumption of mast by hogs was probably a contributing factor to the extinction of the passenger pigeon (Bucher 1992). A single animal has been estimated to consume about 530 pounds of acorns per month. Although the diet of hogs largely consists of plant material, the animals are omnivorous, and many forest-floor invertebrates and small vertebrates are consumed (Peine and Farmer 1990).

Efforts to control wild hog populations in Great Smoky Mountains National Park illustrate the difficulties involved in the eastern United States (Peine and Farmer 1990). Hog control by shooting and trapping has been carried out since 1960, but this technique has created hostility from local residents and animal rights groups on several occasions. From 1986 through 1989, over $1 million was spent to remove 1,327 individuals from a population estimated to lie between 1,000 and 2,000 animals. The cost of removals is thus high, and because of the high reproductive potential of the species, any relaxation of efforts allows the population to recover to a high density.

Ecological Time Bombs

European colonization of North America thus initiated the invasion of exotics. This invasion started slowly, as species deemed desirable were brought to North America, and as hitchhiking species jumped ship in the New World. These introductions were soon abetted by commercial activity and by governmental programs intended to enrich North America in useful plants and animals. For more than two centuries, no one recognized the evils that were escaping from the Pandora's box of Old World exotics.

Many of the introductions to the eastern seaboard in the 1700s and 1800s have proved to be ecological time bombs, species that have only recently begun to exert serious ecological impacts. This means that the fact that an early colonist has not appeared to be especially troublesome does not mean that it will not become a problem exotic. Purple loosestrife is a prime example. It remained a minor invader for 130 or more years after it had been introduced to North America.

Many other examples of ecological time bombs exist among exotics reaching North America during the colonial period. After latent periods of 100 years or more, many of the earliest introductions to the eastern seaboard have recently begun to exert serious ecological impacts. Tatarian honeysuckle, an ornamental shrub introduced in the 1700s, has now become a serious invader of forests in New England (Woods 1993). Norway maple, brought to Philadelphia in the 1700s, is also becoming a common invader of eastern deciduous forests (Webb and Kaunzinger 1993). Other species that were introduced in the 1800s, such as water starwort (Philbrick et al. 1998) and Japanese knotweed (Seiger 1997), may be on the verge of becoming serious invasives.

The legacy of the colonial period is still with us. The freedom that human colonists found in the New World thus went well beyond political and religious freedom. European settlement of North America, along with temperate areas of the Southern Hemisphere, promoted an ecological imperialism (Crosby 1986) under which European plants and animals were introduced freely to new continents. This imperialism quickly evolved into a policy of free biotic exchange, an idea that became institutionalized in North America as a basic principle of free trade. No one questioned the practices of introducing plants and animals from distant regions to North America and Hawaii until crises of agriculture and forestry began to emerge. Only in the past few decades have unrestricted practices that have introduced exotics to North America's wildlands and waters been questioned. North America has been treated as a biotic commons, and we are now beginning to appreciate the tragedy.

Chapter 5

West Coast Bays and Estuaries: Swamping the Natives

> [San Francisco] Bay has lost the distinctive faunal characteristics and the web of community relationships that it had developed since its post–ice age origin, and which distinguished it from the other great estuaries of the world.
>
> —Andrew N. Cohen and J. T. Carlton, *Nonindigenous Aquatic Species in a United States Estuary**

The European green crab appeared in southern San Francisco Bay in 1989 or 1990, probably as larvae in ballast water released by a cargo ship (Grosholz and Ruiz 1995) (Table 5.1). A small but voracious crab, less than 3 inches in carapace width, it is native to the Atlantic coast of Europe from Norway to Portugal. This crab is a classic European colonist, having also invaded coastal waters in South Africa, Australia and Tasmania, and eastern North America (Grosholz and Ruiz 1996). Since its appearance in San Francisco Bay, it has been spreading north and south along the California coast, apparently by dispersal of planktonic larvae. By 1993, it had invaded Bodega Bay, some 75 miles to the north; in 1994, it was collected at Elkhorn Slough in Monterey Bay. By 1997, its range extended from Coos Bay, Oregon, to Monterey Bay, California (Grosholz and Ruiz 1997). Because of patterns of ocean currents along the coast, its rate of spread is faster toward the north than toward the south. No obstacle appears likely, however, to limit its spread until it has colonized much of

---

*Andrew N. Cohen and J. T. Carlton (1995). *Nonindigenous Aquatic Species in a United States Estuary: A Case Study of the Biological Invasions of the San Francisco Bay and Delta*. U.S. Fish and Wildlife Service, Washington, DC.

Table 5.1. Scientific names and native regions of exotics that are serious invaders of bays and estuaries in western North America.

PLANTS	
European cordgrass, *Spartina anglica*	Europe
Japanese sea grass, *Zostera japonica*	Japan
salt-meadow cordgrass, *Spartina patens*	eastern North America
smooth cordgrass, *Spartina alterniflora*	eastern North America
South American cordgrass, *Spartina densiflora*	South America
MARINE INVERTEBRATES	
Asian clam, *Potamocorbula amurensis*	Asia
Asiatic clam, *Corbicula fluminea*	Asia
Asiatic sea star, *Asterias amurensis*	Asia
Atlantic mud snail, *Ilyanassa obsoleta*	eastern North America
Australian isopod, *Sphaeroma quoyanum*	Australia
Chinese mitten crab, *Eriocheir sinensis*	China
cnidarian, *Blackfordia virginica*	Black Sea
cnidarian, *Maeotias inexspectata*	Black Sea
copepod, *Limnoithona sinensis*	Asia
copepod, *Sinocalanus doerrii*	Asia
eastern oyster, *Crassostrea virginica*	Atlantic Coast
european green crab, *Carcinus maenas*	European Atlantic Coast
Japanese cockle, *Tapes japonica*	Asia
Japanese mussel, *Musculista senhousia*	Asia
macoma clam, *Macoma balthica*	Atlantic Ocean
Mediterranean mussel, *Mytilis galloprovincialis*	Mediterranean Sea
moon jelly, *Aurelia* sp.	Japan
New Zealand sea slug, *Philine auriformis*	New Zealand
oyster drill, *Urosalpinx cinerea*	eastern North America
Pacific oyster, *Crassostrea gigas*	Asia
ribbed mussel, *Geukensia demissa*	Atlantic Coast
softshell clam, *Mya arenaria*	eastern North America
FISH	
American shad, *Alosa sapidissima*	eastern North America
blue catfish, *Ictalurus furcatus*	eastern North America
chameleon goby, *Tridentiger trigonocephalus*	Asia
common carp, *Cyprinus carpio*	Asia
striped bass, *Morone saxatilis*	eastern North America
threadfin shad, *Dorosoma petenense*	eastern North America
yellowfin goby, *Acanthogobius flavimanus*	Asia

the West Coast. This crab has broad temperature and salinity tolerances, and it flourishes in estuaries and sheltered bays with shores ranging from rocky or cobbly to sandy or muddy.

As a generalist predator on marine invertebrates ranging from mollusks to polychaete worms and crustaceans, both native and introduced, the European green crab has the potential to influence the overall structure of intertidal animal communities. At Bodega Bay, California, exclosure experiments have already shown that the crab has reduced densities of several invertebrates of intertidal sand flats, including two species of small bivalves and two small crustaceans (Grosholz and Ruiz 1995). The native mud crab (*Hemigrapsus oregonensis*) is also declining at Bodega Bay, probably as a result of predation and competition from the European green crab (Grosholz and Ruiz 1997). Little doubt remains that the "top-down" influence of this predatory crab will restructure the invertebrate communities of Pacific coast bays. Based on observations of the short-term impacts of a population explosion of the native Dungeness crab (*Cancer magister*) in the 1980s, these impacts on invertebrates are even likely to reduce the populations of many shorebirds that winter along the Pacific coast and feed on invertebrates (Grosholz and Ruiz 1997).

Bays and coastal waters from Baja California, Mexico, to British Columbia, Canada, are experiencing invasions of all sorts of marine and estuarine organisms. Exotic sea grasses, salt-marsh cordgrasses, fish, and a host of marine invertebrates, including even protozoans, are the most serious invaders. For many taxa, the numbers of introduced species now exceed the number of natives. Four species of salt-marsh cordgrasses from three continents have become naturalized in coastal marshes where only a single species of similar life-form once existed. More than thirty species of exotic fish have invaded estuaries and their adjacent inland waters. Hundreds of exotic marine invertebrates are established, and at least one new species is added every year. Considering mollusks alone, thirty species have been introduced to the Pacific coast, fourteen from the North American Atlantic coast and sixteen from elsewhere (Carlton 1992b). Of these thirty species, twelve are widespread. Several of the invading plants and animals have major impacts on the physical environment of estuaries. The flood of exotics is also placing at risk a number of native species.

The routes of invasion are diverse. Development of oyster mariculture has played a major role. Eastern oysters from the Atlantic coast were introduced to various locations on the Pacific coast in 1869 and 1870, immediately following completion of the first transcontinental railroad, and in the 1930s (Nichols et al. 1986). The Pacific oyster was first introduced from Asia to the Pacific coast in 1902, with plantings continuing until the 1970s. Overall, the oyster industry accounts for about twenty of the exotic mollusks introduced to the Pacific coast (Carlton 1992b). Massive translocation of fish began in the late 1800s, as well. In 1873, for example, a shipment of 300,000 fish of ten species was sent to California by railroad for release into streams and lakes (Moyle 1986). Other species probably arrived as fouling organisms on or in the hulls of ships; their spawn resulted in colonization of hard surfaces in the bay. Still others, especially in recent decades, have come in ballast water of cargo ships. Most of these have arrived as planktonic larvae, and these forms are consequently introduced free of many of their natural enemies, facilitating their establishment and population growth (Lafferty and Kuris 1996).

These introductions amount to a massive uncontrolled experiment in ecology and evolution. They are almost unpredictable in nature and impact—"ecological roulette," as some ecologists term the introductions (Carlton and Geller 1993). The sort of new community that this novel assortment of exotic species will form is uncertain. Some of the exotics may even modify physical and chemical conditions of Pacific estuaries in unanticipated ways. Marine ecologists are seriously concerned about how this experiment will affect the rich and extraordinarily valuable marine biota of the Pacific coast.

San Francisco Bay and Delta

In 1996, a report prepared for the U.S. Fish and Wildlife Service characterized San Francisco Bay as "the most invaded aquatic ecosystem in North America." More than 234 nonindigenous species have been recorded in the estuary, which includes San Francisco Bay itself and the Sacramento–San Joaquin Delta to the east (Cohen and Carlton 1998). The rate of establishment is about one new exotic species per year. James T. Carlton, a marine ecologist, calls it an "accidental zoo," noting that the biota consists of a wide range of species that have been brought together suddenly in ecological and evolutionary time. Other marine ecologists (Nichols and Thompson 1985) have even recognized an "introduced mudflat community" composed entirely of introduced species.

Like Florida (see Chapter 9) and the Hawaiian Islands (see Chapter 13), the San Francisco Bay and Delta ecosystem may be particularly vulnerable to exotic invasions (Nichols et al. 1986). The estuary is a relatively young geographical feature, and the time for evolutionary and biogeographic processes to act has been short. The estuary's native aquatic biota is thus relatively impoverished. Low biotic diversity, and the fact that it is the most severely disturbed estuary system on the Pacific coast, make it a welcoming environment for new immigrants.

Introductions to the San Francisco Bay and Delta ecosystem have been both deliberate and inadvertent. Smooth cordgrass was introduced deliberately to San Francisco Bay in the mid-1970s (Daehler and Strong 1994). It has spread aggressively in the 1990s (Daehler and Strong 1996). More recently, European cordgrass from a site in Puget Sound and South American cordgrass from Humboldt Bay, California, were transplanted to San Francisco Bay as part of salt-marsh restoration efforts. As yet, European cordgrass has not spread widely. South American cordgrass, however, is spreading actively and is threatening to invade estuaries farther south in California (Daehler and Strong 1996).

Smooth cordgrass has established five major areas of infestation in southern San Francisco Bay. This species grows to a height of 6.5–8.0 feet, compared to about 20 inches for the native rough cordgrass (*Spartina foliosa*). Smooth cordgrass has a wider tidal range, growing at both higher and lower levels, than native cordgrass. It also invades stands of native cordgrass. Smooth cordgrass stands increase sediment deposition and lead to the buildup of a high mud bench. Furthermore, smooth and rough cordgrasses are apparently hybridizing, with the hybrids being more vigorous than rough cordgrass (Daehler and Anttila 1997). A new, aggressively tillering, dwarf ecotype of smooth cordgrass that can colonize upper intertidal mudflats has also appeared in San Francisco Bay (Daehler et al. 1999). The net effect of exotic cordgrass invasions is thus to expand the area of dense cordgrass marsh at the expense of open mudflats. In the long run, this reduces the area of foraging habitat for most species of shorebirds (Daehler and Strong 1996).

Fish were brought to the San Francisco Bay and Delta both deliberately and inadvertently. In all, thirty exotic fish have been introduced successfully to the bay and delta (Cohen and Carlton 1998). Another species, the yellow perch, was introduced in 1891 but became extinct in the 1950s (Herbold and Moyle 1989). In almost all cases, through the late 1960s, introductions were deliberate; two of the more recent introductions have been accidental. Most species were introduced with the goal of improving commercial and sport fisheries. The yellowfin and chameleon gobies, on the other hand, were probably introduced inadvertently in ballast water of cargo ships from Asia. The bay and delta now possess thirty species of exotic fish and only twenty-seven native species, with the total abundance of exotics also exceeding that of natives. Exotics dominate both the fresh and the brackish water portions of this ecosystem, whereas in the near-marine waters of the outer bay, only five of the fifty-seven species are exotics (Baltz 1991).

The dominant fish species of the bay and delta system is the striped bass, introduced from the Atlantic coast in 1871. The successful establishment of this species was probably aided by the hydraulic gold mining that was occurring in the Sacramento drainage. Striped bass have eggs that are tolerant of heavy silt levels in the water. The native salmon populations were intolerant of the stream changes caused by mining and were decimated during the mining years. The American shad, introduced in the 1870s or 1880s from the Atlantic coast, also has silt-tolerant eggs. Both striped bass and shad became important commercial fisheries species in the early 1900s. The shad has declined considerably since the construction of Shasta Dam in the mid-1940s because of reduced spring water flows required for spawning. Commercial fishing for striped bass ceased in 1935, but the species is still a major sport fish. Other exotic fish include threadfin shad, common carp, and a variety of minnows, shiners, mosquitofish, catfish, panfish, and gobies. The blue catfish, brought to southern California in 1969 as a sport fish, appeared in the delta in 1979, probably by unauthorized translocation from southern California.

Introduced fish have very likely contributed to the decline and even extinction of native fish in the San Francisco Bay and Delta system. At least two native fish, the thick-tailed chub (*Gila crassicauda*) and Sacramento perch (*Archoplites interruptus*), have become extinct in this system, and several others have declined considerably (Herbold and Moyle 1989). The delta smelt (*Hypomesus transpacificus*) is federally listed as an endangered species.

Most introduced species, however, have been marine invertebrates, and most of these introductions appear to have been inadvertent. In the late 1800s, when young eastern oysters were shipped by rail from the East Coast to San Francisco Bay and deposited in artificial reefs to mature, various Atlantic invertebrates were carried along with them (Nichols et al. 1986). Although the oyster itself never became naturalized, many invertebrates that traveled with it did become established in the bay. During the decade of the 1890s, when importation of eastern oysters reached its peak, four Atlantic coast mollusks commonly associated with oyster reefs, including the oyster drill, became established in San Francisco Bay. Oyster plantings also inadvertently introduced at least two parasitic invertebrates to San Francisco Bay (Bjergo et al. 1995).

Many other invertebrate species have arrived in San Francisco Bay in ballast water of cargo ships. Two of the dominant copepod crustaceans in the bay are exotics that probably arrived in ballast water: *Sinocalanus* appeared in 1978 and *Limnoithona* in 1979 (Herbold and Moyle 1989). The abundance of these two species, and the fact that *Sinocalanus* occurs farther inland into the freshwaters of the delta region, have largely displaced several native copepods. Grazing on phytoplankton by these exotic copepods has also reduced phytoplankton average abundance, as measured by the concentration of chlorophyll in the water, in the interior portions of the delta. Primary production in this portion of the delta has correspondingly declined. Two additional exotic Asian copepods have since invaded the bay, also in ballast water (Carlton 1996b).

Some of the most influential marine invertebrate colonists of San Francisco Bay have been clams and snails. The Atlantic mud snail, for example, has displaced a native mud snail (*Certhidea californica*) from most of its original habitat (Race 1982). An entire guild of exotic bivalves, including the Japanese mussel, ribbed mussel, Japanese cockle, softshell clam, macoma clam, Asiatic clam, and Asian clam, exists in the bay. The composition of this mollusk assemblage is continually changing as new species appear. Some of these species were introduced with oyster transplants; others have arrived in ballast water.

The Asiatic clam reached San Francisco Bay from other Pacific coast sites that it first colonized. It arrived in North America near Nanaimo, British Columbia, in the early 1920s. The Asiatic clam is a small species, reaching a maximum length of about 1.6–1.8 inches. It is hermaphroditic and incubates its young in its mantle chamber until they reach a tiny bivalve stage. It spread to the Columbia River in 1938 and invaded freshwaters of the San Francisco delta region in the 1940s and 1950s. Since then, it has invaded freshwaters throughout most of temperate North America (McMahon 1983; Morton 1997). It is now the most widely distributed bivalve in North America, occurring in thirty-five states from Washington and California to New Jersey and Florida.

Asiatic clams prefer well-oxygenated flowing waters, and there they may become enormously abundant, forming beds of thousands of animals per square yard. In the San Francisco delta region, Asiatic clams are most numerous in the Sacramento River and the interior portions of the delta. One striking estimate of their abundance was of about 110,000 clams per square yard on a sand bar of the Delta-Mendota Canal. In Suisun Bay, in the middle portion of the estuary system, young Asiatic clams appear in vast numbers in spring, when freshwater flows carry them downstream. They do not survive the high salinities that develop there later in the year, however. Although they are most notorious as fouling agents in intake pipes and cooling systems of power plants, these filter-feeding clams are competitors with native invertebrates for phytoplankton and particulate detritus in streams and lakes.

Even more spectacular has been the invasion of San Francisco Bay by the Asian clam (Carlton et al. 1990). Asian clams are even smaller than the Asiatic clam, and they reach a maximum length of about just over 1 inch. Asian clams also have planktonic larvae and thus probably arrived as larvae in ballast water of cargo ships from eastern Asia. The first adults were collected in October 1986, in Grizzly Bay. Within one year, the Asian clam had become the most abundant benthic invertebrate in the waters from San Pablo Bay east to the Sacramento–San Joaquin Delta, composing more than 95 percent of total invertebrate biomass (Thompson and Schemel 1991). In many places, the density of these clams exceeded 1,000 per square yard. Within two years, the species had invaded the entire estuary, and densities of more than 8,365 per square yard were recorded in places. The species occupies all types of subtidal and lower intertidal substrates and occurs in salinities ranging from near freshwater to near ocean water.

The Asian clam, as the most recent and now most abundant invader, is likely to influence the species composition of the bivalve assemblage throughout the bay and delta. As an efficient filter feeder on diatoms and other phytoplankton, the Asian clam will also influence the planktonic system of the estuary. From 1988 through 1990, for example, the summer phytoplankton biomass in northern areas of the bay was less than a tenth, and annual primary production less than a fifth, the values that prevailed prior to the clam's invasion (Cloern and Alpine 1991). In turn, the Asian clam is prey for species higher in the food chain, including some crabs and diving ducks.

A second group of highly influential exotics in San Francisco Bay comprises predatory marine invertebrates. The introduction and spread of the European green crab, described at the beginning of this chapter, is almost mirrored by several other predatory invertebrates. The Chinese mitten crab, an exotic established earlier on the Atlantic coast of North America, showed up in 1994. It may have reached San Francisco Bay as larvae in ballast water or on algae used for packing of bait worms shipped from New England. In Germany, where this species was introduced in the 1930s, its population exploded, with masses of crabs "migrating up the main rivers, piling up against dams, climbing spillways and swarming over the banks onto shore, sometimes wandering onto streets and entering houses" (Stevens 1996, p. B8). The New Zealand sea slug, a predatory gastropod mollusk, appeared in San Francisco Bay in 1993, and in 1994 in Bodega Bay. It feeds primarily on small bivalve mollusks, both native and introduced. Recently, as well, two species of small jellyfish, or cnidarians (Mills and Sommer 1995), and a new form of moon jelly (Greenberg et al. 1996) have also appeared in the bay. These species feed on zooplankton, and if major population explosions occur, as has occurred in the Black Sea, the impact on open-water food chains could be serious. Two other pelagic crustaceans, species of mysid shrimp that feed on zooplankton, appeared in the bay in 1992 (Modlin and Orsi 1997).

Other Pacific Coast Bays and Estuaries

Other estuaries up and down the West Coast have suffered invasions of similar nature, but involving fewer species. In Puget Sound, three exotic cordgrasses have invaded estuary areas: smooth cordgrass, European cordgrass, and salt-meadow cordgrass. In Willapa Bay, south of Puget Sound, smooth cordgrass, which invaded the bay in about 1911 or slightly earlier, had colonized 1,200 acres by 1990 and was spreading at a rate indicating that it could take over the entire 30,000 acres of intertidal habitat in another fifteen years. It might therefore eliminate native cordgrass and greatly reduce the area of open mudflats. It has also spread north to the Gray's Harbor estuary (Stiller and Denton 1995). Farther south, South American cordgrass has become established in Humboldt Bay, California, and is spreading vigorously by both vegetative reproduction and germination of seed (Kittleson and Boyd 1997). The climatic range it occupies in South America suggests that it will probably prove invasive in estuaries farther south.

Willapa Bay was also the first North American site for another exotic marine flowering plant, Japanese sea grass. Sea grasses, which occur in shallow marine and estuarine waters, influence many features of intertidal ecology by stabilizing the sedimentary substrate, providing attachment sites for small invertebrates, creating shelter from predators for small invertebrates and fish, and furnishing detrital and dissolved organic foods for other organisms. Japanese sea grass, first noted in 1957, probably arrived in shipments of Pacific oysters (Posey 1988). It has spread rapidly and now occurs from southern British Columbia to Oregon. In Asia, it occurs as far south as Vietnam, and thus it is likely to find favorable habitats much farther south on the North American coast.

Japanese sea grass occupies a mid-intertidal range, higher than that of native sea grasses. Invasion of bay areas by Japanese sea grass thus appears to have the general effect of enabling the normal invertebrate community associated with native sea grasses to reach higher intertidal levels (Posey 1988). Many mid-intertidal invertebrates present before the invasion increased in density, but some species decreased. Ultimately, the main effect of this exotic sea grass will probably be to favor species associated with native sea grass beds and disfavor those, including perhaps some shorebirds, that are associated with open mudflats or sand flats.

Coos Bay, Oregon, has been another center of exotic invasions. In the South Slough National Estuarine Research Reserve in Coos Bay, for example, at least thirty-two introduced marine plants and animals exist (Carlton 1989). Two fish, the striped bass and the American shad, and one mollusk, the softshell clam, were deliberately introduced. The remaining species, ranging from marine algae and Japanese eelgrass to marine invertebrates of six phyla, were introduced through oyster mariculture or as fouling organisms or ballast water stowaways on ocean vessels. The potential for additional introductions in ballast water is emphasized by the fact that 367 taxa of marine organisms have been identified in the ballast water of ships arriving in Coos Bay (Carlton and Geller 1993). Sampling of ballast water from ships arriving from Japan between 1987 and 1991, for example, revealed thirty-three species of ciliate protozoans known as tintinnids, some of which were not known to be from eastern Pacific waters (Pierce et al. 1997). These ciliates are so abundant in ballast water that many of the known North American species are probably cryptogenic exotics that arrived before marine biologists had examined the native fauna of the Pacific Coast.

Farther south still, San Diego Bay is another heavily invaded system (Crooks 1997). At least fifty-eight exotic marine plants and animals have been reported from the bay proper and nearby bays and lagoons. These include sixteen species of crustaceans, thirteen tunicates, eight mollusks, and eight polychaete worms. The rate of invasion is increasing, with 43 percent of all exotics having been discovered since 1970.

Two species that have invaded San Diego Bay illustrate ways in which exotics are influencing the physical environment of Pacific estuaries (Crooks 1997). The Australian isopod reached San Francisco Bay in the mid-1800s, probably in the fouling communities on the hulls of ships that brought miners to the California gold rush. It has spread to many other bays and lagoons along the Pacific coast. It appeared in San Diego Bay in the 1920s. The isopods burrow in the vertical mud banks of tidal channels in enormous densities, often creating more than 8,365 holes per square yard. The riddled banks tend to collapse, undercutting the vegetated surface of the salt marsh, which eventually slumps into the channel. The result is widening of the channels and loss of salt marsh that is home to numerous plants, invertebrates, birds, and mammals of the upper intertidal zone. In the San Diego area, measured widening of channels equals about half a yard in eighteen months. Similar loss of marsh habitat has been observed in other California estuaries, particularly Elkhorn Slough in Monterey Bay.

A second exotic with a powerful effect on the intertidal environment in the San Diego area is the Japanese mussel. This mussel, with a life span of only about two years, is a fast-growing form that reaches a maximum size of about 1.3 inches (Crooks 1996). This species first appeared in Puget Sound, where it was probably introduced in the 1920s with Japanese oyster (*Crassostrea gigas*) stock (Crooks 1998). From there, it spread to San Francisco Bay in the 1940s and reached southern California by the 1960s. For the next quarter century, its abundance fluctuated considerably, but in the late 1980s its population exploded. In 1995, intertidal population densities reached 22,587 individuals per square yard, and shallow subtidal beds averaged an astonishing 148,906 individuals per square yard (Crooks 1997). The densest populations tend to occur in low-diversity areas of central Mission Bay, suggesting that the species is an opportunistic colonizer of disturbed sites (Dexter and Crooks, submitted). These populations, in which the individual mussels and other objects are tied together by byssal threads, transform open mudflats into dense, thick mats of mussels and many other marine invertebrates, much like the mussel beds that occur on rocky substrates of the open coast. These mats effectively stabilize the surface of mudflats. Although such beds are biotically rich, they may inhibit the establishment and growth of eelgrasses, especially where eelgrass beds are patchy or stressed (Reusch and Williams 1998). Eelgrass beds support a distinctive marine invertebrate fauna.

These San Diego Bay examples demonstrate that some exotics exert their effects because they act as "ecosystem engineers," species that either destroy or create physical environments that influence many other species (Jeff Crooks, personal communication). One can only imagine what extensive ecosystem engineering another San Diego area exotic, black mangrove (*Avicennia germinans*), would have accomplished had an incipient population not been destroyed. Open mudflats and stands of cordgrass might have been replaced by a dense jungle of shrubby mangroves to which native marsh and estuary animals were ill adapted.

General Impacts of Exotics

How little we know about the true status of many of the common marine organisms in Pacific coast estuaries is highlighted by the case of the bay mussel. Recent electrophoretic analyses have revealed that the bay mussel of the southern Pacific coast, once considered to be the European bay mussel (*Mytilus edulis*), is really the Mediterranean mussel (McDonald and Koehn 1988). Bay mussels from the northern Pacific coast (*Mytilus trossulus*) are related to those of the Baltic Sea and eastern Canada. The Mediterranean mussel is clearly an exotic, and it was probably introduced to southern California in the early 1900s, perhaps to San Diego Bay, where "*Mytilus edulis diegensis*" was collected and described in 1907. The bay mussel of the northern Pacific coast, *Mytilus trossulus*, was described in 1850 from specimens collected in Oregon. It is considered to be a native North American species, but it may well be a cryptogenic exotic. When it invaded, the Mediterranean mussel probably displaced *Mytilus trossulus* from the southern Pacific coast (Heath et al. 1997). Not surprisingly, mussel invasions are still occurring, as genetic analyses of populations along Vancouver Island, Canada, reveal nonindigenous forms and hybrids between these forms and *Mytilus trossulus* (Heath et al. 1997).

The influence of exotic marine plants and animals may have contributed to the extinction of some species, however, as yet, this loss of species is probably minor. Exotics have changed the species composition of most of the original estuarine communities of San Francisco Bay, however, and have begun to alter the physical habitats of estuaries and salt marshes in many other Pacific coast bays. Biodiversity at the level of communities and ecosystems has thus been reduced (Carlton 1996b).

Protecting Bay and Estuary Ecosystems

What can be done about this massive transformation of coastal ecosystems by exotics? The potential for biological control of exotic marine and estuarine species has been little explored. One reason that few efforts have been made is that many marine and estuarine exotics, as well as potential the control agents, have larval stages that are distributed widely by ocean currents (Lafferty and Kuris 1996). Local populations of a problem species thus would not necessarily trigger a local response by a control agent. An effective control relationship would have to involve interaction of the problem species and control agent throughout the oceanic region involved.

Nevertheless, for species such as the European green crab, relatively specific parasites do exist (Lafferty and Kuris 1996), suggesting that a classical biological control relationship might be possible. Another intriguing possibility is the promotion, or even subsidization, of a fishery aimed at the green crab itself. Exploitation of this species might help keep its density at a reduced level. Only a few decades ago, for example, various sea urchins were considered to be pests of kelp, a marine plant with a valuable commercial harvest. Now, a multimillion-dollar fishery exists for these same sea urchins, which are exported to Japan, and the sea urchin harvest itself needs to be regulated.

Future invasions are almost inevitable, given the difficulty of controlling dispersal of marine organisms by ocean shipping and the intense efforts being made to expand mariculture of varied species, many of them exotics (Carlton 1992b). Another predatory invertebrate, the Asiatic sea star, for example, has invaded Tasmania, Australia, and appears likely to make its way eventually to the Pacific coast of North America, most likely as a ballast water immigrant (Lafferty and Kuris 1996). Like a number of other starfish, the Asiatic sea star might exert a "keystone" impact, in which its predation almost completely changes the composition of the community of native intertidal invertebrates.

What is happening to Pacific coast estuaries is also happening to a significant degree to estuaries of the Atlantic and Gulf coasts of North America. As discussed in Chapter 6, ballast water discharges are also introducing many species of exotic invertebrates and fish to the Great Lakes. These invasions are increasing diversity, but at the expense of productivity in the short term, and possibly of native biodiversity in the long run. The lessons evident from the Pacific coast are relevant to many other aquatic ecosystems in North America.

Perhaps it is impossible to prevent all invasions of North American estuaries. Much can be done, however, to reduce the frequency of such invasions. Inventing an effective technology of ballast-water management is absolutely critical. Equally critical is formulation of an oversight plan to assure that ships utilize this technology. Maricultural activities must also be monitored thoroughly to assure that unwanted exotics are not introduced inadvertently. Ways of achieving these goals are discussed in more detail in Chapter 21. San Francisco Bay and several other Pacific coast estuaries are well on the road to showing the ecological and economic price that will be paid for failing to halt these estuarine invasions.

Chapter 6

Northern Temperate Lakes: Chaos along the Food Chain

> The ecological and economic impacts of exotic species in the Great Lakes have been and will continue to be enormous. Even though only a fraction of the species introduced into the Great Lakes have caused documented impacts, all the species have contributed to the artificial character of the Great Lakes. . . . Introduced species exist at almost every level of the food chain of the Great Lakes and changes brought about by these organisms have damaged the natural community of the Great Lakes.
>
> —E. L. Mills et al., "Exotic Species in the Great Lakes"*

Lake Erie illustrates the long history of human impacts on North American lakes. The first major change occurred as the lake's watershed was deforested following settlement. In northwestern Ohio, the Great Black Swamp was drained between 1854 and 1900. This area was a vast, forested bottomland stretching along the Maumee River from Indiana to the southwestern corner of Lake Erie. Drainage of this swampland flushed enormous quantities of organic silt into western Lake Erie. Deoxygenation, turbidity, and eutrophication of the shallow west end of the lake resulted (Egerton 1987). As human populations increased along the lake, fishing intensified. Commercial fishing led to decline of the lake sturgeon before 1900 and of lake trout at the turn of the century. In the early decades of the 1900s, still other commercial fish declined because of heavy

*E. L. Mills, J. H. Leach, J. T. Carlton, and C. L. Secor (1993). "Exotic Species in the Great Lakes: A History of Biotic Crises and Anthropogenic Introductions." *Journal of Great Lakes Research* 19:1–54.

exploitation. The growth of cities, industries, and farms within the watershed of the lake led to pollution of the lake by sewage nutrients, fertilizers, and toxic chemicals. General eutrophication of the lake, combined with toxic chemical pollution, peaked in the 1960s, creating symptoms so severe that many viewed the lake as "dying." Indeed, at least three important fish became extinct.

Concern in the United States and Canada led to formation of the International Joint Commission, which has overseen efforts to restore the health of Lake Erie and the other Great Lakes. Over $7.5 billion has been spent on pollution abatement around Lake Erie (Makarewicz and Bertram 1991). By the 1980s, dramatic improvements in water quality, oxygen availability, and biotic function were evident. The walleye (*Stizostedion vitreum*) fishery, for example, rebounded as a major sport fish, recovering from its near demise in 1972.

Lake Erie, however, can never be restored to its original biotic condition. Its biota, from phytoplankton and higher aquatic plants to fish at the highest levels of the food chain, abounds in exotics. Exotic zebra and quagga mussels (Table 6.1) cover much of the bottom. Exotic phyto-

Table 6.1. Scientific names and native regions of invasive exotics of the Great Lakes and other northern freshwater lakes.

PLANTS	
Eurasian water milfoil, *Myriophyllum spicatum*	Eurasia
FRESHWATER INVERTEBRATES	
amphipod, *Echinogammarus ischnus*	Europe
antipodean snail, *Potamopyrgus antipodium*	New Zealand
quagga mussel, *Dreissena bugensis*	Russia
spiny cladoceran, *Cercopagis pengoi*	Eurasia
spiny water flea, *Bythotrephes cederstroemi*	Eurasia
water flea, *Eubosmina coregoni*	Eurasia
water flea, *Bosmina maritima*	Eurasia
zebra mussel, *Dreissena polymorpha*	Russia
FISH	
alewife, *Alosa pseudoharengus*	Atlantic Ocean
common carp, *Cyprinus carpio*	Eurasia
Eurasian ruffe, *Gymnocephalus cernuus*	Europe
rainbow smelt, *Osmerus mordax*	Atlantic coast
round goby, *Neogobius melanostomus*	Black and Caspian Seas
sea lamprey, *Petromyzon marinus*	Atlantic Ocean
tubenose goby, *Proterorhinus marmoratus*	Black and Caspian Seas
white perch, *Morone americana*	Atlantic Ocean

plankton and zooplankton populate the open waters. Alewives, rainbow smelt, and white perch dominate intermediate levels of the food chain. Introduced salmon vie with the walleye for top carnivore dominance. Sea lampreys prey on the larger fish. The food web of Lake Erie is an almost haphazard assemblage of species from the Old and New Worlds, continually and chaotically changing.

Human Impacts on the North American Great Lakes

Lake Erie and the other Great Lakes stand as prime examples of the ecological upheavals caused by the activities of humans. These lakes, the world's largest set of interconnected freshwater lakes, are beautiful, biotically rich, and economically valuable. They originally held one of the temperate zone's richest fish faunas. This fauna of about 125 species was dominated by salmonids: Atlantic salmon, lake trout, and a variety of whitefish and ciscos—all fish of major value for sport or commercial harvest. All of the lakes have experienced a succession of severe impacts following European settlement. They are now the epicenter of a new ecological cataclysm that is affecting lakes and other freshwater ecosystems throughout temperate North America: a blitzkrieg of exotics. Mills et al. (1993b) documented 139 exotics in the Great Lakes, and more have appeared since.

Human activities have disturbed all sectors of the biotic food web that leads from aquatic producers to salmonid fish, and even to their predators (Scavia and Fahnenstiel 1988). Some actions have affected the highest levels in the food web, with impacts that have cascaded downward. Others have influenced the lower levels of the food web, creating impacts that climb upward through the food web in complex fashion. The most basic patterns of nutrient cycling and energy flow have been affected, and extinction of many ecosystem members has resulted.

History of Exotic Invasions

While efforts in the 1960s and 1970s to restore the Great Lakes from the impacts of deforestation, overfishing, eutrophication, and toxic chemical pollution were making progress, the invasion of exotic species was increasing in magnitude. The stage for this invasion was set much earlier, however, with the construction of the Erie and Welland Canals in the 1800s. The Erie Canal, constructed between 1819 and 1825, connected the Hudson River with Lakes Ontario and Erie. The Welland Canal, which connected Lakes Erie and Ontario by bypassing Niagara Falls, was completed in 1831. The Welland Canal, with a series of forty locks, enabled both canal boats and schooners to enter the upper Great Lakes. The passage required a day or more to negotiate the 326-foot change in elevation between the lakes. In 1845 and 1887, the Welland Canal was reengineered, and the number of locks was reduced. In 1932, the canal was almost completely rebuilt, the locks were reduced to seven, and their size increased to accommodate oceangoing vessels up to 730 feet long. This renovation set the stage for the development of the St. Lawrence Seaway, which brought large-scale ocean cargo vessels into the heart of the Great Lakes.

By making it possible for canal boats and seagoing ships to move between the Hudson and St. Lawrence Rivers into the Great Lakes, the Erie and Welland Canals facilitated the entry of exotic organisms into the Great Lakes. Many of the early vessels using these canals used sand, soil, or rock as ballast. Ballast dumping introduced a variety of aquatic and wetland plants to the shores of the lakes. Animals used these canals to enter the lakes, as well. A snail (*Elimia virginica*) native to the Hudson River was perhaps the first exotic animal invader of the Great Lakes, entering in the mid-1800s. More ominously, the sea lamprey entered Lake Ontario in the 1830s through the Erie Canal, and it appeared in Lake Erie in 1921 (Mills et al. 1993b). The first lampreys were caught in Lake Michigan in 1936, and by 1946 lampreys had colonized all the upper lakes (Smith and Tibbles 1980). The alewife probably entered Lake Ontario via the Erie Canal in the 1860s (Smith 1970), passed through the Welland Canal into Lake Erie in 1931, and spread through the upper lakes between 1933 and 1954 (Smith 1970). The white perch likely entered Lake Ontario between 1946 and 1948 (Scott and Christie 1963) through the Hudson-Mohawk Canal. It had reached Lake Erie by 1953, either through the Erie Canal or the Welland Canal. It reached Lake St. Clair in 1977, Lake Huron in the early 1980s (Boileau 1985), and Lakes Michigan and Superior by the 1990s (Mills et al. 1993b).

Opening of the St. Lawrence Seaway in 1959 greatly increased the rate of invasion of the Great Lakes by exotics. Nearly 30 percent of the exotics recorded in the Great Lakes have appeared since 1959, the bulk of these being brought to the lakes by transoceanic cargo ships that have carried fresh or brackish ballast water.

Exotics gained early entry to the Great Lakes in other ways, too. The rainbow smelt, escaping from a reservoir that drained into Lake Michigan, appeared in Lake Michigan in 1923, Lake Huron in 1925, Lake Superior in 1930, and Lakes Erie and Ontario between 1931 and 1935 (Christie 1974). Other species, many of them native to freshwaters elsewhere in North America, were introduced deliberately. Common carp were stocked in the lakes beginning in the 1870s. The use and the discard of exotic baitfish have led to the introduction of other fish. Aquarium dumping has introduced several aquatic plants and at least one snail into the Great Lakes. Almost all northern lakes have also been the object of deliberate biotic tinkering with exotics, usually in the name of sport fishery improvement. This aspect of ecosystem disruption is examined in detail in Chapter 14.

Impacts of Exotic Fish

Of the exotic fish introductions to the Great Lakes, that of the sea lamprey has probably had the greatest ecological and economic impact. In Lake Ontario, overfishing, deforestation of the watershed, and predation by the sea lamprey contributed to the extinction of two salmonids, the Atlantic salmon, and lake trout before 1900. In Lake Erie, the lamprey remained a minor problem, largely because of the shortage of cold-water, gravel-and-sand tributary streams that are required by lampreys for breeding. In the upper lakes—Huron, Michigan, Superior—suitable breeding streams are numerous. Once established there, sea lampreys decimated populations of all large native fish (Christie 1974). Lake trout were one of the first species affected. The declines in lake trout catches were rapid and catastrophic. In Lake Huron, lake trout catches declined 97 percent between 1938 and 1954, and in Lake Michigan over 99 percent between 1944 and 1953. In Lake Superior, the decline was 89 percent between 1950 and 1960. Lake whitefish, burbot, and walleye populations were also greatly reduced. The declines of these large species cascaded down the food chain. The smaller species of whitefish, seven species known as ciscos, increased in numbers as their predators declined. Eventually, though, lampreys shifted to them as the larger species became scarce. Even in the face of lamprey predation, the smallest cisco, a species known as the bloater, increased greatly in abundance for a while. With the appearance of the alewife, however, it and other smaller, native, open-water species declined.

Beginning in the early 1950s, major efforts have been made to control the sea lamprey (Dahl and McDonald 1980). Lampreys largely spawn in tributary streams to the lakes themselves, and the young live for a period of several years in these streams. Thus, efforts have focused on eliminating access to spawning streams with dams or electric barriers and on killing larval lamprey by the use of a chemical lampricide, TFM (3-trifluoromethyl-4-nitrophenol). Beginning in 1956, these efforts, carried out by the joint U.S./Canada Great Lakes Fishery Commission, led to substantial recovery of salmonid fish in the 1970s and 1980s. Fisheries ecologists recognize that the lamprey can never by eradicated, and continuing control will be necessary to maintain salmonid numbers (Stewart et al. 1981). Recent cuts in the budget of this commission, however, have led to cutbacks in control effort. This, together with the fact that many lampreys spawn in waters at the mouth of tributary streams, where control practices are not very effective, has enabled lamprey populations to rebound quickly. In northern Lake Huron, in particular, lamprey control is now largely ineffective.

The alewife and rainbow smelt, invading on the heels of the decimating effects of the sea lamprey, filled the niches of native zooplankton feeders such as the lake herring and bloater, especially in Lakes Michigan and Huron. The success of these species was also aided by the absence of large salmonid predators. The surging abundance of the alewife between 1954 and 1966 reduced populations of the largest water flea and copepod species, thus allowing increases in populations of medium-sized plankton species (Wells 1970). During this period, many other native fish declined, possibly as a result of competition for plankton food from the alewife and rainbow smelt (Stewart et al. 1981). For example, these declines included yellow perch and emerald shiners, which had been little affected by sea lampreys. Even when sea lamprey populations were reduced by control efforts, the alewife and rainbow smelt tended to inhibit recovery of the native zooplankton feeders.

In 1967, a massive alewife die-off enabled some recovery of major plankton species, although this was partly offset by an increase in rainbow smelt (Jude and Tesar 1985). With success in sea lamprey control, stocking of several salmonids began in the late 1960s and continued throughout the 1980s (Stewart et al. 1981). These species included native lake trout but also exotic coho and chinook salmon and brown and rainbow trout.

The white perch, an estuarine fish of the North American Atlantic coast, has also influenced the composition of the fish community in the lower Great Lakes. The white perch is a predator on yellow perch, bluegills, other small fish, and the eggs of other fish, including walleye (Boileau 1985). In turn, they are preyed on by larger piscivores, including walleye and northern pike. White perch populations tend to grow rapidly to overpopulation level, with resultant stunting of individuals. Such populations can competitively inhibit those of yellow perch and other panfish.

New Invaders

New species of exotic fish are continually challenging the Great Lakes ecosystem. The Eurasian ruffe was probably introduced into the Duluth,

Minnesota, harbor at the western end of Lake Superior in 1985 (McLean et al. 1992). Almost certainly, this introduction was by discharge of ballast water from a transatlantic cargo ship. A small, slimy, rapidly maturing, perch-like fish that reaches a maximum length of about 8 inches, the ruffe nevertheless has an enormous reproductive capacity. It reaches reproductive maturity in two to three years, and it produces tens to hundreds of thousands of eggs annually (Ogle 1998). In Lake Superior, larger ruffe may spawn up to three times in a given year. As a result, the ruffe is increasing in numbers and spreading rapidly. By 1994, it had become the most abundant fish in the lower St. Louis River, which flows into Duluth Harbor, and had spread east to Thunder Bay, Ontario, possibly in ballast water of a Great Lakes ship. Recently, it has been found in the northwestern part of Lake Huron (Gunderson et al. 1998). Its success in Lake Superior, the most oligotrophic of the Great Lakes, bodes ill for the rest of the lakes. In Europe, its abundance tends to be greatest, compared to species such as perch, in more eutrophic lakes (Ogle 1998). Thus, it may do even better in the lower Great Lakes than in Lake Superior. It is also likely to colonize many other freshwater lakes and streams.

The danger posed to the Great Lakes ecosystem by the ruffe is still uncertain. The ruffe feeds largely on small benthic invertebrates, but it is highly opportunistic and sometimes consumes fish eggs. In Europe, it competes with the European perch (*Perca fluviatilis*) and is therefore a possible food competitor for North American yellow perch (*Perca flavescens*) and other benthic fish (Ogle 1998). It has formidable dorsal spines, and it is apparently avoided by larger predatory fish such as walleye and northern pike. Thus, by displacing the food species of these top carnivores, it could cause major disruption of existing food chains. If the ruffe spread throughout the Great Lakes, the damage to fisheries could exceed $100 million annually (Busiahn 1997).

Still more recent fish invaders are the round and tubenose gobies, which appeared in the St. Clair River, between Lake Huron and Lake St. Clair, in 1990 (Jude et al. 1992). Again, these fish apparently were introduced by ballast water discharge of ships coming from the Black and Caspian Seas region of Eurasia. Both are big-headed, bottom-dwelling, voracious predators on small invertebrates, fish, and fish eggs. They also exclude native fish from spawning reefs. By 1994, gobies had apparently decimated populations of the mottled sculpin (*Cottus bairdi*) and reduced those of the logperch (*Percina caprodes*) in the St. Clair River. By 1995, the round goby had appeared in Lakes Superior, Michigan, Huron, and Erie. So far, the tubenose goby has not spread as rapidly as the round goby. By 1996, however, it may have invaded Lake Ontario (Charlebois et al. 1997).

The round goby now appears to be well established in the Great Lakes. How it will impact food chains is still uncertain. Primarily a benthic feeder, the round goby has a very well-developed lateral line sense organ that enables it to detect prey in turbid or dark water. Round gobies are therefore able to feed at night. They feed on crustaceans, mollusks, insect larvae, small fish, and the eggs and fry of native fish such as darters, logperch, and sculpins. They are reported to prey heavily on zebra mussels (Ray and Corkum 1997)—a possible ecological benefit—and, in turn, are preyed on by smallmouth bass, walleye, rock bass, yellow perch, and other large predatory fish (Charlebois et al. 1997).

Zebra and Quagga Mussels

The Great Lakes are beset not only by exotic fish, but also by exotic invertebrates that occupy positions lower in the food web leading to fish. In the benthic environment, the worst of these exotics, at the moment, are zebra and quagga mussels. Zebra mussels are native to brackish and freshwaters from western Europe to southern Russia and the Ukraine (Ludyanskiy et al. 1993). The first zebra mussels were collected in 1988 in western Lake Erie and Lake St. Clair. They were probably introduced by ballast-water discharge from an ocean cargo vessel in 1985 or 1986. By 1990, however, zebra mussels had colonized all of the Great Lakes and the St. Lawrence River. By 1991, they had invaded the Hudson, Mohawk, Susquehanna, Ohio, Illinois, and Mississippi Rivers. Zebra mussels have shown the most explosive population growth and rapid range expansion of any exotic in North America. By A.D. 2000, they will likely have colonized rivers, lakes, ponds, and estuaries throughout much of the United States and southern Canada (Strayer 1991). Whether they can flourish in warm waters of the United States South and Southwest, where temperatures of 84–88 degrees Fahrenheit occur in summer, is uncertain. Their tolerance of salinity suggests that they will invade estuaries to the point at which salinity exceeds two to three parts per thousand.

Individually, zebra mussels are small, most adults reaching only 1.0–1.4 inches in length. In North American waters, some reach sexual maturity in their first year, and the rest in their second year. Mature female mussels spawn batches of a million or more eggs several times a year. Unlike North American freshwater mussels, which have larval stages parasitic on gills of fish, zebra mussels have free-living planktonic larvae that can be transported widely by flowing water and many casual means. Initial dispersal into the various portions of the Great Lakes system was probably effected by ballast water transport by cargo ships (Johnson and Padilla 1996).

Within large lakes, and in streams flowing from them, dispersal occurs by currents that carry the planktonic larvae. Colonization of isolated lakes appears to result primarily from recreational boating activities, such as the transport of larvae in bait buckets and of adults on boat hulls or on aquatic vegetation entangled with propellers.

A second mussel, christened the "quagga" mussel (the quagga is an extinct form of the zebra that lived in South Africa), has also invaded North American freshwaters (Mills et al. 1993a). It is closely related to the zebra mussel, and it is native to the Dnieper and Bug Rivers of southern Russia. The first quagga mussels were collected in western Lake Erie in 1989, but they probably reached Lake Erie in 1987 in ballast water of cargo ships. By 1991, quaggas had spread into Lake Ontario and the St. Lawrence River. Quagga mussels tend to become slightly larger than zebra mussels, but the two are very similar. Zebra and quagga mussels may hybridize, perhaps increasing the genetic variability in already highly variable forms.

Zebra mussels reach enormous densities, up to 627,400 per square yard or more, completely covering submerged hard substrates (Schloesser and Nalepa 1996). Prior to invasion of lakes by these mussels, large attached invertebrates of this sort were essentially absent from North American freshwaters. They also attach to submerged plants. Their attachment to boat hulls and to plants that become entwined with boat propellers is a major factor in the dispersal of mussels between both connected and unconnected bodies of water. They have become serious industrial fouling organisms, because of their tendency to clog water intake and discharge pipes. Not only are adult mussels enormously abundant, but their larvae form a major element of lake zooplankton. In reservoirs, densities of nearly half a million veligers per cubic yard have been reported.

Quagga mussels complement the distribution of zebra mussels. Quaggas tend to occupy deeper waters than zebra mussels, some occurring as deep as 427 feet (Mills et al. 1993a). In Lake Erie, only the very deepest part of the lake, which becomes anoxic in summer, is uninhabited by quagga mussels (Dermott and Munawar 1993). Whereas zebra mussels are the dominant species in western Lake Erie, for example, both zebras and quaggas are common in the deeper lake basins of eastern Lake Erie and in Lake Ontario. Quagga mussels increase in numbers relative to zebra mussels as depth increases. Deepwater quaggas colonize sandy and silty bottoms, forming masses of individuals attached to one another's shells. In spite of the cold bottom waters, quaggas appear to reproduce actively (Roe and Macissac 1997).

Zebra mussels pose a serious threat to one major group of North

American freshwater invertebrates: unionid bivalve mussels. In areas of heavy zebra mussel infestation in Lake St. Clair, for example, densities of unionids declined from 2.0 per square yard in 1986 to 0 in 1992 (Nalepa 1994). In western Lake Erie, five species of unionids were essentially extirpated by 1991 (Schloesser and Nalepa 1994). All living unionids and their dead shells were fouled by zebra mussels. Some individual unionid shells were covered by 15,000 zebra mussels, a quantity equal to about five times the weight of the live unionid. By 1993, heavy infestation of many species of mussels was noted in the Illinois and Mississippi Rivers (Schloesser et al. 1996). Heavy infestations, leading to unionid mortality, have also been noted in the upper St. Lawrence River (Ricciardi et al. 1996). Zebra mussels thus pose a major threat to survival of many unionids. The smaller species of unionids seem to be at higher risk from such fouling than do larger species, presumably because of their higher surface-to-volume ratios (Hunter et al. 1997). Ricciardi et al. (1998) concluded that densities of 2,500 zebra mussels per square yard lead to the extirpation of native mussels in four to eight years. They estimate that the zebra mussel has created an extinction rate of about 12 percent of native mussel species per decade in the Mississippi River basin.

Zebra mussels compete with unionids for food—both are filter feeders on organic particulates—and attach to unionid shells, reducing the capacity of unionids to burrow or maintain normal position in the lake bottom sediment (Schloesser and Nalepa 1996). Heavy mortality results when infestation levels reach about 5,000 per square yard and about 100 per individual unionid. There is no evidence, however, that zebra mussels prefer unionid shells over other hard substrates (Toczylowski and Hunter 1997).

Zebra and quagga mussels substantially modify the trophic dynamics of lakes. They harvest a large fraction of planktonic production and direct it into benthic food chains (see Chapter 18). The mussels themselves enter food chains in a substantial degree, as well. Several native fish are able to crush their shells and consume them (French 1993). These include the freshwater drum (*Aplodinotus grunniens*), redear sunfish (*Lepomis microlophus*), pumpkinseed (*Lepomis gibbosus*), copper redhorse (*Moxostoma hubbsi*), and river redhorse (*Moxostoma carinatum*), which have teeth and chewing pads adapted to crushing mollusks. In Lake Erie, the drum is now feeding heavily on zebra mussels (Morrison et al. 1997). A variety of other native fish also occasionally consume zebra and quagga mussels. The common carp, an exotic, also appears able to consume these mussels. Thus, it is likely that aquatic food chains will experience a shift in composition toward species like these that are able to exploit the exotic mussel food resource. These mussel-feeding fish also constitute potential biological control agents. But to maximize control by such species would require management practices, such as stocking, that constitute one more human pressure on other native fish in the ecosystem.

Other Exotic Invertebrates

Several exotic zooplankters have invaded the Great Lakes. *Eubosmina coregoni,* first noted in Lake Michigan in 1966, is now one of the dominant zooplankton forms throughout the lakes (Mills et al. 1993b). When *Bosmina maritima* appeared is unknown, but it is now an important member of the winter zooplankton community. Genetic analyses show that European strains of some zooplankton species native to the Great Lakes have also been introduced. Populations of *Daphnia galeata* in the lower Great Lakes, for example, are now predominantly hybrids of European and North American races of the species (Taylor and Hebert 1993).

In the open waters of the Great Lakes, and increasingly in smaller lakes, another zooplankter, the spiny water flea, has become a keystone exotic (Lehman and Cáceres 1993). Native to Eurasia, the predaceous spiny water flea was first found in Lake Huron in late 1984. Either it had been present, but not observed, for some time, or else its colonization of the Great Lakes was explosive. In 1985, it was found in Lake Erie and Lake Ontario; in 1986, in Lake Michigan; and in 1987, in Lake Superior. The spiny water flea is large, reaching, with its long spine, a length of almost half an inch. An inhabitant of the open lake waters, it is a predator on smaller zooplankton. It appears capable of occupying lakes varying greatly in area and depth, water chemistry, and fish community structure. Thus, it is likely that this exotic will ultimately invade lakes and ponds throughout much of North America. In 1998, a second spiny cladoceran, *Cercopagis pengoi,* was identified in Lake Ontario (MacIsaac et al. 1999). Because it is similar in appearance to the spiny water flea, it apparently went unrecognized for some time.

In Lake Michigan, appearance of the spiny water flea triggered a major restructuring of the lake's open-water zooplankton community (Lehman and Cáceres 1993). The major native predatory water flea, *Leptodora kindti,* declined immediately in abundance in offshore waters. The composition of herbivorous water fleas also shifted markedly. Two species of *Daphnia, D. pulicaria* and *D. retrocurva,* virtually disappeared. *Daphnia galeata,* on the other hand, increased in abundance but showed a markedly altered vertical distribution pattern, largely becoming restricted to waters deeper than 65 feet during the day. Another water flea, *Bosmina longirostris,* which was a major prey of *Leptodora,* also increased in abundance, even in surface waters. The long-term effects of the spiny water flea are still uncertain. Consideration of the energy requirements of spiny water flea populations suggests that they exceed the productive capacity of *Daphnia* prey, so that it is probably preying heavily on a variety of other zooplankton forms.

Similar restructuring of the zooplankton was noted in Harp Lake, Ontario, a small lake, 176 acres in area (Yan and Pawson 1997). Detailed surveys of the zooplankton of this lake were available for fifteen years prior to invasion of the spiny water flea. The predatory *Leptodora kindti,* which normally peaks in abundance in midsummer, became restricted to spring and early summer. It essentially disappeared during mid- and late summer, when the spiny water flea peaked in abundance. The abundances of several smaller water fleas declined markedly, and those of a few larger species, including *Daphnia galeata,* increased.

As with fish, new exotic invertebrates are continually appearing in the Great Lakes. In 1991, the antipodean snail, native to New Zealand, appeared in Lake Ontario (Zaranko et al. 1997). This snail, which became established in England in 1859, now occurs throughout most of western Europe. In this country, it appeared in the Snake River in 1987 (see Chapter 7), probably traveling in commercial shipments of trout or trout eggs. More recently, it has invaded the Missouri–Mississippi drainage. Populations of this snail can reach 33,460 per square yard, enough to displace native snails and other surface living invertebrates. By 1995, Lake Ontario populations had spread and had reached densities of over 4,180 individuals per square yard in places.

In 1995, an amphipod crustacean native to the Black and Caspian Seas was detected at a site in the Detroit River, between Lake St. Clair and Lake Erie (Witt et al. 1997). Within a year, this species had become the predominant amphipod in parts of Lakes Huron and Erie and was replacing a native amphipod (*Gammarus fasciatus*) in many places (Dermott et al. 1998).

Exotic Plants

Exotic primary producers have also become established in the Great Lakes and other northern lakes. The plant exotics of the Great Lakes include fifty-nine species of aquatic and wetland vascular plants and twenty-four species of algae (Wiley 1997). At least six species of diatoms from the Baltic Sea region were introduced to the Great Lakes in the 1800s, and a variety of other exotic diatoms and green, red, and brown algae had become established in the 1900s (Mills et al. 1993b). More seriously, Eurasian water milfoil, a rooted, submerged aquatic plant, has become established throughout much of eastern North America. The Eurasian species is closely related to a native North American water milfoil (*Myriophyllum exalbescens*), and similar enough in appearance that its presence was not recognized immediately (Reed 1977). It now appears that true Eurasian water milfoil became established in about 1942 near Washington, DC (Engel 1995). From there, it began an explosive geographical spread, largely dispersed by recreational boating activity. It now occurs throughout eastern and midwestern North America from southern Canada to the Gulf of Mexico, as well as in many far western localities.

Native water milfoil itself can be a troubling weed of shallow lake waters, but massive beds of the Eurasian water milfoil, especially in small, relatively shallow lakes, completely transform the littoral habitat and obstruct recreational boating and swimming. Beds of Eurasian water milfoil obstruct the feeding of bass and other fish-eaters, promoting dense populations of stunted panfish that can forage in the dense plant matrix. Plankton-feeding fish are also displaced by dense mats of water milfoil that invade open-water areas.

Food Web Instability

"Food web chaos" is a phrase that well characterizes the effects of exotic introductions to northern lakes. Stable food web relationships have not existed since European settlement of the lake watersheds. Now, however, chaos is being created by influences that are nearly impossible to counteract. Exotic invasions may interact in a positive feedback manner. Disturbance by exotics, creating nonequilibrium patterns of resource availability, may breed vulnerability of these lake ecosystems to invasion by other exotics.

Unpredictability is a key feature of these invasions. Exotics have struck randomly in all sectors of the food web. In some cases, their influence has cascaded downward from high levels in the web, as illustrated by the sea lamprey. In others, it has climbed the food web "ladder" from low levels, as illustrated by the impacts of zebra and quagga mussels. In the food web, much of the energy flow has been shifted from the pelagic to the benthic sector of the food web. Ecological predictability has been nearly destroyed. The sustainability of valuable fisheries and the survival of endemic groups of lake organisms have been placed in question.

Reducing the Rate of Invasions

The enormous changes in biotic composition occasioned by invaders to northern lakes, especially those arriving in ballast water, demand action to reduce the rate of exotic invasions. One approach, adopted by both Canada and the United States, has been to require transoceanic ships to exchange ballast water in midocean (Locke et al. 1993). This procedure assumes that any transoceanic freshwater species will be discharged or killed, and that ballast water will contain only marine species that cannot live in the Great Lakes or in the Hudson River. Many problems exist, however, with this procedure. In tests of this procedure, even ships that performed midocean exchanges were found to be carrying freshwater-tolerant zooplankton when they arrived in the Great Lakes. Some of these species, to the horror of aquatic biologists, were not already established in the Great Lakes.

Midocean exchange of ballast water is thus, at best, only a partial solution to the problem of transport of exotics. Ballast water exchanges may be incomplete, because of the location of intake openings for ballast water pumps. Dormant stages of some invertebrates may survive ballast water exchange in sediments at the bottom of ballast tanks. Also, ballast water exchange does not resolve the problem of transport of species between freshwater or estuarine locations within North America.

Other technologies might involve filtration or sterilization of ballast water by heat, ultraviolet radiation, or chemical means. Alternative means of ballasting also deserve to be explored. In addition, techniques of reducing the transport of fouling organisms on the hulls of vessels need to be considered. As free-market policies expand, ship commerce threatens to bring hoards of new exotics to the Great Lakes and other North American waters.

Reducing the Impact of Established Exotics

Efforts must also be made to reduce the impacts of established exotics. Controlling populations of the sea lamprey, by methods described earlier, allows a more nearly natural food web of open-water fish to be maintained. Some other exotics, such as the ruffe, that threaten major ecological and economic damage, can also be attacked by traditional means. The ruffe is vulnerable to the lampricide TFM, which can be used to treat areas at the mouth of streams, where this species congregates at certain times. Leigh (1998) has shown that the economic benefits to lake fisheries by a ruffe control program are likely to exceed treatment costs by a ratio of 44 to 1.

Once established in lakes, however, most exotics have little chance of being extirpated or controlled directly. The development of biological control systems for aquatic species is still in its infancy. Most aquatic vascular plants do not have specialist herbivores. Eurasian water milfoil may be an exception, a native North American weevil having a stronger preference for this exotic than for the native water milfoil (Newman et al. 1997). A number of specific parasites have been identified for zebra and quagga mussels, but these do not appear to be capable of substantially reducing mussel populations unless ways are developed to augment their densities considerably (Molloy 1998). Manipulations of higher food chain species, such as fish known to prey on zebra mussels, can also be attempted. Because of the broad diets of species at these higher levels, however, the results of such manipulations are difficult to predict. New biological strategies, such as the use of selectively toxic microbes, produced by artificial selection or genetic engineering, may ultimately allow biological control of some exotics (Molloy 1998).

Management Dilemmas

Management of exotics in these lakes thus poses many dilemmas. Once an exotic has gained a foothold, we now have a choice of letting it run its course or trying uncertain control strategies that hold some risk of backfiring. Here, perhaps, is a case in which the only real option is adaptive management, an approach in which major alternative management options are defined, one is applied, and the effect on population dynamics is monitored and used to refine options for subsequent management. A key question remains: Can adaptive management work in a situation in which many options involve tinkering with the species composition of a complex community?

Chapter 7

Western Rivers and Streams: Pollution That Won't Wash Away

> There is a long and honorable tradition in western culture, dating back at least to the Romans, of tinkering with fish faunas by adding new species. This tinkering is part of a much broader tradition of tinkering with nature, to "improve" on it.
>
> —Peter Moyle et al., "The Frankenstein Effect: Impact of Introduced Fishes on Native Fishes in North America"*

The streams of the San Luis Valley in southern Colorado illustrate the carelessness with which streams and their fish faunas have been manipulated in western North America. These streams include the headwaters of the Rio Grande and mountain streams that flow into a closed basin in the upper valley. An estimated fifty-two species of nonindigenous fish have been introduced to these streams (Zuckerman and Behnke 1986). Although a number of the introduced species have been extirpated, at least twenty-five are established, and the status of seventeen is still uncertain. The result has been widespread replacement of native fishes. Many of the species native to the valley exist only as remnant populations in small headwater streams or thermal waters.

The majority of introductions to the San Luis Valley were deliberate.

*Peter Moyle, Hiram Li, and Bruce Barton (1986). "The Frankenstein Effect: Impact of Introduced Fishes on Native Fishes in North America." Pp. 415–426 in R. H. Stroud (Ed.), *Fish Culture in Fisheries Management*. American Fisheries Society, Bethesda, MD.

Federal, state, and private organizations introduced new species to natural waters for varied reasons. Exotic fish were stocked for sport fishing, as prey for sport fish, for insect control, and for aquatic weed control. Public and private fish hatcheries produced sport fish, most of them exotics, for stocking into streams and lakes, including high-elevation waters formerly fishless, with no consideration of impacts on native fish and other organisms. One of these species is the rainbow trout (Table 7.1), once Colorado's

Table 7.1. Scientific names and native regions of invasive exotics of western rivers and streams.

FRESHWATER INVERTEBRATES	
Asiatic clam, *Corbicula fluminea*	Asia
zebra mussel, *Dreissena polymorpha*	Russia
FISH	
black bullhead, *Ameiurus melas*	eastern North America
blue tilapia, *Tilapia aurea*	Africa
brook trout, *Salvelinus fontinalis*	eastern North America
brown trout, *Salmo trutta*	Europe
California roach, *Lavinia symmetricus*	California
channel catfish, *Ictalurus punctatus*	eastern North America
common carp, *Cyprinus carpio*	Asia
convict cichlid, *Cichlasoma nigrofasciatum*	Central America
cutthroat trout, *Oncorhynchus clarki*	western North America
goldfish, *Carassius auratus*	Europe
grass carp, *Ctenopharyngodon idella*	Asia
largemouth bass, *Micropteris salmoides*	eastern North America
mosquitofish, *Gambusia affinis*	Central America
northern pike, *Esox lucius*	eastern North America
oriental weatherfish, *Misgurnus anguillacaudatus*	Asia
rainbow trout, *Oncorhynchus mykiss*	western North America
red shiner, *Notropis lutrensis*	midwestern United States
Sacramento pikeminnow, *Ptychocheilus grandis*	California
sailfin molly, *Poecilia latipinna*	Mexico
sheepshead minnow, *Cyprinodon variegatus*	eastern North America
shortfin molly, *Poecilia mexicana*	Mexico
smallmouth bass, *Micropterus dolomieui*	eastern North America
threadfin shad, *Dorosoma petenense*	Mexico
wakasagi, *Hypomesus nipponensis*	Japan
walleye, *Stizostedion vitreum*	eastern North America
AMPHIBIANS	
African clawed frog, *Xenopus laevis*	Africa
bullfrog, *rana catesbiana*	eastern North America
DISEASES	
whirling disease, *Myxobolus cerebralis*	Europe

state fish, but a species not native to Colorado. At least five species were introduced carelessly in sport fish stockings made by the Colorado Division of Wildlife, and two of these have become established. Interbasin water transfer projects were undertaken with no concern about the possibility of introduction of species to new drainage systems. From time to time, the Colorado Division of Wildlife salvaged fish from lakes or reservoirs that were being drained and transplanted these fish, and any other organisms in the same water, to waters elsewhere in the state. This was often done without knowledge of what species were involved or whether releases were made in the same or different drainages. Exotic fish were also brought into the valley for fish farming and for the breeding of aquarium fish. At least fifteen species from breeding facilities for aquarium fish have escaped or been released into streams, and several of these have become established.

Threats to River and Stream Ecosystems

Western North America has a distinctive, although not very rich, stream fauna. About 170 species of freshwater fish occur in the western United States, compared to about 600 species east of the Rocky Mountains (Dowling and Childs 1992). Of these 170, nearly 105 are federally listed as endangered, threatened, or candidate species for listing. In the western United States and Canada, 23 species of freshwater fish have become extinct, and an additional 9 species have become extinct in Mexico (Miller et al. 1989). Exotic fish introductions are implicated in about 38 percent of these extinctions. Continentwide, but especially in the West, exotics are one of the greatest threats to the species that survive. Of the 54 species of fish listed as endangered or threatened under the U.S. Endangered Species Act, 29 are impacted by exotic fish (Williams and Nowak 1993).

The disruption of western stream faunas by exotics is part of a larger pattern of human transformation of western streams and their watersheds. Most rivers have been dammed to control flooding and create impoundments for water storage, hydropower production, and irrigation. Canal systems were constructed to move water within and between basins. Shaft and placer mining operations polluted streams with toxic chemicals and silt. Waters were diverted for urban and agricultural use. In the course of these activities, natural stream habitats were destroyed and altered. Free-flowing streams were impounded and seasonal flows regulated. Major sections of large rivers were converted into reservoirs. Networks of irrigation and drainage canals were created. In the era of massive water project development from the 1930s to 1960s, it is not surprising that public agencies and sportsmen's groups dealt with this wholesale change of stream habitats by introducing species perceived to be better adapted to the new conditions (Dill and Cordone 1997).

The number of fish species or hybrid forms introduced to North America or to areas outside their natural North American range is astounding. For the United States, a recent compilation (Nico and Fuller 1999) indicates that such introductions number about 524 species and hybrids. Of these, 185 are international introductions, and 317 are U.S. species translocated to areas outside their native range. A surprising fraction, over 4 percent, of these introductions are of hybrid forms. In the western states, the numbers of introductions range from 38 for Wyoming to 114 for California. Not all of these introductions survive, but exotic species make up 35–59 percent of the fish faunas in these states (Moyle et al. 1986). At least 67 species in the Pacific states, and 53 in California alone, are exotic (Moyle 1986; Dill and Cordone 1997), so that the scope of the problem is clearly enormous. Although some of the introduced fish have provided improved sport fishing in reservoirs and other artificial habitats to which native species were not well adapted, very few of these exotics are without some detrimental impacts in natural waters. In western North America they have altered the native fish community, and often the total ecosystem function, of almost every stream and river.

More ominously, exotic fish are often accompanied by other exotic organisms that can cause additional biotic disturbance of stream and river ecology. Several diseases and parasites of fish, to which many exotics are resistant but many natives vulnerable, have invaded North American streams (Allan and Flecker 1993). Activities centered on introduced exotic fish, such as recreational fishing and boating, may also introduce other kinds of exotics. Exotic invertebrates, their numbers as yet small, are also beginning to invade stream invertebrate communities, which are also low in diversity in western streams and rivers.

History of Fish Introductions

The history of tinkering with fish faunas of western streams is long. Intercontinental introductions of exotic fish to North America began early in the 1800s (Courtney and Kohler 1986). Both the United States and Canada experienced two major periods of introduction in the latter half of the 1800s and the 1960s (Crossman 1984). In the late 1800s, introductions were largely in response to the perceived decimation of native species by overfishing. In the 1960s, species thought to be resistant to pollution were introduced widely.

Introduction of Old World fish began in eastern North America, but quickly moved west. Common carp were introduced to the Hudson River in 1831 (see Chapter 4). By 1872, carp had been introduced to freshwaters of the Sacramento Delta (Herbold and Moyle 1989), and by 1880, they were being raised in the San Luis Valley of Colorado (Zuckerman and Behnke 1986). By 1890, carp had been introduced to streams and lakes throughout North America (Moyle 1986). The goldfish was another early introduction to North American waters, having been stocked in the Hudson River in 1843 (Radonski et al. 1984). Goldfish are now widespread in natural and altered waters, even being a common inhabitant of stock tanks on western rangelands. Brown trout, a prized European sport fish, were also an early introduction. This species was first introduced to the United States in Michigan in 1883 (Courtenay et al. 1984) and to Newfoundland, Canada, in the late 1880s (Courtenay and Kohler 1986). It quickly spread by stocking throughout Canada and the United States. This fish is still reared in many state fish hatcheries for stocking in trout streams. A number of species, particularly the threadfin shad and the Asian wakasagi, have been introduced to localities in California to serve as a food fish for introduced trout (Moyle 1976). Several other European sport fish were introduced to eastern streams and lakes but survive in only localized situations (see Chapter 14).

A number of other fish, including the grass carp and several species of tilapia, have been introduced more recently from outside of North America for aquatic weed control (Taylor et al. 1984). Several tilapias have been introduced in the southeastern United States (see Chapter 9). In western North America, the blue tilapia was introduced to Arizona in 1975, and it has now become the dominant fish in parts of the lower Colorado River (Courtenay and Kohler 1986).

Aquarium fish from Mexico and from other continents are increasingly finding their way into natural waters in the United States. Shortfin and sailfin mollys, for example, have been released into southwestern streams, where they have contributed to the decline of several localized endemic fish (Minckley and Deacon 1968). In California and Oregon, as well as in parts of the eastern United States, the oriental weatherfish has recently become established in streams (Logan et al. 1996). Weatherfish might prey on smaller fish and fish fry and are a potential vector for several parasites and a virus disease.

Many exotic fish introduced into western rivers and streams, however, are native to other regions of North America. Most of these introductions have been in the name of improving sport fishing. The most frequently introduced species include rainbow and brook trout, largemouth and smallmouth bass, walleye and northern pike, and a variety of catfish and panfish (Boydstun et al. 1995). Rainbow trout, for example, have been introduced to waters outside their native range in forty-seven states, largemouth bass to locations to which they were not native in forty-one states. Other widespread deliberate introductions have been of mosquitofish for aquatic insect control. Native to western North America, mosquitofish have been introduced to locations to which they are not native in twenty-five states.

Impacts of Introduced Fish

Exotic fish have major impacts on both stream habitats and stream communities. Common carp, for example, are notorious for their modification of the aquatic habitat. Carp destroy beds of aquatic plants, disrupt bottom sediments and their benthic invertebrate populations, and create turbid water conditions (Taylor et al. 1984; Moyle et al. 1986). They thus transform aquatic habitats, in which rooted aquatic plants are important producers and invertebrate-feeding fish are important consumers into phytoplankton-based systems in which carp act as scavengers. In the Eel River in California, the California roach has also modified the stream habitat. Introduced about 1970, it has substantially changed benthic algal and invertebrate populations (Power 1990).

The fish communities of most river systems in western North America have also been heavily impacted by exotic fish introductions, but probably none more than the Colorado River. Biotically, the Colorado River has greater endemicity in its fish fauna than has any other North American river (Minckley and Deacon 1968). Loss of the river habitats to which these species were adapted, and competition from exotic species introduced into the reservoirs, have endangered this fauna almost in its entirety. Nine species of larger native fish, including eight species of river chubs and the Colorado pikeminnow (*Ptychocheilus lucius*), are now designated as threatened or endangered (Hickman 1983). In the Glen Canyon–Grand Canyon region of the river, eight species of native fish once occurred. Now, nineteen species of exotics are present. Three of the eight original natives have been extirpated in this region, leaving only five native species with reasonably healthy populations (Carothers and Johnson 1983). In the colder portions of the river, these natives have largely been replaced by introduced trout, and in warmer regions, by channel catfish.

The fish fauna of the Eel River in northern California has been similarly transformed (Brown and Moyle 1997). Some sixteen species of fish have been introduced into a fauna that originally consisted of fourteen species. Many of the exotics come from other streams in interior California. At least ten of the exotics have become well established, while only three of the natives show stable populations. All ten species of native anadromous fish have declined in abundance or disappeared. Four of the ten exotics are restricted to reservoirs, so that flowing water areas are occupied primarily by six exotics and three natives.

The result of these introductions in the Eel River is the creation of several new assemblages of fish that are associated with major stream habitats. One of these assemblages, dominated by the Sacramento pikeminnow, occupies the lower main channel of the river; a second, dominated by the California roach, occupies major warm-water tributaries to the lower main channel. A third assemblage, dominated by native coho salmon and rainbow trout, is concentrated in shaded, cold-water tributaries where deep pools exist. A fourth—not really an assemblage—is a native rainbow trout population, which is confined to high-gradient headwater streams with cold water and only small pools. Thus, communities dominated by native species are now restricted to the cold-water tributaries of the river system, whereas exotics dominate the larger, warm-water portions of the system. The implications of these altered assemblages for recovery of native anadromous fish are unknown.

Farther north, the Columbia River has been less impacted than the Eel by introduction of exotic species, either from elsewhere in North America or from other continents. However, the decline of many species of anadromous salmon has led to programs of hatchery production and release of fry. Major concerns have now arisen about these hatchery programs. Many fisheries biologists, for example, fear that hatcheries might serve as sources of disease that could impact wild populations.

Beyond the threat of disease, considerable evidence now indicates that hatchery fish are contributing to the decline of native runs of the species involved (Hilborn 1992). In effect, hatchery fish are exotic genotypes, selected for hatchery environments, but contributing to competitive displacement of wild fish. Hatchery fish are often not derived from fish stock of the river system into which they are released and thus, genetically speaking, are exotics. On the Nooksack River, Washington, for example, a hatchery program for coho salmon (*Oncorhynchus kisutch*), utilizing fish from other sources, may have driven the local stock to extinction, although the salmon run survives with hatchery fish (Nehlsen et al. 1991). Similar cases exist for several chinook salmon stocks. Of 214 salmon stocks in the northwestern United States, in fact, 104 may have experienced genetic introgression from hatchery fish (Nehlsen et al. 1991).

Evidence is also accumulating that genetic differences exist between fish produced in hatcheries and fish of wild origin, even when fish come from local origin (Hindar et al. 1991). Recent studies in New Zealand have compared life history characteristics and fry-to-adult survival rates of wild-hatched chinook salmon (*Oncorhynchus tshawytscha*) with those of salmon reared in a hatchery for six to twelve months before release into the same streams. Hatchery fish tended more often to mature after only one year at sea and tended to be smaller than wild-hatched fish at ages of two to three years (Unwin and Glova 1997). The survival of hatchery fish was relatively poor, considering their larger size and greater age at release, compared to wild-hatched fish (Unwin 1997). During the twenty-eight-year period when the hatchery was operating, the percentage of wild-hatched fish in runs declined to 34 percent, indicating that the natural spawning component of the population was being displaced by hatchery fish.

Exotics and Extinction of Native Species

Exotic fish have been the primary or major contributing factor to extinctions of many fish in western North America. Many of these forms occupied small, isolated streams, lakes, or ponds that were also affected by many human activities related to ranching and to water diversion for human use. Nevertheless, at least fifteen species or subspecies in the western United States and two species in northern Mexico have disappeared largely because of introductions of exotics that eliminated them by competition, predation, habitat disruption, or hybridization (Miller et al. 1989). These extinctions include two subspecies of cutthroat trout, together with endemic chubs (*Gila* spp.), shiners (*Notropis* spp.), dace (*Rhinichthys* spp.), gambusias (*Gambusia* spp.), and others.

Exotic fish prey on many species of endangered native fish. In the Little Colorado River in Arizona, for example, channel catfish, black bullheads, and rainbow trout are predators on the endangered humpback chub (*Gila cypha*), depleting the largest population of this endangered species (Marsh and Douglas 1997). Predation by exotic fish also appears to be the major limitation to successful recruitment by the razorback sucker (*Xyrauchen texanus*) in Lake Mojave on the Colorado River (Minckley et al. 1991).

Many endemic forms of western North America are thus threatened by exotic fish. The Sonoran topminnow (*Poeciliopsis occidentalis*), for example, has been displaced from most of its original range in Arizona, New Mexico, and Sonora, Mexico, by the mosquitofish (Meffe and Vrijenhoek 1988). A federally designated endangered species, it survives in only a dozen headwater locations in Arizona. In Mexico, it is still common, evidently because mosquitofish have not been stocked extensively in Sonoran streams. The Shoshone pupfish (*Cyprinodon nevadensis shoshone*), thought to be extinct, was rediscovered just in time to prevent its extermination by habitat degradation and competition from mosquitofish (Taylor et al. 1988). The Gila spinedace (*Meda fulgida*), once widely distributed through the Gila River system in Arizona and western New Mexico, has been displaced in much of the lower Gila River by the exotic red shiner (Minckley and Deacon 1968). In Nevada, several local populations of the White River springfish (*Crenichthys baileyi*) have been eliminated or depressed by largemouth bass, convict cichlids, and other exotic fish (Tippie et al. 1991). In the Eel River in California, one of the dominant exotics, the Sacramento pikeminnow, introduced in 1979, appears likely to restrict one of the remaining natives, the threespine stickleback (*Gasterosteus aculeatus*), to very local areas or perhaps extirpate it completely (Brown and Moyle 1997).

Some exotic fish, such as mosquitofish, also prey on larvae of stream amphibians (see also Chapter 14). In the Santa Monica Mountains of southern California, for example, mosquitofish and exotic crayfish were probably responsible for the disappearance of California newts (*Taricha torosa*) from several streams (Gamradt and Kats 1996). Laboratory experiments showed that crayfish ate egg masses of the newts and that both crayfish and mosquitofish were predators on newt larvae.

Hybridization between Exotics and Natives

Still another impact of exotic fish results from interbreeding with closely related, endemic species. Rainbow and cutthroat trout have been introduced to streams throughout the western states. In Arizona, the native trout of the White Mountains was the Apache trout (*Oncorhynchus apache*). Over half a century, the streams to which this species are native have been stocked with rainbow, cutthroat, brook, and brown trout. In addition to suffering from competition from brown and brook trout, hybridization has occurred with both rainbows and cutthroats (Carmichael et al. 1993). Of thirty-one Apache trout populations examined in the late 1980s, twenty showed some degree of hybridization with one or both exotics. In some streams, every fish taken was a hybrid. Pure Apache trout remain in only a few locations. The Gila trout was native to the Verde River in Arizona, where it has been extirpated, and to the upper Gila River in New Mexico (Dowling and Childs 1992), and it is likely to have been affected by hybridization, as well.

Hybridization is the major threat to many endemic forms of the cutthroat trout. In addition to the two subspecies of cutthroat trout that have been driven to extinction with hybridization with rainbow trout, several other subspecies of cutthroats have also experienced hybridization with rainbow trout or with other subspecies of cutthroats themselves (Allendorf and Leary 1988). Hybridization among even subspecies of the cutthroat has been shown to cause developmental problems.

Brook trout have also been introduced to the Klamath and the Columbia river systems, where a closely related species, the federally threatened bull trout (*Salvelinus confluentus*), is native. Hybridization occurs between the two, with the hybrids probably being sterile, and this may have contributed to replacement of bull trout by brook trout in a tributary of the Bitterroot River in Montana (Leary et al. 1993).

Hybridization between exotic and native fish is not limited to trout. In the Guadaloupe River in Texas, introduced smallmouth bass have hybridized with the endemic Guadaloupe bass (*Micropterus treculi*), threatening its survival (Edwards 1979). Several endemic species of pupfish (genus *Cyprinodon*) have also been affected by hybridization. In the Pecos River, Texas, several shiners and minnows have been introduced from drainages in the western Great Plains and have hybridized with native species (Fausch and Bestgen 1997). For example, sheepshead minnows were introduced sometime between 1980 and 1984, probably by release of baitfish (Wilde and Echelle 1997). The result is that populations of the endemic Pecos pupfish (*Cyprinodon pecosensis*) along about 240 miles of the river are now of hybrid origin. This hybridization occurred in less than five years. A similar situation occurred when sheepshead minnows were introduced into the range of the Leon Springs pupfish (*Cyprinodon bovinus*) in Texas. In this case, it was possible to eradicate sheepshead minnows and hybrids before the entire population was affected.

Other Exotics

Fish are not the only exotics to cause problems in western streams. An exotic amphibian, the African clawed frog, has become established in several rivers in California (St. Amant et al. 1973). Whether this frog has contributed to the extirpation of populations of small fish and native frogs is uncertain. In the Santa Clara River, Ventura County, California, however, clawed frogs were found to be abundant in the brackish lagoon at the mouth of the river. In this location, several individuals were found with tidewater gobies (*Eucyclogobius newberryi*), a federally endangered species, in their stomachs (Lafferty and Page 1997). Bullfrogs have been introduced widely throughout western North America, with severe localized impacts on native stream animals (see Chapter 15).

A number of exotic diseases and parasites have appeared in western streams, frequently introduced with exotic fish. One of the most serious new diseases of native fish is whirling disease. Caused by a protozoan parasite, whirling disease was first detected in Pennsylvania in 1956 (Bergersen and Anderson, 1997). It quickly spread through neighboring states, and it now occurs from Michigan to New Hampshire and south to Virginia. By 1965, it had appeared in California and Nevada; by 1996, it had been reported from most areas of the West with self-sustaining trout populations.

The whirling disease parasite has a life history involving two hosts: trout and some other salmonid fish and tubificid worms. Protozoan spores from dead fish enter the stream water and are ingested by the tubifex worms, which live in the muddy bottoms of pools. In the worms, the spores mature and form clusters of a new spore stage. These clusters are equipped with hooks that enable them to attach to internal tissues of fish, such as the gills or surfaces of the mouth and the digestive tract. Once attached, the spores enter the circulatory system of the fish. One infected tubifex worm can introduce thousands of spores into a trout that consumes it. If the spore clusters enter the water, they can also infect trout when water is pumped through the gills. In the fish, the parasite attacks developing cartilage tissues of young fish, eventually disrupting the cartilage tissues associated with cranial balance organs. This leads to a tail-chasing behavior pattern, giving the disease its “whirling” name.

The most severe effects of whirling disease have been on rainbow trout in some western states, particularly Montana and Colorado. In some streams, populations of this species have been reduced by 95 percent. In other areas, such as California, the parasite is widely established but has little impact (Modin 1998). European brown trout, an exotic, are resistant to the disease. So far, eastern brook trout also do not appear to be seriously affected by the disease.

The Asiatic clam may be the vanguard of invertebrate invaders of western rivers and streams (McMahon 1983). Appearing in the lower Columbia River in 1938, this clam colonized the Sacramento Delta in the 1940s and the Gila River drainage in Arizona in the 1950s. From there, it moved into the Rio Grande in the 1960s. In all probability, the spread of this clam has resulted from sport fishing and recreational boating activities that transfer young clams from place to place in bait buckets and bilge water. Spread of the clam has certainly been aided by the transfer of water in major canal systems, such as those operating in central California and central Arizona. In the western United States, this clam now occurs in all states except Montana and Wyoming (McMahon 1983).

The zebra mussel is likely to be the next mollusk to invade western freshwaters. Only four years after it was first found in the Mississippi River, the zebra mussel was present, and often abundant, in the impounded upper portion of the river south of Minneapolis (Cope et al. 1997). As of 1995, the western limit of zebra mussels was still confined to the Mississippi drainage. The climatic relationships of the species indicate that it can potentially occupy waters in most of western North America except for hotter portions of the southwestern United States (Strayer 1991). The ease with which this species is transported overland by recreational boating activity virtually assures that it will invade the western region.

Endangered Rivers and Streams

Western rivers and streams have been some of the most carelessly treated natural ecosystems in North America. Until very recently, little attention has been given to the ecological integrity of these waters. The water itself has been viewed as a resource that is wasted unless it is fully consumed for productive human activity. Many westerners still regard in-stream flows and river discharges into the ocean as having no beneficial purpose. The biological value of river and stream habitats has been viewed almost entirely as the production of commercial and sport fish. Perceived shortcomings of commercial and sport yields have been addressed by quick fixes of stocking and introduction of exotic species. Compounding these actions have been the accidental or thoughtless releases of nongame exotics. Efforts to manage rivers and streams to enhance the survival and productivity of native species have been inadequate and belated.

The result of the deliberate and unplanned introduction of exotics to western rivers and streams has been severe ecological disruption. Moyle et al. (1986) have characterized deliberate introductions of fish to western streams as an example of the "Frankenstein Effect." Count Frankenstein sought to create an improved human being but instead created a deranged being. Similarly, efforts to create improved stream ecology by exotic fish introductions have created a deranged stream ecology. Magnified by the unplanned invasions of western streams by exotic fish and other organisms, this Frankenstein Effect is altering the ecology of western streams and threatening to destroy much of our native stream biota.

Chapter 8

Eastern Forests: The Dark Side of Forest Biodiversity

> [The chestnut blight] is probably the largest single change in any natural plant population that has ever been recorded by man.
>
> —J. L. Harper, *The Population Biology of Plants**

Theodore Roosevelt Island, 88 acres in area, lies in the Potomac River, within the city of Washington, D.C. Now administered by the U.S. National Park Service, the island's name comes from a monument to Theodore Roosevelt that was erected on the island. The forest vegetation of the island was at least partially cleared for farming in the late 1700s and the 1800s and was disturbed at other times until the Park Service acquired the island in 1932. Most of the island has now regrown into a mixed deciduous forest.

Theodore Roosevelt Island illustrates the sorts of exotic invasions that the eastern deciduous forest now faces. A common upland tree, for example, is white mulberry (Table 8.1). This species was probably introduced during the early farming period. Another of the dominant trees in the 1950s, American elm (*Ulmus americana*), suffered heavily from Dutch elm disease in the 1960s. Many trees died or were weakened. The forest understory has been invaded by Japanese honeysuckle and English ivy, probably also holdovers from the farming era. Spreading tangles of these vines can virtually eliminate the native herb layer (Thomas 1980). Japanese honeysuckle also overgrows and kills saplings of the dominant

*J. L. Harper (1977). *The Population Biology of Plants*. Academic Press, New York.

Table 8.1. Scientific names and native regions of exotics that have invaded deciduous forests in eastern North America.

PLANTS	
ailanthus, *Ailanthus altissima*	Asia
Amur honeysuckle, *Lonicera maackii*	Asia
Amur maple, *Acer ginnala*	Asia
autumn olive, *Eleagnus umbellata*	Asia
common barberry, *Berberis vulgaris*	Europe
common buckthorn, *Rhamnus cathartica*	Europe
English ivy, *Hedera helix*	Europe
garlic mustard, *Alliaria officinalis*	Europe
golden rain tree, *Kolreuteria paniculata*	Asia
Japanese barberry, *Berberis thunbergii*	Japan
Japanese honeysuckle, *Lonicera japonica*	Asia
kudzu, *Pueraria lobata*	Asia
multiflora rose, *Rosa multiflora*	Asia
Norway maple, *Acer platanoides*	Europe
Oriental bittersweet, *Celastrus orbiculatus*	Europe
princesstree, *Paulownia tomentosa*	Asia
Tatarian honeysuckle, *Lonicera tatarica*	Eurasia
tearthumb polygonum, *Polygonum perfoliatum*	Asia
white mulberry, *Morus alba*	Asia
white poplar, *Populus alba*	Europe
yellow iris, *Iris pseudacoris*	Europe
PLANT DISEASES	
beech bark disease, *Nectria coccinea*	Europe
butternut canker, *Sirococcus clavigignenti juglandacearum*	Europe
chestnut blight, *Cryphonectria parasitica*	Asia
dogwood anthracnose, *Discula destructiva*	uncertain
Dutch elm disease, *Ophiostoma ulmi*	Asia
white pine blister rust, *Cromartium ribicola*	Eurasia
INSECTS	
Asian long-horned beetle, *Anoplophora glabribennis*	Asia
balsam wooly adelgid, *Adelges piceae*	Europe
gypsy moth, *Lymantria dispar*	Eurasia
hemlock wooly adelgid, *Adelges tsugae*	Asia
winter moth, *Operopthera brumata*	Europe
BIRDS AND MAMMALS	
European starling, *Sturnus vulgaris*	Europe
feral pig, *Sus scrofa*	Eurasia
wild boar, *Sus scrofa*	Eurasia
sika deer, *Cervus nippon*	Japan

canopy trees. English ivy is even more destructive. It climbs into both understory and canopy trees, shading their foliage and eventually killing them. Both vines are favored by increased light, and thus by any disturbance of the canopy, as by the death of American elms. Finally, an exotic yellow iris has invaded open marsh areas, forming dense stands that crowd out arrow-arum (*Peltandra virginica*), a plant whose fruits are an important food of wood ducks during the nesting season. The iris also promotes the establishment of swamp forest trees, hastening the total conversion of open marsh to swamp forest.

The forests of eastern North America appear remarkably resilient, as evidenced by their recovery from near destruction in the wake of European settlement. As settlement spread west, much of the original forest was cut and burned to open up land for farms and plantations. In New England, for example, about 98 percent of the original forest was cleared. The forests that escaped clearing were logged destructively to obtain fuelwood and timber. Only a few scattered patches of uncleared forest escaped severe disturbance. Yet, over most of the eastern forest region, particularly in New England and the Appalachian Mountains, forest cover has been reestablished through ecological succession. Forest cover in New Hampshire today is about 87 percent, for example. The composition of these forests, however, is very different from that of pre-settlement times.

Now, however, eastern forests are being subjected to an increasing variety of anthropogenic influences. These include rising carbon dioxide level, acid deposition, chemical pollution, nitrogen overloading, timber harvesting, fragmentation, outbreaks of native and exotic insects and diseases, and invasion by exotic plants, mammals, and birds. Introductions of exotic diseases, insects, and plants are perhaps the most serious. These began in colonial times, but their number and impact have increased in magnitude with time. New forest tree diseases and insect pests appear every year. Most of these exotics are disease agents or herbivorous insects that are native to related host trees in Europe and eastern Asia. Some North American species lack evolved resistance to these insects and diseases, and many of the exotic insects have escaped their Old World enemies. Consequently, they sometimes run wild. Exotic trees, shrubs, vines, and herbaceous plants from Europe, Asia, and elsewhere are invading forest areas and displacing native species. The most destructive exotics have come from Europe and Asia.

In these eastern forests, the effect of invasive exotics together with other forms of stress and disturbance can be synergistic. In other words, the total degree of disturbance due to the different interacting exotics and stresses is more than additive. This synergism can be seen clearly on

Theodore Roosevelt Island. When the native forest structure is disrupted by disease, insect damage, or human disturbance, exotic trees, shrubs, and vines are able to invade. Human settlement has compounded these stresses by fragmenting the forests and increasing the ratio of edge to interior habitat. The opportunities for invasion of the forest interior by exotic plants from outside areas are thus increased.

Exotic Tree Diseases

The most serious invaders of the forests of eastern North America have been tree diseases and insect pests from Europe and Asia (Liebhold et al. 1996; Niemela and Mattson 1996). Many of the major tree diseases and insect pests of North American forests are exotics. At least 20 serious diseases of native woody plants are exotics. Nearly 400 exotic insect pests of woody plants have been identified in the United States. Of the 70 most serious forest insect pests, 19 are exotics. Three quarters of the insect exotics are from Europe, a region of similar climate and closely related biota.

White pine blister rust was the first major tree disease to invade North America. It was apparently brought to several locations on white pine nursery stock from Europe in the 1890s and early 1900s. This rust is capable of infecting most members of the five-needled group of pines, including eastern white pine (*Pinus strobus*), western white pine (*P. monticola*), southwestern white pine (*P. strobiliformis*), sugar pine (*P. lambertiana*), whitebark pine (*P. albicaulis*), and bristlecone pine (*P. aristata*). It was first noted infesting eastern white pines in Pennsylvania in 1905, but within just a few years, it had spread through most of the northeastern states. Shipments of white pine seedlings to western North America from France introduced the rust to this region in about 1910.

This rust has a life history involving stages both on trees of the white pine group and on shrubs of the genus *Ribes*—currants and gooseberries—which are widely distributed in North America. In an effort to stem the disease on pines, massive campaigns were undertaken in the 1930s to eradicate shrubs of the genus *Ribes* in regions where white pines were important timber trees. These campaigns had little or no success in reducing damage to pines.

The most destructive tree disease to strike North America was the chestnut blight. The American chestnut (*Castanea dentata*) was once the most ecologically and economically important hardwood tree of eastern North America, as well as one of the most widespread. Its range extended from northern Georgia and Alabama north to central Maine and west to the Mississippi River. Old-growth trees reached heights of 100 feet and diameters of 6 feet or more. It is said to have constituted about a quarter of the biomass of the deciduous forests within its range. Chestnut mast was consumed by many native mammals and birds. Until its demise, it was one of the most important timber trees of the eastern forests. Chestnut supplied wood for uses ranging from construction lumber to furniture manufacture.

Chestnut blight, a fungal canker disease, was introduced to New York prior to 1904, when it was first noted in New York City. The fungus probably entered North America on Chinese or Japanese chestnut trees introduced as urban ornamentals. The fungus invades the bark, spreads through the xylem and phloem tissues, eventually girdling and killing the tree. Once established in New York, the fungus spread at an astounding rate of about 25 miles per year. Within twenty years, it had killed almost all mature chestnut trees in New England, and in another two decades, it had spread throughout the southern and western range of the chestnut. The chestnut blight also attacks the chinquapin (*Castanea pumila*) with lesser intensity. The blight fungus survives as an essentially harmless infection of various oaks, hickories, maples, and other woody plants. In the case of the American chestnut, the blight quickly kills the aboveground portion of the tree, but the roots are resistant to the disease. Thus, although the American chestnut has essentially been eliminated as a member of the forest canopy, sprouts continue to arise from the root systems of former trees. These sprouts grow into small saplings, which are eventually killed. The American chestnut may well survive for centuries by this tenacious mode of vegetative regrowth.

The rapid disappearance of the American chestnut led to a major change in the composition of the oak and chestnut forests of the Appalachian region. Oaks, and on some sites hickories and tulip poplar (*Liriodendron tulipfera*), filled the gaps left by the death of chestnuts (Spurr 1980). On the Allegheny Plateau to the west, loss of the chestnut was compensated by a variety of early successional trees and species that occurred in mature stands with the chestnut (Spurr 1980). The loss of the American chestnut as a mast-producing tree doubtless had a significant effect on forest wildlife (Campbell and Schlarbaum 1994).

Dutch elm disease comes close to chestnut blight in its ecological and economic destructiveness. The American elm (*Ulmus americana*), the species most severely affected, has an even broader range than the American chestnut. The American elm occurs throughout the eastern deciduous forest region. This species is most abundant on moist lowland or floodplain soils and was extensively used as an ornamental tree in urban areas. The value of trees lost to this disease is estimated to be several billions of dollars (Sinclair and Campana 1975). Dutch elm disease seriously affects slippery elm (*Ulmus rubra*), another widespread species, and other native American elms are also somewhat susceptible.

Dutch elm disease probably originated in Asia, where elms resistant to the disease are found. It received its name because of its decimation of European elms in Holland in the early 1900s. Dutch elm disease was introduced to Ohio in imported European elm veneer logs in about 1930. This fungal disease is transmitted from tree to tree by two species of bark beetles, one of which is a European exotic introduced to North America at about the same time as the disease itself.

Within about forty years, Dutch elm disease killed most mature American elms and has greatly reduced the abundance of this species. In 1944, for example, Champaign-Urbana, Illinois, had about 14,000 American elms gracing its streets. By 1972, only forty living trees remained (Sinclair and Campana 1975). Because American elms begin to produce seed when relatively young, the net effect of the disease in forests has been to change the demography of the tree from an overstory dominant to that of a short-lived understory species.

The death of American elms has altered plant and animal composition of the affected forests. The bird composition of many forest areas was considerably altered during the height of the disease. In Trelease Woods, a research area of the University of Illinois, for example, elm deaths increased populations of a number of woodpeckers but also of exotic birds such as the European starling and English sparrow (Kendeigh 1982). Forest composition also shifted. As the canopy elms have declined in numbers, other trees of moist bottomland soils have increased in abundance.

Several other important trees of the eastern forests are suffering from exotic diseases, and their long-term prospects are uncertain. Butternut canker, an exotic fungal disease, has virtually eliminated this species from the Appalachians, including Smoky Mountains National Park (Schlarbaum 1994), as well as from much of the upper Midwest. White ash (*Fraxinus americana*) is being attacked by a number of exotic fungal blights. Dogwood (*Cornus florida*) is now suffering high mortality throughout much of its range because of dogwood anthracnose, an exotic disease that appeared in Ohio in 1990 (Daughtrey et al. 1996). On the Cumberland Plateau in Tennessee, a study in 1995 found that no fruit was produced by dogwoods in either moist cove or drier upland forest, and that seedlings and saplings had experienced very heavy mortality (Hiers and Evans 1997). Dogwood appears likely to disappear from these forests. Because dogwood plays an important role as a calcium "pump" that brings this mineral from deep soil levels to the forest floor, loss of this species could have major repercussions for calcium nutrition of many forest plants and animals.

Exotic Forest Insects

Exotic forest insects compound the ecological problems of tree disease. Of these insects, the gypsy moth is probably the most destructive. The gypsy moth was brought to North America in 1869 from France by an amateur lepidopterist who was interested in breeding silkworms. Within a few years, it escaped into the wild in Medford, Massachusetts. The moth became established, causing local outbreaks that stimulated efforts to eradicate it. In the case of this European biotype, females are flightless, a feature that probably limited the initial rate of spread of this insect. In about 1905, however, the gypsy moth began a gradual spread that still continues, now at a rate of about 13 miles per year (Liebhold et al. 1992). By 1991, it had invaded about 200,000 square miles of the northeastern states and eastern Canada. Its range now extends west to Michigan and south to Virginia, with outliers of infestation in the Midwest, Rocky Mountains, and Pacific Northwest.

The gypsy moth, an insect widespread in Eurasia, feeds on many woody plants, with oaks being preferred. Sweetgum (*Liquidambar styraciflua*), widely distributed in the southern United States, and quaking aspen (*Populus tremuloides*), transcontinental in the northern states and southern Canada, are also highly preferred species (Liebhold et al. 1997). The distribution of highly preferred hosts, in fact, extends far beyond the present range of the moth. Several important understory shrubs, such as hazelnut (*Corylus* spp.) and serviceberry (*Amelanchier* spp.), are also preferred hosts. When preferred food species have been exhausted, gypsy moth larvae will feed on a wide range of other hardwoods and conifers.

Because of the distribution of its preferred hosts, the gypsy moth will probably spread throughout the forested region of eastern North America. Incidents of long-distance dispersal, usually as a result of egg masses transported on cars or trucks, suggest that it will eventually reach parts of western North America where oaks and other broad-leaved trees occur. An Asian strain of the gypsy moth appeared near port areas in Vancouver, British Columbia, and Portland, Oregon, the result of egg masses brought here on Russian cargo vessels. This biotype of the species has females capable of flight, so that its establishment might lead to very rapid spread of the moth throughout western North America. These initial invasions are thought to have been eradicated.

Defoliating outbreaks of the gypsy moth tend to occur in areas where at least 20 percent of the forest consists of oaks or other preferred species of trees (Liebhold et al. 1997). Defoliation weakens trees, increasing their susceptibility to other insects and diseases (Liebhold et al. 1996). Repeated defoliation can kill up to 90 percent of the preferred host trees, as well. Defoliation favors the increase of nonpreferred tree species, slowly altering the composition of affected forests. Defoliation also reduces forest transpiration and increases water runoff. By increasing litter and litter decomposition, it adds to nitrogen losses in stream outflow.

Outbreaks of the gypsy moth have varied impacts on birds and mammals. Gypsy moth outbreaks can reduce mast production by oaks (Gottschalk 1989). Oak mast is an important food of several forest mammals, including white-footed mice (*Peromyscus leucopus*) and white-tailed deer (*Odocoileus virginianus*), which play major roles in the transmission of Lyme disease to humans (Jones et al. 1998). Forest-edge birds may be benefited by gypsy moth outbreaks, but reproductive success of forest-interior birds may be reduced (Thurber et al. 1994). Defoliation of canopy trees reduces suitable nesting sites for canopy birds. It also encourages increased growth and fruiting of understory shrubs, which in turn attracts increased numbers of terrestrial, omnivorous mammals that can prey on ground-nesting birds. Studies with artificial nests suggest that predation on nests is almost twice as great in defoliated forests as in undefoliated stands. Furthermore, the greatest intensity of predation is at or near the ground surface. Many birds breeding in defoliated forest areas thus may not be achieving replacement breeding success.

Another exotic tree pest, the hemlock wooly adelgid, is a sap-feeding homopteran related to scale insects, whiteflies, and aphids (Watson 1992). The insect positions itself on small branches and twigs, inserts its sucking mouthparts into the tree's vascular tissues and feeds on the phloem fluid. Once in place, the adelgid covers itself with a soft, white material, from which its "wooly" appellation comes. Severe infestations drain the tree of its nutrients, and these may inject toxic substances that disrupt hormonal and vascular processes. Adelgids are dispersed in part by wind, but birds and mammals such as deer also transport eggs and the "crawler" stages of the insect on their feathers or fur (McClure 1990).

The hemlock wooly adelgid is native to eastern Asia, where it is a harmless associate of species of Asian hemlocks (Watson 1992). It invaded the Pacific Northwest in the 1920s, its host becoming the western hemlock

(*Tsuga heterophylla*). By 1953, it had spread to eastern North America, attacking the eastern hemlock (*Tsuga canadensis*) and Carolina hemlock (*Tsuga caroliniana*). It now occurs from Massachusetts to North Carolina (Orwig and Foster 1998).

The areas of most concentrated damage by the hemlock wooly adelgid are in central Virginia and coastal New England. In southeastern New York and Connecticut, tree mortality has been heavy. In some Connecticut stands, the density of hemlock has been reduced by over 95 percent (Orwig and Foster 1998). In 1988–1989, in addition, it was detected in Shenandoah National Park, Virginia. There, many old-growth hemlocks had been killed and others severely defoliated. Hemlocks are a unique component of eastern North American forests—long-lived, evergreen, canopy trees of unique life-form. Their presence is strongly associated with that of several birds, including the barred owl (*Strix varia*), the great horned owl (*Bubo virginianus*), the ruffed grouse (*Bonasa umbellus*), and several songbirds. If this insect is not brought under control, eastern and Carolina hemlocks might be reduced to minor components of the eastern forests.

A close relative of the hemlock wooly adelgid, the balsam wooly adelgid, attacks North American firs of the genus *Abies.* It appeared in New England in 1908, arriving on European nursery stock. It quickly spread through the balsam fir (*Abies balsamea*) forests of New England and eastern Canada. Moving south through the Appalachian Mountains, it has virtually eliminated the southern form of balsam fir and is now contributing heavily to mortality of Fraser fir (*Abies fraseri*). Here, the adelgid is one of several factors, including acid deposition, that are creating stress for trees in high-elevation forests. These southernmost spruce-fir forests, rich in endemic species, are an archipelago of isolated communities with biotas related to those of the boreal forests of the northern United States and Canada.

In Great Smoky Mountains National Park, Rabenold et al. (1998) found that twenty years of adelgid infestation had eliminated almost all large Fraser firs and had promoted increased mortality of red spruce (*Picea rubens*) by windthrow. In the more open, drier stands, breeding bird populations declined by 41 percent. Six species typical of mature spruce-fir stands declined by more than half, and several species typical of brushy, early successional habitat increased in abundance. In other mountain areas in the southern Appalachians, where smaller areas of spruce-fir habitat exist, the declines of bird species typical of mature stands has been even greater.

Many other exotic, herbivorous insects pose threats to eastern forests.

Beech bark disease, for example, is induced by a scale insect that was accidentally brought to northern New England on nursery stock of European beech, probably in the 1920s. Damage to the outer bark by the scale permits fungi to invade the inner bark, weakening or killing the tree. Beech bark disease has spread slowly through the northeastern states, killing 50 percent or more of the beeches in many places. Winter moth, introduced to Nova Scotia in the 1930s, is a defoliator of various hardwood trees, particularly red oak (*Quercus ruber*). By the 1950s, it had become a serious pest in parts of eastern Canada. Now, it appears to be largely under biological control by two introduced parasites. Recently, Asian long-horned beetles have become established in the New York City and Chicago areas. Between 1996 and 1998, over 2,400 infested trees have had to be removed in New York City, at a cost of over $1 million (Haack et al. 1997). Because the beetle attacks maples, elms, and a variety of other hardwood trees, it poses a serious threat to eastern forests.

Invasive Plants

Largely in response to disturbance and stress, several species of deciduous trees from Europe and Asia have become frequent invaders of eastern forests. One of the most serious, ecologically, is the Norway maple. Norway maples were first brought to North America in 1756 as urban ornamentals and are now one of the most common street trees in the East. Norway maple is very shade tolerant and is able to invade undisturbed, closed-canopy deciduous forests. It is now a common tree in many eastern forests.

At the Drew University Forest Preserve in New Jersey, studies have shown the ecologically destructiveness of Norway maple. Norway maple had invaded the Drew Forest by 1915 (Webb and Kaunzinger 1993). It now constitutes over a quarter of the density of stems 1 inch or greater in diameter, and it is the third most important tree from the standpoint of total basal area (Wyckoff and Webb 1996). Norway maple seedlings are far more abundant than those of beech and sugar maple, the two native canopy dominants, beneath the canopies of beech and maple. Norway maple saplings are far more common beneath beech than those of any other tree, and beneath sugar maple, they are equal in abundance to those of sugar maple itself. Beneath Norway maple, the only saplings present are those of Norway maple. Furthermore, the richness of understory plants is significantly lower beneath Norway maple canopies than beneath those of beech and sugar maple.

The ailanthus tree, or tree-of-heaven, was brought to the eastern states from France in about 1784. Originally from China, it had been brought to France in 1751. By the early 1800s, it was planted widely as an ornamental because of its rapid growth, tolerance of polluted urban sites, and attractive appearance. Soon, however, its less desirable features emerged: the offensive odor of flowering male trees and the invasive nature of the tree in urban and nonurban habitats.

Made famous as "the tree that grew in Brooklyn," ailanthus is an opportunistic invader of disturbed sites. Mature trees have a prodigious capacity for production of winged seeds, which are distributed widely by wind. Seedlings quickly establish a deep taproot and grow rapidly, overtopping young native tree seedlings. The leaves and bark of ailanthus contain allelopathic chemicals that inhibit germination and growth of other plants. Spreading from waste places and disturbed forest edges, ailanthus invades openings in forest stands in the eastern states, where it can crowd out native herbaceous and woody plants.

Several other trees have become naturalized in eastern forests, although their ecological impacts have been little studied. White poplar and white mulberry have become established throughout much of eastern North America. Princess tree and golden rain tree occur widely from New York to Florida and west to the Mississippi River. Amur maple, a small tree with potential to become a major weed, occurs in the wild in several areas in Illinois and Missouri (Ebinger 1996).

A number of exotic vines have also invaded forest edges and openings in eastern North America. From these starting points, they often overgrow understory vegetation, suppress tree reproduction, and even smother tree canopies. The heavy load of these vines in tree crowns also makes the trees more vulnerable to wind and ice damage. The most pernicious of these vines are Japanese honeysuckle, English ivy, oriental bittersweet, and, most recently, tearthumb polygonum. Kudzu, especially in the southern states (see Chapter 9), is another.

Japanese honeysuckle was planted throughout most of the eastern United States "by those who do not value the rapidly destroyed indigenous vegetation" (Fernald 1950). This pernicious vine invades thickets and forest edges, eventually colonizing openings within forest areas. Comparisons of Japanese honeysuckle with native vines have shown that in northern states Japanese honeysuckle retains leaves for a greater portion of the year. It begins new leaf expansion earlier in the spring, gaining as much as two months of growth before insect herbivores become active. In southern areas, Japanese honeysuckle is evergreen. In addition, it is better able than native honeysuckles to compensate for herbivory (Schierenbeck et al. 1994).

Tearthumb polygonum, the most recent viny invader, is an annual. Nevertheless, its stems can climb to heights of 15 feet or more in shrubs and trees. Introduced to Pennsylvania in the 1930s, it is spreading rapidly in the middle Atlantic region. How serious an invader of forests it will prove to be is still uncertain.

Shrubs, as well, are serious deciduous forest invaders. These include Tatarian and Amur honeysuckles, autumn olive, common buckthorn, Japanese barberry, common barberry, and multiflora rose. These shrubs were all introduced deliberately to North America, and their planting was often actively promoted by governmental wildlife and conservation agencies. Most have bird-dispersed fruits, which is a major contributor to their invasiveness. Several, including common buckthorn and both species of shrub honeysuckles, derive part of their success from their extended vegetative season (Harrington et al. 1989). They leaf out early in spring and retain their leaves longer than most native shrubs.

Tatarian honeysuckle, one of the first ornamental shrubs brought to North America, was introduced before 1800. It was also planted widely as a wildlife cover and food plant. Now, however, it has proved to be an aggressive invader of northeastern forests, spreading from edges and openings into stand interiors (Woods 1993). Tatarian honeysuckle can form a nearly continuous shrub layer in forests that had only a sparse growth of native understory shrubs. In Vermont, the density and diversity of herbs, as well as the density of tree seedlings, declined with increasing cover of Tatarian honeysuckle (Woods 1993).

Amur honeysuckle was introduced to North America in the late 1890s as an urban ornamental (Luken and Mattimiro 1991). Now naturalized in forest and shrubland areas in at least twenty-three eastern states, it is a particular problem in forest nature preserves in Ohio and Kentucky. It produces abundant, bird-dispersed fruits. It also seems to be almost completely free of herbivory, suggesting that it has escaped herbivores that exploited it in Asia (Hutchinson and Vankat 1997). Near Oxford, Ohio, Amur honeysuckle has invaded many areas of disturbed deciduous forest (Hutchinson and Vankat 1997). In ninety-three forest stands, the cover of this exotic shrub averaged 25 percent and reached over 95 percent in some stands that had been invaded for more than twelve years. Amur honeysuckle reduces the diversity and density of tree seedlings, and thus it opposes successional reestablishment of closed-canopy forest. The cover of native herbs is also reduced by Amur honeysuckle.

Common buckthorn, a large shrub introduced as a hedge plant, has invaded deciduous forests from Quebec west to Minnesota and south to Virginia and Missouri. Escaping into farm fencerows and pasture, it has now become a common invader of deciduous forest understories. In the MacArthur Woods Forest Preserve in Lake County, Illinois, for example, the understory consists almost entirely of buckthorn, with a density of almost 200 shrubs per acre (Osmund 1997).

In New Jersey, New York, and eastern Pennsylvania, Japanese barberry has become a serious invader of the understory of closed forests, including many nature preserves (Ehrenfeld 1997). Introduced in 1875 as an ornamental and hedge plant, this shrub has gradually spread inward from the edges of forest stands. Its foliage is unpalatable to deer, which has contributed to its widespread use in landscaping in the Northeast, and may also be a factor promoting its invasion of forest areas with white-tailed deer. A recent survey (Ehrenfeld 1997) reported this species to be present, and often abundant, in forty-three protected forest areas in the New York–New Jersey area.

Common barberry and multiflora rose, like buckthorn, were introduced as hedge plants. Multiflora rose was also promoted by wildlife agencies as a shelter and food plant for quail and other wildlife. These shrubs are established throughout the northeastern U.S. and southern Canada, and they are common invaders of forest edges and openings.

Among herbaceous plants, garlic mustard has become the most widespread and serious invader of the forest floor community. Garlic mustard was probably brought to North America by immigrants from Europe; the first record of the species was on Long Island, New York, in 1868. It was used as a flavoring herb in cooking and was thought to have value in treating ulcers and gangrene. The plant spread rapidly, and it is now found throughout eastern forests, especially on calcareous soils, from southern Canada to Tennessee and Virginia. It has also become established in the Pacific Northwest (Nuzzo 1991).

Garlic mustard, a shade-tolerant biennial, invades broadleaf forests and quickly becomes the understory dominant. Disturbance of the forest floor, such as by vehicular activity or even trampling by white-tailed deer, promotes its establishment, but disturbance is not necessary for invasion of garlic mustard. As a biennial or winter annual, garlic mustard can form a dense canopy before native spring herbs emerge. It produces large quantities of seed. Within ten years of its invasion of a forest stand, it can become the forest floor dominant (Nuzzo 1991). The dense canopy of garlic mustard can choke out not only these native herbs, but also vines and the seedlings of shrubs and trees.

Forest fragmentation contributes to the invasion of almost all of these trees, vines, shrubs, and herbaceous exotics. As the forest is reduced to small patches, the relative extent of edge habitats increases, providing greater opportunity for the invasion of exotics. Few analyses of this potential influence have been carried out. In Indiana, small old-growth woodlots surrounded by agricultural fields were found to be relatively resistant to invasion by exotic herbs, vines, and shrubs. Most of the exotics present were not shade tolerant, however. Access of shade-tolerant exotics to these stands may, in fact, have been prevented by the open farmland habitats that surrounded them (Brothers and Spingarn 1992).

Other Exotics

With such profound changes in eastern forest vegetation, it is perhaps surprising that so few exotic birds and mammals have become serious problems. The European starling is the most serious exotic bird in the eastern forests, because of nest-hole competition with native species (see Chapter 16). Sika deer, wild boar, and feral hogs (see Chapters 14 and 16) are also problem species in forest areas of the East and Southeast.

Forest Restoration

Fully mitigating the impacts of exotics on the eastern forests is unrealistic, given the massive changes that have occurred and the continuing invasion of exotics of many sorts. But partial restoration may be possible. For the chestnut and American elm, efforts have centered on developing biological control systems and in finding resistant genotypes. American and Chinese chestnuts are being crossed, for example, to introduce resistance genes against chestnut blight to the American tree. In an effort to inactivate the chestnut blight fungus, a genetically engineered virus that outcompetes the fungus is also being tested.

At a U.S. Department of Agriculture laboratory in Glenn Dale, Maryland, several resistant American elm genotypes have been isolated (Line 1997). The two most resistant genotypes, named "New Harmony" and "Valley Forge" after their homes, have been cloned in large numbers and made available for sale by nurseries. Even these genotypes, however, are not completely resistant. These first plants are almost all destined for urban planting. Reestablishing wild, resistant elms over the full range of the American elm is another matter. Even if planted in numbers, American elms require fifty or more years to mature. Restoration must also establish populations in which the resistant genotypes predominate over remnant trees with high susceptibility to Dutch elm disease. Because elms in the wild are outcrossing plants that reproduce mainly by seed, interbreeding with sensitive strains will cause the purity of the resistant genotypes to decline over several generations.

Efforts are being made to develop biological controls for several of the major insect pests of deciduous forest trees. The gypsy moth is subject to epizootics of a nuclear polyhedrosis virus, which appeared in outbreak populations of the moth in the late 1800s. These epizootics reduce gypsy moth populations to low densities that persist for a number of years. In 1989, a second pathogen, a fungus (*Entomophaga maimaiga*), appeared, stimulating interest in its possible use as a biological control agent. Concern arose, as well, about whether or not it might interfere with the ability of the virus to reach an epizootic state and decimate populations of the moth (Hajek 1997). In any case, it appears that the North American gypsy moth is accumulating pathogens from which it might initially have been free.

These efforts are encouraging. Protecting and restoring the composition of eastern forests, however, must recognize the complex interrelations that exist between all major abiotic and biotic stresses. Stresses due to pollution, changing atmospheric composition, and global warming may exacerbate the impacts of exotic tree diseases and insects (Mattson 1996). In turn, these may increase invasions of exotic woody and herbaceous plants, which thrive on disturbance. Although the resilience of the eastern forests appears substantial, at some point changes in composition may translate into loss of ecological function, such as reduced productivity and impaired nutrient cycling. These sorts of ecosystem impacts are considered in more detail in Chapter 17.

The complexity of the changes now occurring in eastern North American forests demonstrates that simple species-by-species remediation cannot protect and restore these ecosystems. Nothing short of a total ecosystem approach to management of the eastern forests can alleviate the impacts of exotics and the stresses that are promoting their invasions.

Chapter 9

Florida and the Gulf Lowlands: Hostile Ecosystem Takeovers

> A precious and irreplaceable part of Florida's, and the nation's, heritage is disappearing. Plants, animals, and entire ecosystems that took tens of millions of years to evolve are at risk. What is being gained in their place? A hodgepodge of species found in other parts of the world. . . . Florida is being homogenized, and everyone, for all time to come, will be the poorer for it.
>
> —E. O. Wilson in Simberloff et al., *Strangers in Paradise**

A new profession has emerged in urban areas of southern Florida: the large reptile trapper (Dalrymple 1994). In Broward County, for example, a 20-foot reticulated python had to be removed from under a house, where it had resided for some time, apparently feeding on small urban mammals and household pets (Table 9.1). At least five other species of large boas and pythons sometimes appear in urban and suburban areas, along with three species of large monitor lizards. One trapper of large reptiles has reported that he removes more than forty large boas and pythons annually. The spectacled caiman, another large, exotic reptile, is locally established in Dade and Palm Beach Counties. Most of the egg-laying reptiles appear unlikely to establish breeding populations in southern Florida, but numerous reports indicate that the boa constrictor, a livebearer, does reproduce successfully. What danger these exotic reptiles pose is uncertain, but they serve to emphasize the truly enormous and diverse

*E. O. Wilson (1997). In D. Simberloff, D. C. Shultz, and T. C. Brown (Eds.). *Strangers in Paradise*. Island Press, Washington, DC.

Table 9.1. Scientific names and native regions of invasive exotics that have become established in Florida and the Gulf Coast.

PLANTS	
air potato, *Dioscorea bulbifera*	Africa, Asia
Asian climbing fern, *Lygodium microphyllum*	Asia
Australian paperbark, *Melaleuca quinquenervia*	Australia
Brazilian pepper, *Schinus terebinthifolius*	Brazil
Burma reed, *Neyraudia reynaudiana*	Asia
carrotwood, *Cupaniopsis anacardioides*	Australia
Chinese tallow, *Sapium sebiferum*	Eastern Asia
cogon grass, *Imperata cylindrica*	Southeast Asia
horsetail casuarina, *Casuarina equisetifolia*	Australia
hydrilla, *Hydrilla crassipes*	Sri Lanka
Japanese climbing fern, *Lygodium japonicum*	Japan
kudzu, *Pueraria montana*	China
latherleaf, *Colubrina asiatica*	Hawaii
papaya, *Carica papaya*	Central America
skunk vines, *Paederia foetida* and *P. crudasiana*	Asia
suckering casuarina, *Casuarina glauca*	Australia
torpedo grass, *Panicum repens*	Europe
tropical soda apple, *Solanum viarum*	South America
water hyacinth, *Eichhornia crassipes*	South America
water lettuce, *Pistia stratiotes*	South America
wetland nightshade, *Solanum tampicense*	Central America
FRESHWATER INVERTEBRATES	
Asiatic clam, *Corbicula fluminea*	Asia
goldenhorn marissa, *Marissa cornuaurietus*	South America
red-rimmed melania, *Melania tuberculata*	Asia
spiketopped applesnail, *Pomacea bridgesi*	Brazil
MARINE INVERTEBRATES	
wood-boring isopod, *Sphaeroma terebrans*	Indo-Pacific
INSECTS AND LAND MOLLUSKS	
bromeliad weevil, *Metamasius callizona*	Mexico
citrus blackfly, *Aleurocanthus woglumi*	Asia
giant African snail, *Achatina fulica*	Africa
prickly pear moth, *Cactoblastis cactorum*	West Indies
red fire ant, *Solenopsis invicta*	South America
tortoise beetle, *Chelymorpha cribaria*	West Indies
FISH	
black acara, *Cichlasoma bimaculatum*	South America
blue tilapia, *Oreochromis aureus*	Africa
flathead catfish, *Pylodictus olivaris*	Mississippi Valley
Mayan cichlid, *Cichlasoma uropthalmus*	Central America
oscar, *Astronotus ocellatus*	South America
peacock bass, *Cichla ocellaris*	South America

pike killifish, *Belonesox belizianus*	Central America
spotted tilapia, *Tilapia mariae*	Africa
walking catfish, *Clarias batrachus*	Southeastern Asia
AMPHIBIANS AND REPTILES	
boa constrictor, *Boa constrictor*	South America
brown anole, *Anolis sagrei*	West Indies
Cuban treefrog, *Osteopilus septentrionalis*	Cuba
greenhouse frog, *Eleutherodactylus planirostris*	Bahamas
marine toad, *Bufo marinus*	Tropical America
red-eared turtle, *Pseudemys scripta*	United States
reticulated python, *Python reticulatus*	Asia
spectacled caiman, *Caiman crocodilus*	South America
BIRDS AND MAMMALS	
armadillo, *Dasypus novemcinctus*	North America
black-hooded parakeet, *Nandayus nenday*	South America
Eurasian collared dove, *Streptopelia decaocto*	Eurasia
feral hog, *Sus scrofa*	Eurasia
monk parakeet, *Myiopsitta monarchus*	South America
nutria, *Myocastor coypus*	South America

nature of the invasion of nonindigenous species in Florida and in the Gulf Lowlands.

Florida and the Gulf Lowlands are second only to Hawaii in the magnitude of invasion by nonindigenous species. Many of these exotics are well adapted to the altered conditions that humans have created and are outcompeting native species. In southern Florida, especially, the spread and increase in abundance of exotics have reached crisis status. Exotic species are invading every major ecosystem. More ominously, they are converting large areas of natural ecosystems into new ecosystem types dominated by exotics.

Vulnerability to Exotic Invasions

The invasion center for exotics in this region is south Florida. Why is this? Three features make south Florida vulnerable: its insular nature, an equable climate, and the massive disturbance of the landscape that has accompanied urban-agricultural development (Simberloff 1994, 1997). Complementing these features is the massive inflow of exotic species through air and sea commerce (Gordon and Thomas 1997).

Although the Florida peninsula is broadly connected to the southeastern United States, it is still much like an island. Peninsular Florida, bounded on three sides by water, is in a sense bounded on the north by freezing weather, a condition as inhospitable to many subtropical species as an ocean. In addition, although we think of south Florida as a lush subtropical region, its isolation from main areas of the tropics has left it an impoverished biota. Also, high sea levels during Pleistocene interglacial periods isolated parts of Florida as real islands. The Florida Keys today show such an isolation pattern.

Florida's climate is permissive. Temperatures are mild to warm, and rainfall is plentiful. Freshwater lakes and streams are abundant, as are coastal marshlands and marine habitats. Thus, a wide range of terrestrial and aquatic organisms find available habitats.

Physical disturbance of the land surface and alteration of hydrology have created additional favorable conditions for exotic species, both terrestrial and aquatic. Agriculture, phosphate mining, urban development, and construction of canals for water management and transportation have all created welcoming habitats for exotics.

South Florida, too, is the major entry point for many kinds of organisms. Miami has long been a port of entry for agricultural and horticultural plants and for pet and aquarium animals. As a result, most of the problem exotics of Florida were introduced deliberately. Major seaport facilities in Florida and the Gulf Coast have also been the entry points for many unintentional introductions.

Nonindigenous species are a major component of Florida's biota. Florida has about 2,523 species of native plants and an estimated 925 exotic species that have become established in the wild (Frank and McCoy 1995a). Thus, about 27 percent of the wild flora of the state is exotic. Because perhaps 25,000 other species are in agricultural or horticultural cultivation, the potential for future plant invasions is enormous. For example, most of the 60 or so species of fig trees (genus *Ficus*) that have been planted ornamentally do not produce fertile fruit because of the absence of specific wasp pollinators. But wasp pollinators for several species have recently become established, giving these trees the potential to reproduce and invade wildlands. Non-indigenous animals are also prominent in Florida's biota, especially in south Florida. A thousand or more arthropods and more than 100 vertebrates have become established in Florida.

Invasive Land Plants

Many of the established exotic plants are still very restricted in distribution or are nonaggressive in their growth patterns. Others grow and spread unrestrained, causing massive disruption of native ecosystems. The area invaded by exotic plants is greatest in south Florida, where 17.6 percent of the acreage of natural areas was dominated by exotic species, compared to 9.2 percent in central Florida and almost zero in northern Florida (Gregg 1994).

Terrestrial plant invaders include trees, vines, grasses, and broad-leaved herbaceous plants. Most serious are several tropical trees. These include Brazilian pepper, Australian paperbark, casuarinas, Chinese tallow, and carrotwood. These trees often form dense, single-species stands. They constitute the most serious exotic threat to open marshlands and coastal shrub and strand communities.

Brazilian pepper is probably the most widespread woody exotic in Florida. It is an evergreen shrub or small tree that was introduced as an ornamental plant. A pioneer species of disturbed land, its small berrylike fruits are dispersed widely by fruit-eating birds. Brazilian pepper is quick to invade disturbed areas such as canal and ditch banks and idle farmland. About 800,000 acres are now infested by pepper trees (Jones and Doren 1997). Mature stands of these trees have lower density and diversity of native birds than do native pinelands and forest edges (Jones and Doren 1997). Brazilian pepper also invades undisturbed coastal marshland, shrubland, and maritime hammocks. In coastal marshes, it eventually forms a low, closed-canopy forest that destroys foraging areas for herons, egrets, and other waterbirds. In maritime hammocks, it can form a dense understory, crowding out native shrubs and tree seedlings. Increasingly, it is invading the inland portions of mangrove swamps. By 1994, an estimated 100,000 acres of mangroves in Everglades National Park had been heavily invaded by Brazilian pepper. These areas are rendered unsuitable as waterbird rookeries and as breeding habitat for native mangrove songbirds. Once Brazilian pepper has become established in a mangrove habitat, its eradication is logistically almost impossible because of the swampy terrain and densely tangled structure of the vegetation.

The Australian paperbark tree is probably the most damaging exotic species in south Florida wetlands. A member of the eucalyptus family, the paperbark is evergreen and produces attractive spikes of white flowers, sometimes flowering five times a year. The name paperbark comes from its thick, spongy bark, the outer layers of which peel off in large sheets. A fast-growing tree of swampy land in its native Australia, it was introduced to Florida in the early 1900s as an ornamental and as a species that would help "drain the Everglades" by transpiring large amounts of water from wetlands. The altered hydrology resulting from the construction of drainage canals and dikes, however, created a habitat ideal for the growth and rampant spread of this tree.

Australian paperbark has proved to be strongly invasive on every type of wetland soil, from freshwater to brackish (Bodle et al. 1994). A mature tree produces millions of seeds, which are shed in response to stresses such as cold, drought, fire, or, unfortunately, herbicide treatments. The tree's resinous leaves make its foliage flammable, while the bark protects the trunk and branches from fire damage. New branchlets sprout quickly from the trunk and branches after fire. Seeds germinate quickly in moist soil, especially in disturbed areas, enabling the plant to colonize ditch banks, lake margins, moist pastures and wet prairies, open sawgrass marshes, cypress swamps, and the inland margins of mangrove swamps (Myers 1983; Bodle et al. 1994). In just twenty-five years, a sawgrass marsh can be converted into a closed-canopy stand of paperbarks, beneath which few other plants can survive. Paperbark also invades the ecotone between pine forest and cypress swamp, displacing pond cypress (*Taxodium ascendens*) (Myers 1984; Ewel 1986). Drier habitats, such as pine flatwoods and dry shrublands, are also invaded to a lesser degree.

Paperbark now infests almost 500,000 acres of Florida south of Lake Okeechobee. In 1992, the South Florida Water Management District estimated that 26,000 acres consisted of paperback stands with 95 percent or more tree cover. Without control, the spread and population growth of paperbark could enable it to take over all remaining suitable habitat in about thirty years.

Chinese tallow is a fast-growing, deciduous tree that reaches a height of 30–35 or so feet (Bruce et al. 1997). The tree was introduced to the United States in the late 1700s, perhaps by Benjamin Franklin, who reportedly sent seeds to a plantation owner in Georgia in 1772. Later, it was promoted for possible production of oil from its seeds, as an ornamental tree and as a flowering tree useful for honey production. The U.S. Bureau of Plant Industry, an agency of the Department of Agriculture, introduced the tree near Houston, Texas, in the early 1900s. It soon escaped, first colonizing riparian areas and then spreading to uplands. The seeds of Chinese tallow are apparently dispersed primarily by flowing water, but some birds consume and probably disperse them, as well. The tree has a poisonous, milky sap, and various tannins and phenols in its decaying leaves are reported to be toxic to many aquatic invertebrates. This accumulated litter seems to facilitate germination of the Chinese tallow itself. It has been claimed that the litter is toxic to other woody plants, but no experimental evidence for this exists.

In Florida and Texas, Chinese tallow has invaded riparian borders, hardwood hammocks, bottomland forests, and coastal prairie. It now occurs in coastal areas from south Texas to Florida and north to North

Carolina. In Texas, it is a particularly serious invader of the Gulf coastal prairie, an endangered ecosystem type. Near Houston, Texas, aerial photographs document the invasion of coastal prairies, beginning in about 1958, and their conversion to closed-canopy tallow woodland in twenty to twenty-five years. In some locations, tallow has displaced native grasses and has formed a nearly closed tree canopy in less than ten years. The long-term fate of these woodlands is uncertain, because young individuals of several native trees are found in many stands. Possibly, these species will increase in relative abundance and increase tree diversity in tallow-invaded areas (Bruce et al. 1995). Nevertheless, Chinese tallow has permanently transformed large areas of prairie to woodland.

Two casuarinas, or Australian pines, the horsetail casuarina and the suckering casuarina, are established in Florida. Nearly 400,000 acres have been invaded by these trees. Horsetail casuarinas tend to colonize open, sandy habitats, especially on shores and barrier islands. They quickly invade sites disturbed by human activity or storms, but they also invade sandy areas with undisturbed strand vegetation. Casuarinas can completely displace native plants, so that some ecologists predict that, without control, they could completely eliminate native coastal vegetation in south Florida. On sandy coasts and barrier island beaches, where stands of horsetail casuarina have replaced herbaceous strand vegetation, the beach profile has been altered. Gently sloping beaches have been replaced by upper beaches, stabilized by casuarina roots, that drop off sharply to the wave-wash zone. Such beaches are unsuitable for nesting by loggerhead and green sea turtles, and possibly also by the American crocodile. Constriction of the beach habitat is especially serious in the Florida Keys, where the original beaches were already narrow. Horsetail casuarina is not tolerant of prolonged flooding and so is largely restricted to well-drained sites. Suckering casuarina, however, is more tolerant of wet conditions and invades sites such as sawgrass marshes and hardwood hammocks.

Two much more recently established trees are also becoming problem species. Carrotwood, an evergreen tree that grows to about 30 feet in height, was introduced to Florida in 1980 as an ornamental plant. It is now invading disturbed, low-lying areas on both east and west coasts of south Florida. Maritime hammocks in these areas are now often fringed by dense thickets of carrotwood. It can germinate and grow beneath Australian pines, compounding the effect of these invaders on native vegetation. The full nature of the threat that this plant poses is still uncertain, but it seems likely to be invasive in fresh and estuarine wetlands throughout all of Florida (Oliver 1992). Papaya, cultivated extensively in south Florida, has been dispersed into the seed banks of many Everglades hammocks. Following Hurricane Andrew in 1992, papaya germinated and grew rapidly in many damaged hammocks, creating an exotic canopy in only four months (Horvitz et al. 1995).

A host of viny plants, including kudzu, air potato, Asian and Japanese climbing ferns, and skunk vine, has also invaded native vegetation in Florida and neighboring states, smothering the canopies of trees and shading out understory species. These exotics pose the most serious threat to native forest vegetation.

Kudzu, a deliberate introduction to the southeastern United States, infests some 2 million acres of forest from Alabama and Tennessee to the Carolinas and Florida. In Florida, it has spread through the north and central regions and threatens to invade the Everglades region. Kudzu tends to invade disturbed areas and forest edges, where it overgrows and smothers all native vegetation. It is tolerant of long periods of flooding, and thus it is well adapted to invade the hardwood hammocks of the Everglades.

Air potato, not a potato at all, is a high-climbing vine that produces aerial clusters of potato-like tubers. Introduced to Florida in 1905, it is now an aggressive invader of tropical hardwood hammocks (Schmitz et al. 1997). It climbs into tree canopies, shading out the foliage of both trees and understory plants. Hurricane damage to trees, such as that caused by Hurricane Andrew in 1992, enables the air potato to develop such extensive canopy foliage that recovery of the trees is inhibited.

Skunk vine, Asian and Japanese climbing ferns, and latherleaf are other climbers that are widely established in Florida. In south and central Florida, skunk vine, its seeds dispersed by birds, has invaded almost all hardwood hammocks. In Withlacooche State Forest, in west-central Florida, skunk vine has eliminated almost all native understory plants, and it and air potato have become major canopy invaders. Asian climbing fern also tends to invade tropical hardwood hammocks and cypress domes, killing canopy trees and shading out understory natives. The climbing ferns can smother the canopies of shrubs and trees up to 30 or so feet in height, forming thick mats over their foliage. Japanese climbing fern has invaded northern Florida. Both climbing ferns are spreading and are likely to invade most of the state. Latherleaf, a viny, sprawling shrub, appeared in the Florida Keys in the 1950s. It invades and overgrows native plants of coastal dunes and marshes, maritime hammocks, and mangrove swamps. Similar to the climbing vines, it overgrows the canopy of invaded vegetation, shading out native species.

Several coarse, exotic grasses are also serious invaders of native vegetation. Cogon grass and Burma reed are tall grasses that are highly flammable, and they promote a regime of frequent fire that favors their own spread. Cogon grass was brought to Alabama and Mississippi in the early 1900s and was subsequently planted in Florida for soil stabilization and as a pasture forage. It forms dense, monospecific stands that crowd out all other plants and become fire-prone once the shoots die. Its deep rhizomes survive fire and resprout quickly. Cogon is the most widespread invader of forest habitats in Florida and can crowd out almost all native plants in the understories of pine flatwoods. In pastures, cogon can become dense enough to exclude gopher tortoises. Torpedo grass, another invader, forms dense stands along wetland borders, crowding out native species. Torpedo grass is now one of the dominant plants in marshes of Lake Okeechobee.

Tropical soda apple and wetland nightshade are two spiny, noxious weeds that have recently invaded open habitats in Florida (Fox and Bryson 1998). Tropical soda apple now occurs on more than 1.2 million acres of upland pasture. Wetland nightshade has colonized wetter habitats in six counties in southwest Florida. Whether these plants will become major threats to natural communities is not yet known.

Exotic Land Animals

Terrestrial habitats are also invaded by numerous animals. About 1,000 exotic insects are established in Florida, or about 8 percent of the 12,500 insect species estimated to occur in the state (Frank and McCoy 1995b). In the United States, Florida is second among U. S. states only to Hawaii in the percentage of exotic insects. These exotics represent many different orders and families, but ants, mayflies, and weevils are the most numerous (Frank and McCoy 1995b). Most of these exotics have entered Florida unintentionally. Some have arrived by overland dispersal from neighboring states or by aerial dispersal from Caribbean islands. Others have reached Florida as stowaways on air or sea shipments of plants. Miami International Airport is the entry point for most living plants brought into the continental United States, and most such plants carry some living insects such as eggs, larvae, or adults.

About forty-two exotic arthropods have deliberately been introduced for the biological control of insects and weeds. In some cases, these are exotics brought in to control exotics. An exotic flea beetle (*Agasicles hygrophila*), for example, has become an effective control on alligatorweed (*Alternanthera philoxeroides*), formerly a pernicious aquatic weed.

Most of the unintentional insect invaders are pests of agriculture and horticulture, but some have invaded native vegetation. Of the latter, the red fire ant is the most deleterious. This ant entered the United States near

Mobile, Alabama, between 1933 and 1945 (Vinson and Greenberg 1986). Since then, it has spread west to Texas, north to North Carolina, and south through Florida. It is likely to invade much of the Pacific Coast. News reports in 1999, in fact, indicated that a major infestation had been discovered in Orange County, California. The red fire ant has two social forms: one in which each colony has a single queen ant, and a second in which each colony has many queens. The single-queen colonies are strongly territorial, a behavior that limits their density. The multiple-queen colonies are not territorial and may reach high densities. In extreme cases, newly invaded areas may possess 2,500 colonies per acre. Once established, a density of 20 to 30 large colonies per acre is normal. The densities of fire ants in North America, however, are about four to seven times those normal for their range in South America, apparently because they have escaped limitation by many of their natural enemies in South America (Porter et al. 1997).

The red fire ant invades disturbed areas, including lawns and pastures and also less disturbed habitats such as pine flatwoods, savannas, and wetland borders. Red fire ants displace most native ants, including native fire ants. They are primarily predaceous and reduce the general diversity of native ground-dwelling arthropods. They also invade ground nests of birds, such as bobwhite quail, and those of reptiles, such as the American alligator (*Alligator mississippiensis*) and the six-lined racerunner (*Cnemidophorus sexlineatus*), preying on eggs and newly hatched young (Allen et al. 1998). Drees (1994) found that red fire ants attacked nestlings, causing a 92 percent reduction in nesting success of waterbirds nesting on the ground and in low shrubs on spoil islands in Galveston Bay, Texas. Red fire ants may also be responsible for declines of birds such as the common nighthawk, ground dove, and eastern meadowlark along the Gulf coast.

Other recent insect invaders are threatening endemic plants in south Florida. The bromeliad weevil was apparently brought in on imported bromeliads (Frank and Thomas 1994). It has become naturalized and is threatening populations of several native bromeliads. The prickly pear moth has been used effectively in Australia and on several small West Indian islands as a biological control on exotic prickly pear cacti. Unfortunately, this moth has spread on its own from the West Indies to the Florida Keys, where it now endangers endemic native species of prickly pear cacti. Likewise, the tortoise beetle, a leaf-feeding species that probably entered Florida on imported sweet potato stock, now threatens two endangered species of morning glories.

Four species of frogs and toads have invaded Florida. The marine toad, the world's largest toad, is almost invulnerable to predation as an adult, because of toxic secretions from its parotid glands. The Cuban tree frog secretes a noxious slime that deters predation by birds and snakes. These species may displace native toads and frogs and thus alter predation relationships in food chains that include these species. The greenhouse frog does not have a tadpole stage, but it develops directly from the egg to a tiny adult. If this species displaced other native frogs, aquatic food chains that involve tadpoles would thus be affected.

Some thirty-two species of exotic reptiles, mostly lizards, are now at home in south Florida. Exotic reptiles make up an astonishing 38 percent of the total reptile fauna. Some nineteen species of geckos, West Indian anoles, and other tropical lizards now breed in south Florida; exotic lizards outnumber native species. The brown anole may be a competitor of the native green anole, a widespread native species. As noted earlier, the boa constrictor may be an established breeder in the Miami area.

Eleven species of exotic birds breed in Florida, several of these, such as rock doves, house sparrows, and European starlings, being continentwide exotics. Many more species have escaped and survive in the wild but have not established breeding populations. The Eurasian collared dove, introduced to the Bahamas in 1974 by escapes from a commercial aviary, has now colonized all of Florida, and it is spreading north and west (Smith 1987). It now breeds through the Gulf Coast region, as well as in the Texas panhandle, eastern New Mexico, and southeastern Colorado. The population of this dove is clearly exploding, and it may prove to be the most aggressive exotic invader to North America since the cattle egret. Monk parakeets are now widespread in southern Florida, and the black-hooded parakeet and several other small parrots have localized breeding populations in southern Florida.

About twenty-two species of nonindigenous mammals and four feral domestic species have become established in Florida. These range from squirrel, green, and rhesus monkeys to axis and sambar deer, and even elk. Aside from rats and mice, the nine-banded armadillo and feral hog are probably the most widespread exotics. Armadillos were first introduced near Miami in the early 1900s. They spread throughout the state from subsequent introductions in east-central Florida and Alabama, eventually meeting animals that entered the Florida Panhandle from farther west (Layne 1997). Feral hogs, however, are Florida's most detrimental exotic mammal. Hogs are predators on ground nests of sea turtles, gopher tortoises, crocodiles, and shorebirds. Their rooting disturbs ground-cover vegetation and the litter layer. Hogs also carry diseases such as viral pseudorabies and bacterial brucellosis, making them a threat to other large native mammals, as well as to livestock. Hogs, for example, are one of the prey of the endangered Florida panther. Nutrias occur in many southeastern coastal marshes and are often a disrupting factor in marsh ecology (see Chapter 14).

Invasive Freshwater Plants

Aquatic habitats have their own set of exotic plant and animal invaders. Hydrilla, an aquatic plant rooted in the bottom, and water hyacinth and water lettuce, floating aquatics, grow in such profusion that they form dense mats over the surface of lakes and streams. These mats block out light and reduce oxygen supply to deeper waters, suffocating benthic life. As a result, they almost completely alter the composition and ecosystem function of these freshwaters, thus creating immense problems for boating and other human activities.

Hydrilla was introduced to Florida waterways in the 1950s by the discard of tropical aquarium plants into streams and canals (Joyce 1992). Hydrilla can grow under lower light intensity, more so than most native aquatic plants, and can occupy the deep-water areas of lakes and streams. It quickly became a major aquatic weed throughout Florida and the Gulf states. About 75,000 to 100,000 acres of canals and lakes are now choked by this plant.

Orange Lake, near Gainesville, provides a case history of the impact of hydrilla on Florida lakes. Prior to 1974, when hydrilla first appeared, this 13,340-acre lake possessed 6,670 acres of deep, open water. By 1977, only about 280 acres of open water remained, nearly the entire lake being choked by dense hydrilla mats (Joyce 1992). The impact on sport fishing at Orange Lake was estimated to run into millions of dollars.

Water hyacinth, introduced to the Gulf states in 1884 at an exposition in New Orleans, quickly became an exotic plant scourge of southern waterways. In just four years it had invaded freshwater areas from Texas to Alabama, and in two years more, it reached Florida. By the 1950s, it had covered 120,000 acres in Florida, and by 1972, its coverage was estimated to be 200,000 acres. Water hyacinth can completely cover lakes and streams, shading and reducing oxygen levels in the water, killing native submerged vegetation, and reducing fish populations. Intensive control efforts—mechanical, herbicidal, and biological—have reduced the extent of heavily infested waters to about 2,600 acres.

Water lettuce, present in Florida since at least the 1700s, is probably derived from South American stock. In South America, a variety of host-specific insects are associated with water lettuce, whereas none occur in

Florida (Cordo et al. 1981). How water lettuce reached Florida is uncertain. During the 1980s, water lettuce increased in distribution, triggering intensified control efforts with mechanical and herbicidal procedures. Now, about 2,500 acres are heavily infested with this species.

Invasive Freshwater Animals

Introduced freshwater invertebrates, especially mollusks, impact planktonic and rooted plant producers, native herbivores, and native detritivores. The result is basic alteration of the dynamics of lower levels of food webs in streams, ponds, and lakes. Among these exotics, the Asiatic clam is the most widespread and serious invader. Apparently introduced into British Columbia in the 1920s (see Chapter 5), this small clam reached Florida in the early 1960s, and it now occurs in almost all watersheds. The Asiatic clam develops dense populations that filter out much of the suspended organic matter and plankton that support food chains of native species. The clams and their shells alter bottom conditions, reducing populations of native mud-bottom invertebrates, including native mussels.

Several large freshwater snails, brought in by aquarists and aquaculturalists, have become established in Florida's waters, The spiketopped applesnail is a close relative of the native Florida applesnail, which it tends to displace. The Florida applesnail is the primary food of the snail kite, a state and federal endangered species, which does not feed on the spike-toothed form (Warren 1994). Loss of the Florida applesnail in the Everglades could thus lead to disappearance of the kite. Another snail, the red-rimmed melania, has spread throughout the Florida region. It, and two other melania species, can reach enormous densities and displace a variety of native snails. The goldenhorn marissa, widespread in south Florida, feeds voraciously on rooted aquatic plants, both native and exotic.

About twenty-eight species of alien freshwater fish are permanently established in Florida. Another seven or so species have bred temporarily, only to disappear during cold winters. An additional forty-four species have been found living, but not breeding, in natural waters (Courtenay 1997). Some of these exotics are game species, and some are harvested commercially on a limited basis. Several are considered threats to native fish. The flathead catfish is a voracious predator on other fish and invertebrates. A sport fish native to the Mississippi valley (see Chapter 15), it has invaded the Appalachicola and Escambria Rivers in the Florida Panhandle. It is a threat to several small fish endemic to these coastal drainages, particularly the snail and spotted bullheads (*Ameiurus brunneus* and *A. serracanthus*, respectively). The pike killifish, introduced near Miami in 1957, is an able predator on small fish, particularly the native mosquitofish (*Gambusia holbrooki*). The pike killifish is now spreading to the west and north (Thomas Lodge, personal communication). The ecological threats of other exotic fish are uncertain. The blue tilapia is a plant feeder that reaches large size and can substantially reduce the abundance of aquatic macrophytes. The spotted tilapia and walking catfish are also widespread in freshwaters of south Florida. Several Central and South American cichlids, particularly peacock bass, oscar, black acara, and Mayan cichlid, have invaded undisturbed waters in Everglades Park and other protected areas (Courtenay 1997). All feed on invertebrates and small fish, and they are likely to impact populations of native fish.

As massive as the representation of exotic fish is, few native fishes have yet been endangered. In part, this may reflect simply the lack of data on native fish communities prior to the exotic onslaught. On the other hand, none of the many introductions made in the name of fisheries improvement can be said to have enhanced Florida fisheries (Courtenay 1997).

Among aquatic reptiles, the red-eared turtle has invaded Florida from farther north. It apparently hybridizes with the closely related cooters (*Chrysemys floridana*) and red-bellied turtles (*Chrysemys nelsoni*). The spectacled caiman has localized breeding populations in the Miami area. The impact of this large predator could be great, should it become widespread.

Invasive Marine Organisms

In the marine environment, Florida is home to many species—so-called cryptogenic species—that have probably been introduced to tropical and subtropical regions worldwide by centuries of ship traffic. These include many fouling and wood-boring invertebrates whose geographic origins are cloaked in obscurity. The wood-boring isopod is one such species. It was probably introduced to the western Atlantic from the Indo-Pacific region by early wood-hulled ships. In Florida, it causes damage to the prop roots of red mangrove trees. Some biologists have suggested that this damage limits the invasion of deeper waters by red mangroves. Others, however, have suggested that their effects might, in fact, stimulate root branching and thus be beneficial (Odum et al. 1982). In any case, major deterioration of red mangrove stands does not seem to be occurring where the wood-boring isopod is active.

New Exotic Ecosystems

Plant and animal exotics seem to get along well together in many instances, thus forming their own new ecosystems (Schmitz et al. 1997). Forests of casuarinas now exist along formerly treeless coasts. In many places these stands are developing an understory of carrotwood, one of the few plants that can grow in the litter beneath casuarinas. Similarly, Brazilian pepper and Australian paperbarks form forest communities in formerly treeless marshes. In the Everglades, these stands are populated by exotic Cuban tree frogs, greenhouse frogs, and brown anoles. Other exotics, such as cogon grass, are likely to transform woodland communities into exotic grasslands.

Meeting the Challenge

The appearance of other exotics in Florida and the Gulf Coast region is almost inevitable. Thousands of exotic species, not yet established in the wild, already exist on farms, in urban plantings, or as pets. More are still being brought into the region, legally and illegally. Other species are present in neighboring states and are actively expanding their distribution. Still others are almost certain to be introduced by marine transport activities.

To meet the challenge of these exotics to native biodiversity, all types of control will have to be mobilized in Florida and in the Gulf Coast region. Early detection of new exotics is a key strategy. Successful eradication campaigns, using a variety of techniques, have been mounted against incipient infestations of species such as the citrus blackfly in Key West and the giant African snail in the Miami area. These successes highlight the importance of monitoring to detect new exotics and of the potential of prompt action to eradicate them.

Mechanical, chemical, and biological techniques can all play a useful role. The earliest attempts to control problem aquatic weeds were mechanical. Direct cutting and weeding can be effective approaches in areas of localized infestations of many exotic plants. Chemical pesticides can be used effectively, in a sense as the lesser of evils. Herbicides, for example, have been used extensively in efforts to control water hyacinth, hydrilla, Brazilian pepper, and Australian paperbark.

Biological control, however, offers the greatest opportunities for control of exotics in Florida and the Gulf states. Many exotics have few close relatives in North America, and biological control agents are thus likely to be specific to them. The Australian paperbark and Brazilian pepper, for example, have no close North American relatives but do have hundreds of insect associates in their native regions (Jones and Doren 1997). A number of insects from the native regions of Australian paperbark and Brazilian pepper are under study. A parasitic fly (family Phoridae) from Brazil is also being tested as a biological control for the red fire ant. Discovering safe and effective biological control agents, however, requires extensive study and prolonged testing, and most efforts are still in their early stages.

Some biological control successes have already been achieved. Several aquatic weeds have effectively been reduced with introduced insects. These include alligatorweed, water hyacinth, water lettuce, and hydrilla (Center et al. 1997). Combining biological and chemical control agents may also be effective, as in the case of a fungal agent and herbicide for killing Australian paperbark (Rayachhetry and Elliott 1997).

Preventing the takeover of Florida and the Gulf Coast by exotics presents an enormous challenge. Efforts must be made to reduce the rate of entry of new exotics. Those that do arrive and become established must be identified and eradicated or contained. Widely established exotics must be treated with an integrated control strategy that utilizes the varied, but safe, techniques available. A strong commitment by the citizens of this region, through state and federal action, must underpin these efforts.

Chapter 10

Plains and Intermontane Grasslands: Exotics at Home on the Range

> Others arrived from the West, and found thousands of square miles of ready-made seed bed prepared by the trampling hoofs of range livestock. In such cases the spread was often so rapid as to escape recording. . . .
>
> —Aldo Leopold, "Cheat Takes Over"*

Ranchers of the southern plains are continually searching for the perfect forage grass: long in its growing season, tolerant of drought, productive under high stocking rates, and effective in animal growth. Many ranchers in Oklahoma, Texas, and New Mexico believed they had found it in King Ranch bluestem, variety *songarica* of Caucasian bluestem (Table 10.1). Imported from Russia in 1929, Caucasian bluegrass is a perennial bunchgrass that eventually becomes a sod-forming grass in mature stands. It is a warm-season grass that establishes readily from seed, has a long growing season, tolerates heavy grazing, and is rated a fair forage in quality. King Ranch bluestem has been seeded into pastures throughout much of the southern plains. In western Oklahoma, about 2 million acres are now in King Ranch bluestem.

The trade-off: King Ranch bluestem is extremely invasive, taking over highway rights-of-way to the exclusion of native plants and invading pastures of native grasses. Spreading by rhizomes, it is difficult to control

*Aldo Leopold (1941). "Cheat Takes Over." *The Land* 1:310–313.

Table 10.1. Scientific names and native regions of invasive exotic plants of western North American grasslands.

PLANTS	
Boer love grass, *Eragrostis chloromelas*	Africa
buffel grass, *Pennisetum ciliaris*	Africa
bull thistle, *Cirsium vulgare*	Europe
Canada thistle, *Cirsium arvense*	Europe
Caucasian bluestem, *Bothriochloa ischaemum*	Asia
cheatgrass, *Bromus tectorum*	Eurasia
common crupina, *Crupina vulgaris*	Europe
crested wheatgrass, *Agropyron cristatum*	Eurasia
Dalmatian toadflax, *Linaria dalmatica*	Eurasia
diffuse knapweed, *Centaurea diffusa*	Eurasia
halogeton, *Halogeton glomeratus*	Eurasia
Kentucky bluegrass, *Poa pratensis*	Europe
leafy spurge, *Euphorbia esula*	Europe
Lehmann love grass, *Eragrostis lehmanniana*	Africa
musk thistle, *Carduus nutans*	Eurasia
Russian knapweed, *Centaurea repens*	Eurasia
Russian thistle, *Salsola kali*	Asia
smooth brome, *Bromus inermis*	Europe
spotted knapweed, *Centaurea maculosa*	Europe
yellow star thistle, *Centaurea solsticialis*	Europe
yellow sweet clover, *Melilotus officinalis*	Europe

with either herbicides or mechanical tilling. In regions where native plants struggle to survive in roadsides, cemeteries, and native-grass pastures, King Ranch bluestem is becoming recognized not as a superb forage, but as a noxious weed. In Texas, for example, it has been identified as one of the threats to survival of a federally endangered plant, south Texas ambrosia (*Ambrosia cheiranthefolia*).

The legacy of ranching and grain farming in the plains and the intermontane regions of western North America is the continuing invasion of native grasslands and shrubsteppe by a host of annual and perennial grasses and broadleaf weeds from Eurasia. Adapted to grazing and disturbance by long evolution in the Old World, and often freed from their natural biological control agents, dozens of exotic plants are running rampant. In many cases, these plants are altering basic ecosystem processes, such as fire frequency and nutrient cycling patterns. These alterations often create positive feedbacks, favoring additional exotics. They also promote new stable states, with exotic species differing in life-form from native dominants, in many places. The evidence is now strong that these invasions are causing impoverishment and shifts in species composition of both native plant and vertebrate animal communities.

Grassland Ecosystems

Grasslands, often termed prairies, originally occupied several regions of western North America. These grasslands were dominated by perennial grasses, in some areas almost entirely bunchgrasses, which grow in dense clumps, or in other areas by a mixture of bunchgrasses and sod-forming species. The most extensive grasslands occupied the Great Plains. In the eastern and southern plains, tall-grass prairie predominated, typified by tall-growing species such as big bluestem (*Andropogon gerardi*). In the central plains, tall-grass prairie gave way to midgrass prairie, with dominant species of intermediate height, such as little bluestem (*Schizachyrium scoparium*). Still farther west lie the short-grass prairie, dominated by low-growing bunchgrass and sod-forming grasses, exemplified by buffalo grass (*Buchloë dactyloides*).

Grasslands, known as the Palouse prairie, and mixed grass and shrub communities, termed the shrubsteppe or sagebrush steppe, also occurred in the northern intermontane region—interior British Columbia, the Columbia Plateau, the northern Great Basin, and western Wyoming. Most of the plains and foothills of the more moist, northern sector of this region were originally covered by bunchgrass prairie, dominated by Idaho fescue (*Festuca idahoensis*) and bluebunch wheatgrass (*Agropyron spicatum*). Farther south, and in drier habitats, cold desert shrubs such as basin sagebrush (*Artemisia tridentata*), rabbitbrush (*Chrysothamnus nauseosus*), and bitterbrush (*Purshia tridentata*) were the dominant plants, although bunchgrasses were common in the spaces between shrubs.

On the borders of the Chihuahuan and Sonoran Deserts in southern Arizona, southern New Mexico, and northern Mexico, desert grasslands flourished. These grasslands were also dominated by bunchgrasses, including plains lovegrass (*Eragrostis intermedia*) and various species of grama grasses (*Bouteloua* spp.), three-awn grasses (*Aristida* spp.), dropseeds (Sporobolus spp.), and many others. In California, valley grassland originally covered most of the central valley and the southern coastal zone (see Chapter 12).

Most of the tall-grass prairie and the deep-soil Palouse prairie have been converted to cropland. Nevertheless, in midgrass and short-grass prairie regions, and in intermontane and desert grassland regions, large areas of grassland still survive as grazing lands and protected grasslands. Now, however, these grasslands are experiencing massive waves of invasion by exotic grasses and forbs. On the Great Plains, for example, between 30 and 60 percent of all plant species are exotics. Exotic invasions began with the first ranching and farming enterprises, but have proceeded kaleidoscopically to today, and new species are continually appearing and spreading.

Invaders in the Northern Intermontane Region

The grasslands of the northern intermontane region were the first to experience severe impacts of exotics. European settlement brought cattle ranching to this region in the 1850s, imposing a new ecological pressure on grasslands and shrublands. The bunchgrasses were at once prime forage plants and species intolerant of trampling. Originally, only sparse populations of ungulates, including elk and pronghorn, but not bison, occupied the region (Mack and Thompson 1982; Billings 1990). Unlike the dominant grasses of the Great Plains, most of which are low, sprawling plants that spread by rhizomes and are little affected by trampling, those of the Columbia region were nonrhizomatous bunchgrasses (Mack 1989). Bunchgrasses produce dense sets of shoots, or tillers, at their base. These tillers, growing upright, are vulnerable to grazing by animals, which may destroy their flowering capacity and eliminate the only means of reproduction: seeds. Trampling by large ungulates can also fragment the clump bases of bunchgrasses. Seedlings of many of the bunch grasses are also sensitive to grazing and usually die, even if grazed only once. Trampling also disturbs the soil surface and creates germination sites that are ideal for exotic annual grasses and weeds. Furthermore, the surface of the soil between clumps of bunchgrasses and shrubs is stabilized largely by a crust of lichens, algae, and fungi—so-called cryptogamic or microphytic crusts (West 1990). This crust protects the soil against erosion and, because its components include nitrogen-fixing microorganisms, is a source of fixed nitrogen. Grazing by cattle quickly consumed the prime forages, destroyed the bunchgrass clumps by trampling, and pulverized the cryptogamic crusts.

With penetration of railroads through the region in the 1880s, homesteading of the region began in earnest, and grain farming proliferated. By 1890, more than 2 million acres of land was farmed. Thus, by the turn of the century, much of the Columbia Plateau was severely overgrazed or cultivated for wheat and other grains. Within just twenty-five years of the first wave of settlement, noxious weeds were becoming a serious problem (Mack 1986). In less than fifty years, the region's native ecosystems had been degraded, with many areas almost beyond recovery.

With farming came a host of exotic annual plants, in many cases brought in as contaminants in seeds of grains and legumes. By the 1880s and 1890s, several species of exotic bromegrasses had become established in the intermontane region. Broad-leaved weeds had also invaded. Russian thistle, native to Eurasia, arrived in South Dakota in 1877 as a contaminant in flax seed. By 1900, it had spread westward to the Pacific coast. It now occurs throughout western North America. Not a true thistle, but a member of the goosefoot, or chenopod, family, Russian thistle is a globular, prickly-leaved weed that may develop a diameter of 1.0–1.5 yards. As it matures, a quarter million or so seeds ripen. The quintessential "tumbleweed," Russian thistle is an annual that dies at maturity, its mature shoot separating neatly from the root. The shoot is then free to disperse its seeds as the wind tumbles the dead globe across the land. It is quick to colonize disturbed soil, including cropland and areas of overgrazed range. In the late 1800s and early 1900s, Russian thistle became the most pernicious weed of grain farms in the plains and western states. Usually, however, if disturbance ceases, sites suitable for germination gradually disappear, and Russian thistle declines in abundance. Even small areas of disturbance by digging animals, however, can enable this species to survive at low density in an area, once it has invaded.

The worst was still to come. The most aggressive of the bromegrasses, cheatgrass, first became established near Spence's Bridge, British Columbia, in about 1899 (Mack 1981). A winter annual, preadapted to the winter moisture regime of the interior Northwest, cheatgrass is a hardy and prolific grass. Its specific name, *tectorum,* means "of roofs" and refers to its Eurasian habit of growing on the sod roofs of houses. On the Columbia Plateau, it found disturbed soils and winter moisture to be ideal. It spread throughout the region, following railroad lines that aided in the dispersal of its seeds in seed grain, livestock feed and bedding, and the dung of transported animals. There were even deliberate introductions. Depletion of native grasses led ranchers and even agricultural scientists to seek new grasses to replace the lost natives.

Over the vast region of the Columbia Plateau and northern Great Basin, from British Columbia south through Washington and Idaho to Oregon, Nevada, and Utah, cheatgrass has been the keystone species in the restructuring of grassland and arid shrubland ecosystems. Cheatgrass occurs as a weed in grain fields, giving it the common name "cheat" because it cheats grain farmers of their expected wheat yields. By 1930, cheatgrass had spread throughout the region, colonizing disturbed land and invading shrublands by occupying the bare soil areas between shrubs. These stands of cheatgrass burn readily (Billings 1990), and in shrub-dominated areas of dry rangeland the result is disappearance of the shrubs, most of which do not resprout after fire. The dominant shrub of the shrubsteppe is basin sagebrush (*Artemesia tridentata*), a nonsprouter. Fire thus acts as a positive feedback for spread and increase in abundance of cheatgrass. Since the mid-1940s, hundreds of thousands of acres of rangelands have burned annually in the intermontane region. In this region, nearly 62 million acres are at risk of being converted to monocultures of cheatgrass.

Several broadleaf annual weeds followed cheatgrass. Yellow star thistle, a winter annual native to southern Europe, was first noted in western North America in the late 1800s (Roché et al. 1994). Not until the 1950s and 1960s, however, did it spread and become a major rangeland pest. By the 1990s, it had occupied over 9.4 million acres of the intermontane West, from California and Nevada north to Washington and Idaho. Over 8 million of these acres are in California, and nearly half a million in Idaho.

Yellow star thistle favors deep soils on warm slopes. It germinates in the fall and allocates most of its growth to a leaf rosette with a deep taproot during the winter and spring. In mid- to late summer, when other plants have become dormant because of the hot, dry conditions, yellow star thistle flowers and matures its seed (Roché et al. 1994). Like cheatgrass, it appears unable to invade healthy perennial grassland, but where the grass cover has been disturbed by heavy grazing, it quickly becomes dominant. Its spiny flower heads inhibit grazing by livestock, and it also is toxic to some animals.

Another weedy annual, halogeton, also a member of the goosefoot family, invaded the intermontane region in the 1930s, first appearing near Elko, Nevada (Mack 1986). Halogeton stores quantities of oxalate compounds that make it toxic to livestock. Within twenty-five years, halogeton had spread over 10 million acres of arid rangeland in the northern intermontane region, especially on saline soils, and thousands of sheep had succumbed to its toxins.

Still more recently, in 1969, common crupina, an annual plant of the sunflower family, appeared in Idaho (Thill et al. 1985). From Idaho, common crupina has spread widely through the intermontane region and the Pacific Northwest, now infesting about 50,000 acres of grassland (Roché et al. 1997). It has the potential to become a very serious weed in the Pacific Coast states, as well as in parts of eastern North America (Patterson and Mortenson 1985).

Weedy annuals set the stage for invasion of the northern intermontane region by several exotic perennial grasses and biennial and perennial broad-leaved plants Among the grasses, crested wheatgrass, in dry sites, and smooth brome and Kentucky bluegrass, in moist sites, are common exotics. Leafy spurge, various species of knapweeds, and Dalmatian toadflax are the principal broadleaf biennial and perennial weeds. Several true thistles are also invasive weeds throughout the western grassland region, especially in riparian areas and on moist soils. Among these is Canada thistle (see Chapter 4). Bull thistle and musk thistle are other species that form dense stands of tall weeds.

Among the grasses, crested wheatgrass is perhaps the most serious. Crested wheatgrass was seeded into millions of acres of degraded rangelands and shrubsteppe habitat in the name of range improvement and restoration. Crested wheatgrass can also sometimes spread into undisturbed grassland (Marlette and Anderson 1986). Once established, it resists invasion by native plants.

Leafy spurge is just one of many spurges brought to North America from southern Russia or the Ukraine, probably as a contaminant of grain seed. The form invading North America may actually be a hybrid between leafy spurge proper and a second Old World species, *Euphorbia waldsteinii.* Arriving in the East in about 1827, leafy spurge spread westward, appearing in the Red River Valley of North Dakota in the 1880s. It has now become one of the most pernicious weeds of rangelands and natural grasslands in the northern Great Plains. The most serious infestations of leafy spurge are in Manitoba, Saskatchewan, North and South Dakota, Nebraska, Montana, and Wyoming. About 7.4 million acres of range- and grassland in Canada and the United States are heavily infested. A deep-rooted perennial, able to store large reserves of nutrients underground, leafy spurge grows vigorously and produces abundant seed. Seeds are dispersed by grazing animals in their coats, especially by sheep, and in feces (Olson et al. 1997). The seeds germinate profusely on areas of soil disturbed by livestock grazing, and the plants form dense stands that can reduce range productivity by 50 to 75 percent. Leafy spurge also has a milky, bitter sap that makes it irritating, or even toxic, to cattle and possibly to native ungulates. This sap also gives the plant another common name: wolf's milk.

In natural prairies, leafy spurge stands tend to be associated with areas of soil disturbance, such as paths, vehicle tracks, and firebreaks. Where leafy spurge is abundant, other exotic species, such as smooth brome and Kentucky bluegrass, also tend to be abundant. Dense stands of leafy spurge also crowd out most native plant species (Belcher and Wilson 1989).

The knapweeds are a group of noxious weedy biennials and perennials that have invaded the northern intermontane region. At least fourteen species of these exotics have become established in western North America, and four have become widespread and abundant (Roché and Roché 1988, 1991). Knapweeds, growing 2–3 feet tall, are weeds in grain fields and disturbed sites, such as roadsides and railroad rights-of-way. Most tend to invade rangelands that have been grazed heavily enough to suppress native perennial grasses, and here they may form dense stands that crowd out most other plants. Many of these weeds were introduced as contaminants in alfalfa seed brought to this country from Europe and Asia.

Spotted knapweed first appeared on Vancouver Island, British Columbia, in 1905 and was introduced to Montana in the 1920s. Spotted knapweed is a herbaceous biennial or perennial with a strong taproot. It was apparently introduced to the Midwest in seed of alfalfa and clover. In summer, it can form dense stands of shoots over 3 feet in height. Now it infests about 7.2 million acres in the western states, mostly in Montana and Idaho. In western Montana, spotted knapweed is the most serious rangeland weed. Based on its climatic behavior in the Old World, spotted knapweed could invade more than 38 million acres in the intermontane region (Chicoine et al. 1985).

Two other related species, diffuse and Russian knapweeds, have also become widespread grassland invaders. Diffuse knapweed was first recorded from alfalfa fields in Washington State in 1907. Now it infests about 3.2 million acres in the western United States, mostly in Idaho, Washington, and Oregon. Diffuse knapweed might ultimately infest 12.2 million acres. Russian knapweed was introduced to western North America in 1928 as a contaminant in alfalfa seed that was widely sold. It now occurs on about 1.4 million acres in most western states, with its most serious infestations being in Idaho.

The knapweeds degrade the habitats invaded and promote conditions favoring their dominance. Some, such as spotted and diffuse knapweeds, can invade ungrazed native grasslands (Lacey et al. 1990; Sheley et al. 1997), but grazers, which tend to avoid these weeds and concentrate on native grasses and forbs, can speed the takeover of grassland by these exotics. Their seeds remain viable in the soil for a decade or more, so that once established they are difficult to eliminate by any treatment. As grasses are replaced by these weeds, the ground surface becomes less protected against erosion, and soil loss increases.

Dalmatian toadflax, a herbaceous perennial with showy, light yellow flowers, was brought to North America as a garden ornamental in 1894 (Vujnovik and Wein 1997). In the 1920s and 1930s, it became naturalized throughout much of the United States and Canada, particularly in the western states and provinces. It invades a wide range of open habitats, spreading both by seed and by vegetative growth. Although it is a particularly aggressive invader of disturbed sites, it also invades undisturbed grasslands and sagebrush habitats. It is a detrimental rangeland weed because it contains alkaloids and glycosides that are toxic to livestock.

Impacts of Exotics in the Intermontane Region

Ecologically, the result of these invasions has been the conversion of ecosystems dominated by native perennial grasses and shrubs to systems dominated by exotic grasses and broad-leaved weeds, in some cases annuals, in others perennials. Dominance of annuals such as cheatgrass leads to major changes in nutrient cycling patterns. Deep-rooted grasses and shrubs are replaced by shallow-rooted annual grasses that draw nutrients and moisture from surface soil layers only. Fire and decomposition volatilize nutrients or release them to the surface, where they are vulnerable to erosion. Furthermore, as colonists of disturbed soils, cheatgrass and other exotic annual grasses have largely displaced native annual grasses that acted as colonizing species. The perennial grasses and weeds tend to form monocultures of plants largely unpalatable to native herbivores and unsuitable for many smaller vertebrates.

At the most basic level, that of the soil, exotic annual grasses and broad-leaved weeds have reduced the infiltration of water, increased runoff, and accelerated erosion. Shallow-rooted annual grasses and perennials such as the knapweeds simply do not hold the soil. On a spotted knapweed site, for example, water runoff was 56 percent greater than on a nearby native bunchgrass site, with erosion being 192 percent greater. This equates to a reduction in soil moisture availability for the deep-rooted perennials and thus acts as a positive feedback favoring exotic annuals.

Exotic plant invasions are also threatening native plant diversity. Cheatgrass, in particular, forms monospecific stands that exclude almost all other plants. Knapweeds can also reduce native plant diversity. In Glacier National Park, for example, spotted knapweed first established along roadways but then spread into native grasslands. As knapweed colonies spread into grasslands, many native plants become locally rare (Tyser and Key 1988).

Native animals, as well as livestock, have been affected. Transformation of the shrubsteppe by exotics may lead to either decreased or increased numbers of small mammals. When shrubsteppe habitat is converted to cheatgrass, for example, both the diversity and the density of small mammals, including least chipmunks, kangaroo rats, various species of mice, and Townsend's ground squirrels, in most cases decline. In Colorado and Wyoming, however, rangeland infested by Russian knapweed had three times the density of rodents that existed in native rangeland. Larger animals are also affected. Deer and elk often feed on the rosettes of knapweeds early in winter and spring. In central Idaho, elk and deer were common on winter range infested by spotted knapweed, and deer especially fed heavily on spotted knapweed rosettes (Wright and Kelsey 1997). A study in Montana, however, found that the use by elk (*Cervus canadensis*) of range dominated by spotted knapweed was only about 2 percent as great as the use of native bunchgrass range (Beck 1993). Herbicidal control of spotted knapweed in Montana also showed a 47 percent increase in biomass of native grasses palatable to native ungulates (Rice et al. 1997).

Many native birds of shrubsteppe habitat on the Columbia Plateau are declining, at least partly because of the impacts of exotics. Few native birds use cheatgrass habitat (Rich 1995). When sagebrush communities are replaced by cheatgrass, the number of breeding bird species drops from 8–12 to only 1–2. Total breeding bird density declines from 150–328 per 100 acres to about 40 per 100 acres. The birds that decline or disappear include several sagebrush specialists, such as the sage grouse, sage thrasher, sage sparrow, and Brewer's sparrow. Other species that decline regionally as sagebrush habitat disappears are the loggerhead shrike, rock wren, gray flycatcher, vesper sparrow, green-tailed towhee, spotted towhee, lark sparrow, and western meadowlark. The only passerine species present in cheatgrass monocultures are likely to be the horned lark and the western meadowlark, but even these generalists are much less common than in sagebrush habitat. Burrowing owls and long-billed curlews, on the other hand, increase in numbers as cheatgrass takes over. Both of these species hold large territories and benefit, to some degree, by the elimination of shrubs.

In the Great Basin, sagebrush specialist songbirds also decline in diversity when shrubsteppe habitat is converted to crested wheatgrass (Bradford et al. 1998). In Saskatchewan, where stands of crested wheatgrass tend to be more open than native mixed-grass prairie, the diversity of songbirds and the abundance of certain species, such as Sprague's pipit (*Anthus spragueii*), may be reduced by conversion of prairie to crested wheatgrass stands (Sutter and Brigham 1998).

Declines in rodent and jackrabbit populations due to the spread of cheatgrass may be translating into decline of golden eagle (*Aquila chrysaetos*) populations in the intermontane region. In 1997, fewer golden eagles than ever before were observed in fall migration by HawkWatch observers at the Grand Canyon, Arizona, observation station. Fewer golden eagles than normal were also recorded in 1997 at several other hawk migration lookouts in the western states (Anonymous 1998).

Where shrubsteppe is transformed to stands of cheatgrass or dense stands of crested wheatgrass, the reptile fauna is also impoverished (Rich, in preparation). Some species, such as the sagebrush lizard (*Sceloporus graciosus*), are dependent on the presence of sagebrush shrubs. Others, such as the horned lizard (*Phrynosoma douglassi*), require a habitat with bare ground between shrubs and grass clumps.

Exotics in the Western Great Plains Region

Other grassland regions of North America have also suffered. In the western Great Plains, many of the invaders have been perennial grasses, most of these the result of deliberate introductions. Smooth brome, crested wheatgrass, and several other exotic wheatgrasses have been seeded into grazing land for supposed range improvement. The exotic wheatgrasses tend to interbreed with the twenty or so native species of wheatgrasses (Bock and Bock 1995). They also spread into areas of disturbed soil, and, once established, are slow to be outcompeted by native vegetation. Many of these grasslands have also been invaded by Kentucky bluegrass, leafy spurge, and yellow sweet clover.

In Theodore Roosevelt National Park in southwestern North Dakota, for example, only 13 percent of mixed-grass prairie sites in which plant composition was sampled were free of exotic species (Larson et al. 1997). The abundances of several of these exotic species—Kentucky bluegrass, leafy spurge, smooth brome, and Japanese brome—were negatively related to the diversity of native plant species. Riparian areas were most heavily invaded by exotics, especially yellow sweet clover, which forms tall, dense stands.

Few studies have been conducted on the impact of these exotics on native birds and mammals of the western Great Plains. In Manitoba, bird communities tended to be richer in native mixed-grass prairie than in grassland dominated by smooth brome, Kentucky bluegrass, and leafy spurge. Seven of the eight species recorded in native prairie also occurred in exotic grassland. The presence of one species, Baird's sparrow (*Ammodramus bairdii*), was significantly related to a low-growing juniper that was present in native prairie, but not in exotic grassland. Sprague's pipits (*Anthus spragueii*), upland sandpipers (*Bartramia longicauda*), and western meadowlarks (*Sturnella neglecta*) were other species significantly correlated with plant species of native prairie, whereas grasshopper sparrows (*Ammodramus savannarum*) and clay-colored sparrows (*Spizella pallida*) were correlated with the abundance of Kentucky bluegrass and smooth brome, respectively. Thus, exotic species caused a substantial shift in composition of the bird community.

In Theodore Roosevelt National Park, infestations of leafy spurge reduced the use of preferred habitats by several ungulates (Trammel and Butler 1995). Grassland use by bison (*Bos bison*) was reduced by 83 percent in infested areas. Use of creeping juniper (*Juniperus horizontalis*) and little bluestem habitat by deer (*Odocoileus* spp.) was also reduced by up to 70 percent. Leafy spurge apparently reduced the availability of preferred native food plants, but it may have caused animals to avoid infested areas because of taste aversion.

Exotics in the Southern Plains and Desert Grasslands

In the southern portions of the Great Plains and the southwestern desert grasslands, deliberate introductions of perennial range grasses have also created exotic problems. As noted at the chapter outset, a variety of Caucasian bluestem has been distributed widely in the southern Great Plains as a pasture grass. This species not only flourishes in areas where it has been seeded, but also aggressively colonizes roadsides and other ungrazed areas that are the repositories of native plant diversity in this region. Buffel grass is another exotic introduced in many parts of the southern plains. Both of these grasses are considered threats to native plant diversity.

In desert grasslands, a number of love grasses have been introduced for erosion control and range improvement. Lehmann love grass and Boer love grass have displaced the plains love grass, a native, over large areas. Lehmann lovegrass was introduced to the southwestern United States in the early 1930s after an extensive search for grasses to rehabilitate degraded desert grasslands in Arizona, New Mexico, and Texas (Humphrey 1994). For this purpose, it has been successful, in that it forms dense, single-species stands of perennial grass. These stands are fire-prone during dry summer periods, and burning favors increased dominance of the species (Anable et al. 1992). It also produces enormous quantities of tiny seeds—6 million per pound—that are distributed easily by wind and water. Mature plants may produce two crops of seed annually. Its fire tolerance and reproductive capacity, plus the fact that cattle selectively graze native grasses during their primary growth and reproduction period in summer, have allowed it to displace native grasses, even on moderately grazed areas. Lehmann love grass now covers about 350,000 acres of rangeland in Arizona, although it was seeded on only about a third of this area.

In addition, stands of exotic love grasses are low in diversity of native plant and animal species (Bock et al. 1986). In Arizona, areas seeded with exotic love grasses lost over 60 percent of their native grass cover, over 50 percent of their herb species, and almost 60 percent of individual shrubs. The dominant insect group, grasshoppers, declined in numbers by 44 percent. Three of the four grassland breeding birds declined significantly in abundance. Of the five principal birds wintering in grassland, four declined significantly and one showed no change. Among small mammals, three species declined significantly, three showed no change in abundance, and one increased. The bird species, Botteri's sparrow (*Aimophila botterii*), and the mammal, the hispid cotton rat (*Sigmodon hispidus*), that were favored by love grasses are species typical of floodplain habitats where sacaton grass (*Sporobolus wrightii*), a form similar in growth pattern to love grass, is the natural dominant. Diversity of both native plants and animals was thus substantially reduced in areas where love grasses became dominant. As the authors of a report on the Arizona studies noted, ". . . plantations of these African aliens are biologically sterile, compared with undisturbed native grassland" (Bock et al. 1986, p. 462). Furthermore, although they are efficient soil holders, the exotic love grasses are now regarded as being poor forage for cattle.

In the Southwest desert, buffel grass shows similar properties. Native to the deserts of northwestern Kenya, buffel grass was brought to the Southwest by the U.S. Soil Conservation Service in 1946. It was also planted widely in Sonora, Mexico. In northern Sonora and southern Arizona, buffel grass produces dense stands that are prone to fire, a factor to which most native desert plants are intolerant. Buffel grass is thus displacing cacti and desert shrubs over large areas.

Controlling Invasive Exotics

Containing or suppressing these exotics, many of which are still spreading, is an enormous challenge. Herbicides alone do not provide a solution. Certain herbicides cannot be used against exotic grasses because they also kill native grasses. Herbicides are usually too expensive to use against broadleaf weeds infesting hundreds of thousands of acres. Furthermore, unless herbicide treatment completely eliminates a weed species, its recovery is likely and repeated treatments become necessary. Perennial knapweeds, which have deep rhizomes that survive when shoots are killed, are particularly difficult to control with herbicides. Herbicides may also kill native broadleaf plants, as well as have other side effects.

Efforts are being made to develop control strategies for exotic rangeland grasses and forbs that combine herbicide use with other techniques, including intensive grazing pressure, and competition from perennial grasses. Whitson and Koch (1998), for example, found that in Wyoming cheatgrass can be reduced and native grasses encouraged by repeated herbicide treatments in combination with intensive grazing. Herbicidal treatments combined with seeding of several perennial grasses and grazing by sheep were effective in reducing leafy spurge and Russian knapweed in Wyoming (Bottoms and Whitson 1998; Ferrell et al. 1998). Unfortunately, some of the most effective grasses in these treatments are themselves exotics, and their use may not be appropriate where protection of native plant biodiversity is desired.

Biological control can also contribute to control of many prairie and rangeland weeds (Louda and Masters 1993). The establishment of biotic enemies that are specific to noxious weeds can directly reduce their populations or, at least, remove the competitive edge they have over native species (also see Chapter 18). The goal of biological control is reduction, but not elimination, of the exotic weed so that its biotic enemies are also able to survive and exert continuing control. Thus, finding and introducing appropriate natural enemies are considered to be the most profitable strategies for controlling exotics in western rangelands.

Biological control programs have been initiated for several of the most pernicious broadleaf weeds of the northern plains and intermontane region. For leafy spurge, for example, the most promising control agents appear to be a series of six flea beetles (genus *Aphthona*) that feed on the plant's leaves and roots (Hansen et al. 1997). Three other species of insects have also been identified as potential biological control agents: a gall-forming fly (appropriately named *Spurgia esulae*), a root-boring moth (*Chamaesphecia hungarica*), and a root-boring beetle (*Oberea erythrocephala*). These nine insects have been released widely, and at least one species is established in 148 counties in 18 states.

The knapweeds have a rich fauna of specialized arthropods in their Eurasian home areas, so that biological control agents are a good possibility. A host of insects, including gall-forming flies, root-feeding beetles, flower-feeding moths, and gall-forming mites, is under study as possible biological controls of spotted and diffuse knapweeds. Gall-forming nematodes, mites, and wasps, together with a leaf rust, are under study as possible control agents for Russian knapweed.

Biological control is an approach that can often be combined with herbicide use, grazing pressure, and plant competition. In North Dakota, for example, herbicidal treatments combined with biological control are more effective in controlling leafy spurge than these are techniques alone (Lym 1998), and programs that combine several or all of these approaches are likely to be even more effective.

Many of these control programs are still in their early stages, so that the prospects of widespread control cannot yet be assessed. The fact that some exotic plants have been almost fully controlled by biological agents suggests that additional successes are likely, however (see Chapter 20). The variety of biotic agents that have been identified for many problem exotics is impressive, which gives hope that integration of biological control with other approaches can suppress many of the exotic plagues of plains and intermontane grasslands.

Chapter 11

Western Floodplains: Disturbing the Disturbance Regime

> Unless management practices are implemented that control the spread of Russian-olive and salt cedar . . . the riparian community of the Rio Grande will likely become dominated by exotic shrubs over the next 50 to 100 years as the current overstory cottonwoods die out.
>
> —William Howe and Fritz Knopf, "On the Imminent Decline of Rio Grande Cottonwoods in Central New Mexico"*

Bosque del Apache National Wildlife Refuge spreads over nearly 90 square miles of the Rio Grande floodplain south of Socorro, New Mexico. The term "bosque," Spanish for "woodland," denotes the open forest of Fremont cottonwood (*Populus fremontii*), Goodding willow (*Salix gooddingii*), and peach-leaf willow (*Salix amygdaloides*) that originally covered much of the Rio Grande floodplain in central New Mexico. This woodland supported an understory of several native shrubs, such as coyote willow (*Salix exigua*), seepwillow (*Baccharis glutinosa*), false indigobush (*Amorpha fruticosa*), and New Mexico olive (*Forestiera neomexicana*). It hosted the most diverse array of wildlife of the interior Southwest.

Perhaps 90 percent of the *bosque* habitat in New Mexico has been

*William Howe and Fritz Knopf (1991). "On the Imminent Decline of Rio Grande Cottonwoods in Central New Mexico." *Southwestern Naturalist* 36:218–224.

destroyed or greatly modified by human activities and the invasion of exotic species. At Bosque del Apache, saltcedar and other alien trees threaten to replace nearly all native species (Table 11.1). Virtually impenetrable stands of saltcedar now cover much of the floodplain close to the river channel. Saltcedar invaded the refuge area in the late 1930s but expanded explosively in abundance in the early 1940s. The last significant reproduction of cottonwoods occurred in 1942, and most of the large trees that have survived date from about this year (Howe and Knopf 1991). Construction of Cochiti Dam, near Santa Fe, in 1975 eliminated the spring floods necessary for cottonwood reproduction along the middle Rio Grande (Crawford et al. 1996). Beneath the large cottonwoods, an understory of saltcedar now crowds out young cottonwoods and many of the native shrubs. But an even greater threat exists if fire occurs. For example, over four days in June 1996, a wildfire swept through more than 6,000 acres of *bosque* habitat, including 4,100 acres in the refuge. Over half of the refuge's cottonwood *bosque* was burned, and the mature cottonwoods were killed. Left to natural processes, ecologists predict that this entire burned area will develop into a monospecific saltcedar jungle. The cost of combatting this possibility during the first five years of recovery has been put at $5 million.

The invasion of Bosque del Apache is one example of the massive replacement of native riparian vegetation by exotic trees, shrubs, and weeds that is now occurring in the western United States. This replacement highlights the critical role of natural disturbance in maintaining

Table 11.1. Scientific names and native regions of the invasive exotics of western riparian habitats.

PLANTS	
ailanthus, *Ailanthus altissima*	China
Bermuda grass, *Cynodon dactylon*	Europe
Johnson grass, *Sorghum halapense*	Africa
musk thistle, *Carduus nutans*	Europe
Osage orange, *Maclura pomifera*	United States
Russian olive, *Elaeagnus angustifolia*	Eurasia
saltcedar, *Tamarix chinensis*	Eurasia
saltcedar, *Tamarix parviflora*	Eurasia
saltcedar, *Tamarix ramosissima*	Eurasia
Siberian elm, *Ulmus pumila*	Asia
teasel, *Dipsacus fullonum*	Europe
white mulberry, *Morus alba*	Europe
white sweet clover, *Melilotus alba*	Europe
yellow sweet clover, *Melilotus officinalis*	Europe

many diverse and productive natural ecosystems. Any increase, decrease, or qualitative change in the disturbance regime is likely to reduce or eliminate native species adjusted to the prevailing regime and consequently will favor exotic invaders (Hobbs and Huenneke 1992).

Throughout western North America, human activities have altered the pattern of flooding, fire, and animal activity in riparian zones. The result is an accelerating invasion of exotics and displacement of native plants and animals. Invasion of exotic plants is, however, only the most recent phase of the accelerating sequence of ecological alteration of southwestern rivers that has followed European settlement (Crawford et al. 1996). Prior to human intervention, these rivers experienced major, but highly variable, spring flood flows, the result of snowmelt in the mountains (Molles et al. 1998). Their channels shifted and meandered across their floodplains (Rood and Mahoney 1990). Flooding created microhabitats required for germination of cottonwoods and willows, and the meandering of channels produced a complex mosaic of discontinuous stands of trees of differing ages.

Human Modification of Riparian Ecosystems

Settlement of the valleys by farming peoples initiated change in the vegetation and hydrology of the river systems. In Arizona and New Mexico, Indian farming peoples settled the floodplains of major rivers between about 600 and 1,000 years ago, clearing land for crops. Sophisticated irrigation systems were developed in central Arizona, whereas the New Mexico pueblos initially relied largely on natural flood irrigation. In most places, however, the impact of these activities was small.

With Spanish settlement, ranching and irrigated farming expanded. Large areas of the floodplains were cleared. The 1800s brought an explosion of change. Fur trappers decimated beaver populations in the mountain streams that fed southwestern rivers, and hundreds of thousands of dams that moderated spring flood flows were lost. Following the Mexican War, acquisition of the region by the United States brought large-scale ranching, logging, and mining that disturbed the watersheds and promoted runoff and erosion. Sedimentation and waterlogging of floodplains of the lower rivers resulted. In the 1900s, dams, levees, and drainage ditches were constructed to regulate the flows of most rivers. Spring floods were contained, and the channels were confined. Elimination of flooding also reduced decomposition of litter and increased the accumulation of flammable material beneath the floodplain forest (Ellis et al. 1998). *Bosque* fires became more frequent and intense.

These changes set the stage for the current invasion of exotics (Busch and Scott 1995). The most serious invaders were several species of trees: three saltcedars and Russian olive. These trees were deliberately introduced but have become naturalized with a vengeance. Their success is due largely to the regulation of flow in western rivers. After spring flooding, native cottonwoods and willows adapt and establish themselves quickly. They produce abundant wind-dispersed seed in late spring and early summer, just in time to colonize moist sand and mud bars left by the spring floods. Because they are not tolerant of shade, and their seeds have only a few weeks of viability, these are essential conditions for germination and early growth of trees of the *bosque* woodlands. On most streams, dams and levees have largely eliminated the intense spring floods and have confined flows to narrow main channels, where any cottonwood and willow seedlings that do appear are quickly swept away.

Similar requirements for cottonwood establishment exist on the Missouri River in Montana. Most plains cottonwoods (*Populus deltoides*) along the river west of Fort Peck Reservoir date from years of large flood flows (Busch and Scott 1995; Scott et al. 1997). These floods created bare areas of alluvial soil on higher floodplain benches, where seedlings were not scoured away by lower floods. Altering the river's flood regime to eliminate large floods means that cottonwood seedlings can establish only on the lowest banks and bars, where they are soon swept away. The result here also has been invasion of exotic riparian trees, in this case Russian olive.

In Arizona, recent floods illustrate the importance of flooding for establishment of native trees (Stromberg 1997). Major floods occurred on the Hassayampa River in the winters of 1993 and 1995. These floods created a scour zone 55 yards wide that was ideal for the germination of Fremont cottonwood and Gooding willow. Both natives germinated earlier than saltcedars and were able to overtop both the exotics and native shrubs. A new generation of native trees was thus able to establish.

Saltcedars and Their Ecological Impacts

Three species of saltcedars have become serious invaders in western riparian areas. Saltcedars are not cedars at all but angiosperms with linear leaves that appear needle-like. The genus *Tamarix*, named after the Tamaris River in Spain, contains about ninety shrubs and trees native to dry regions of Europe, Asia, and North Africa. In the mid-1800s, eight species of saltcedars were introduced to the western United States (DiTomaso 1998). Some were planted for windbreaks or to stabilize stream banks; others, as ornamentals. Several of these species became naturalized by the late 1800s. Saltcedars now occur from Montana to Texas and west to California. *Tamarix ramosissima* and *T. parviflora* are widespread but most common in Arizona and California. *T. chinensis* is most common in Arizona, New Mexico, Texas, and Oklahoma (DiTomaso 1998).

Invasion of riparian environments began along major rivers such as the Colorado, Salt, Gila, Rio Grande, and Pecos, especially where dams had moderated the intensity of spring floods. In the lower Colorado River valley, riparian cottonwood woodland originally covered about 5,000 acres along the river. Many of the large cottonwoods were cut for fuel in the 1800s, when steamboats plied the river. When the riverboat era ended in the early 1900s, the cottonwood stands began to recover. Saltcedar apparently invaded sometime between 1910 and 1920, however, and spread rapidly over the next few decades. By the 1920s, saltcedars had invaded most of the major river systems in the Southwest. In the 1930s and 1940s, they spread rapidly, in the Green River of Utah advancing upstream by about 12.5 miles annually (Graf 1978).

Since the 1940s, however, saltcedars have continued to spread and have invaded tributary streams, desert washes, and desert springs. Now, saltcedar dominates 44 percent of the riparian zone of the lower Colorado, and less than 500 acres of pure cottonwood habitat remain (Hunter et al. 1988). Invasion has been even greater in New Mexico. Along the middle Rio Grande, 64 percent, and in the middle Pecos, 93 percent of the riparian zone is dominated by saltcedar. By the late 1980s, about 1.5 million acres of riparian habitat in the southwestern United States had been taken over by saltcedar (DiTomaso 1998).

All three saltcedars are large, multistemmed shrubs that generally grow to heights of 16–26 feet. These species are perhaps the epitome of pernicious woody weeds. Saltcedars grow rapidly, achieving a height of 10–13 feet in their first year. Their wide-spreading root system, which produces adventitious shoots, enables them to form dense stands on the banks and floodplains of permanent streams (DiTomaso 1998). Their deep roots also enable them to reach the water tables beneath the channels of temporary streams and desert washes. The deeper roots can penetrate to water tables at depths of 5–20 feet. Saltcedars may begin reproduction in their first year, producing fruits that can be dispersed widely by both wind and water (DiTomaso 1998). Mature shrubs may produce half a million tiny seeds annually. Unlike cottonwoods and willows, which concentrate their seed production in early summer, saltcedars produce seeds throughout the summer. These seeds germinate quickly when wetted and colonize soils wetted by summer storms or floods.

The physiology of saltcedars also gives them a competitive edge over native riparian species. Established seedlings are able to tolerate severe moisture stress for more than four weeks and recover normal function if moisture becomes available (Devitt et al. 1997). Older established plants are hardy, tolerating desert heat, winter cold, saline soils, and prolonged flooding. Under drought conditions, saltcedars are able to extract water from unsaturated soil, which native willows cannot (Busch et al. 1992). They also possess higher water-use efficiency than do native willows and cottonwoods (Busch and Smith 1995). During drought years, and in the later portion of the growing season, saltcedars thus are able to maintain greater leaf area per unit of stem than can native riparian species (Cleverly et al. 1997). In saline soils, saltcedars exude salt crystals from their leaves, creating even more saline surface soil conditions that inhibit germination and growth of native plants (DiTomaso 1998). Buried stems produce adventitious roots, enabling the plants to tolerate sand and silt deposition due to floods. As a result of these adaptations alone, saltcedars are eventually able to outcompete native riparian woody plants over a period of decades and create a monospecific stand.

Saltcedar also promotes fire in riparian communities in which it was originally rare (Crawford et al. 1996). Saltcedar shrubs produce multiple basal stems and accumulate numerous dead stems and small branches. Because of the lack of flooding on regulated rivers, standing and fallen debris is not flushed away. Mature saltcedar stands thus contain large quantities of highly flammable fuel, so that they burn easily. After a fire, however, they are quick to resprout, as well as being tolerant of increased salinity and boron content of postfire soils (Busch and Smith 1993). Within ten to twenty years, a stand once again loaded with flammable fuel typically develops, so that in a few fire cycles saltcedar can easily displace willows and cottonwoods that do not resprout after fire.

Dense saltcedar stands transpire large volumes of water. One estimate is that a dense stand of saltcedar transpires 9 acre feet of water per acre per year (Kerpez and Smith 1987). Compared on a unit leaf area basis, saltcedar and native riparian shrubs do not differ greatly in water loss. The greater foliage area of dense stands of saltcedar, however, enable them to evaporate water at a rate 1.6–2.0 times the evaporation that would occur from an open-water surface (Sala et al. 1996). They thus desiccate stream courses in arid regions, promoting conditions that only they can tolerate. Such dense stands thus represent an economic as well as an ecological liability in regions where stream water is subject to human use.

The change in vegetation structure has had profound impacts on riparian animal communities, especially those of birds. The impacts of saltcedar invasion on riparian bird communities differ from place to place, and their overall significance is controversial. In general, however, saltcedar is poorer habitat than native riparian communities for birds.

Along the lower Colorado River, where single-species stands of saltcedar form wide belts across the riparian zone, the diversity and density of birds are lower throughout the year in saltcedar than in native riparian vegetation (Hunter et al. 1988). Saltcedar stands are lower in insect abundance than other types of riparian vegetation, and they also lack mistletoe plants, which are important to many frugivorous birds. The loss of riparian cottonwood woodlands has placed regional populations of several breeding species at risk of extinction, including the ferruginous owl (*Glaucidium brasilianum*), yellow-billed cuckoo (*Coccyzus anericanus*), summer tanager (*Piranga rubra*), Wied's crested flycatcher (*Myiarchus tyrannulus*), and yellow-breasted chat (*Icteria virens*). These species, most of them midsummer nesters, apparently depend on the multilayered canopy structure that creates sheltered nesting sites in native cottonwood woodland. Although adequate insect food seems to be available in midsummer in saltcedar stands, the low stature of these stands does not moderate the extreme summer temperatures of the lower Colorado valley.

Along the lower Colorado, overall, saltcedar use ranks well below cottonwood-willow, honey and screwbean mesquite communities, and native desert wash vegetation (Hunter et al. 1988). The one major exception to this pattern is that young, dense stands of saltcedar 10–20 feet tall are heavily used by doves for nesting. Also, old saltcedar stands with dense canopies at heights of 33–66 feet do become attractive nesting areas for many nesting birds, such as summer tanagers. Communities like this are scarce, however, because of the tendency of saltcedar stands to burn before such old growth develops. An experimental study of vertebrate response to clearing of saltcedar and its replacement by native species on a 50-acre area on the lower Colorado showed that bird diversity increased significantly (Anderson and Ohmart 1985).

Along the Colorado River in the Grand Canyon, where *Tamarix ramosissima* is the dominant species, riparian vegetation was originally only weakly developed because of the intense scouring of the banks by spring floods. Since the construction of Glen Canyon Dam, scouring has been weak, and a dense riparian community of saltcedar and some native shrubs developed—a dense shrub vegetation that did not exist previously (Brown and Trosset 1989; Brown 1990). Eight species of birds make significant use of this vegetation for nesting: willow flycatcher (*Empidonax traillii*), Bell's vireo (*Vireo bellii*), yellow warbler (*Dendroica aestiva*), common yellowthroat (*Geothlypis trichas*), yellow-breasted chat (*Icteria *virens*), blue grosbeak (*Guiraca caerulea*), hooded oriole (*Icterus cucullatus*), and northern oriole (*Icterus galbula*). Some of these species, particularly Bell's vireo, common yellowthroat, and yellow-breasted chat, make heavy use of low, dense stands composed largely of saltcedar. The rest tend to favor taller, patchier vegetation in which native species are more prevalent. Thus, the new exotic-dominated vegetation of the river provides habitat for many birds. On the other hand, it is probably not as favorable a habitat as riparian vegetation composed entirely of native shrubs and trees. Along the Virgin River in southern Utah, in support of this idea, willow flycatchers are believed to have declined because willow vegetation has been replaced by saltcedar.

The situation along the lower Rio Grande valley in Texas is similar to that of the lower Colorado. Replacement of cottonwood-willow vegetation by saltcedar has been extensive. Many of the bird species that have declined in the lower Colorado valley have also declined here, but to a lesser extent (Hunter et al. 1988).

In the Bosque del Apache area of the central Rio Grande in New Mexico, where the dominant saltcedar is also *Tamarix ramosissima*, comparisons have been made of bird use of cottonwood and saltcedar habitats (Ellis 1995). This survey revealed that avian diversity, particularly in spring and summer, was greatest in cottonwoods. Hummingbirds, woodpeckers, and trunk-foraging species such as the white-breasted nuthatch (*Sitta carolinensis*) were largely restricted to cottonwood areas. Neotropical migrants also showed a preference for cottonwood areas, especially during spring and summer. Although some species, especially insectivores, were common visitors, or even breeders, in saltcedar areas, only two were found in saltcedar and not in cottonwood areas: the ring-necked pheasant (an exotic itself!) and the greater roadrunner. Both are open-habitat birds that use saltcedar mostly for cover.

Along the Pecos River in New Mexico, and perhaps along portions of the Rio Grande in southern New Mexico, cottonwood and willow woodland were originally of limited extent. Here, too, invasion of saltcedar has created a new, dense riparian shrubland. As a result, riparian bird populations have apparently increased, compared to those prior to saltcedar invasion (Hunter et al. 1988). Species such as crissal thrashers (*Toxostoma crissale*), yellow-billed cuckoos, yellow-breasted chats, painted buntings (*Passerina ciris*), and blue grosbeaks have probably spread and increased in abundance along the Pecos. Again, though, it is unclear that these species are benefited more by saltcedar than they would be by native riparian vegetation.

Although few studies have been done, it appears that riparian mammals have been less affected by saltcedar invasion than have birds. Some larger mammals, however, are negatively impacted by saltcedar (Kimball Harper, personal communication). Neither mule deer (*Odocoileus hemionus*) nor beaver (*Castor canadensis*) browse saltcedar. For mule deer, the dense saltcedar stands may limit access to the river itself. For beaver, the high concentration of roots in the stabilized bars makes it difficult to excavate nest burrows. In the middle Rio Grande valley of New Mexico, one small rodent, the white-footed mouse (*Peromyscus leucopus*), was abundant both in cottonwood and in saltcedar stands bordering the river (Ellis et al. 1997). White-throated woodrats (*Neotoma albigula*), however, were present in cottonwood but absent in saltcedar stands. Several other species of mice, typical of desert upland habitat and human residential areas, were also found in saltcedar areas, but perhaps only because the saltcedar study areas were close to these other habitats. Along the lower Colorado, small mammals such as mice, kangaroo rats, and woodrats were not markedly different in abundance in saltcedar than in native riparian community types. Experimental clearing of saltcedar and its replacement by native shrubs, however, did lead to an increase in rodent density (Anderson and Ohmart 1985).

Stands of saltcedars also modify river channels by stabilizing sandbars and causing the fusion of islands with the main banks. This has contributed to the decline of several native fish of the Colorado River, including humpback chub (*Gila cypha*), bony-tailed chub (*Gila elegans*), razorback sucker (*Xyrauchen texanus*), and Colorado pikeminnow (*Ptychocheilus lucius*). The eddies and backwaters associated with sandbars and islands are the breeding and nursery areas for these native fish.

Russian Olive and Other Riparian Invaders

Russian olive is a small, drought tolerant tree native to Eurasia. As its name implies, it is a member of the olive family and has a large-seeded, olive-like fruit that is dispersed by birds and small mammals. In the late 1800s, Russian olive was introduced to the Great Plains as a windbreak tree. Subsequently, it was planted widely throughout the plains states, western states, and western Canadian provinces (Knopf and Olson 1984). Planting of the species was actively subsidized by soil conservation, forestry, wildlife, and highway agencies for erosion control, windbreaks, wildlife habitat improvement, and highway landscaping. In Oregon, for example, the State Department of Fish and Wildlife distributed seedlings without cost to private landowners. Many other states made seedlings available at nominal cost. This practice still prevails in many areas, and many nurseries, even those specializing in native plants, sell Russian olives.

By the 1920s, Russian olive had become naturalized in Utah and Nevada, and over the next three decades it appeared in the wild throughout most of the West. In some places its colonization was very rapid. In Canyon de Chelly, Arizona, for example, it first appeared in 1964, and within ten years it had become one of the canyon-floor dominants. A survey published in 1986 found that it is now naturalized in all seventeen western states (Olson and Knopf 1986a). It is also established in Alberta, Saskatchewan, and Manitoba.

Although it is not as aggressive a colonizer as saltcedar, Russian olive is relatively shade tolerant and can germinate and grow beneath canopies of cottonwood and other native trees and shrubs (Shafroth et al. 1995). Thus, even without disturbance of the native vegetation, Russian olive gradually invades and displaces native riparian species. Unlike saltcedar and cottonwood, which are favored by physical disturbance of stream floodplains, as by flood scouring, Russian olive is favored by river control, which reduces disturbance that accompanies flooding. Russian olive, because of its drought tolerance, also tends to invade areas beyond the outer edge of native riparian woodland, broadening the woody riparian zone. Like saltcedar, dense stands of Russian olive stabilize floodplain terraces and increase alluvial deposition, constricting the river channel and causing it to deepen. Eventually, this creates a higher, drier alluvial terrace on which cottonwoods cannot establish.

Extensive pure stands of Russian olive border the channels of many rivers in the Great Plains, the intermontane region, and the Southwest. Major areas of infestation exist in the eastern Dakotas, south-central Kansas, central New Mexico, northeastern Utah, the Snake River plain of Idaho, and portions of the Columbia River basin in Washington and Oregon. In South Dakota and Nebraska, without control, Russian olive is predicted to become the dominant riparian tree. One ecologist has suggested that most of eastern South Dakota might become a Russian olive forest (Olson and Knopf 1986b). In other locations, such as central New Mexico, Russian olive becomes codominant with saltcedar, creating a complex riparian woodland made up primarily of exotics. Russian olive also invades the borders of prairie marshland and in some cases replaces marshland altogether.

Along the Platte River in Nebraska, Wyoming, and Colorado, where riparian woodland was originally very restricted in distribution, Russian olive adds one more step in the transformation of the riparian landscape (Knopf and Scott 1990). Riparian woodlands of cottonwood and other native trees that colonized the floodplain in response to stream regulation and irrigated farming (Johnson 1994) are now being replaced by Russian olive.

Russian olive also tends to promote an understory growth of exotic grasses and forbs. Near Milliken, Colorado, for example, one study found that the undergrowth of a Russian olive stand consisted mostly of cheatgrass, Canada thistle, and stinging nettle, all exotics (Knopf and Olson 1984). In Utah, cheatgrass, mustard, and alfalfa, again all exotics, were the understory dominants of a Russian olive stand (Knopf and Olson 1984).

Russian olive provides food or shelter for many species of birds and mammals in winter, the primary reason for its promotion by fish and wildlife agencies. Many fruit-eating birds and mammals, such as robins (*Turdus migratorius*), evening grosbeaks (*Cocothraustes vespertinus*), fox squirrels (*Sciurus niger*), and raccoons (*Procyon lotor*), feed on the fruits of Russian olive. Other species use stands of this species for winter cover and summer nesting. One of the few studies comparing quantitative mammal and bird use of Russian olive stands versus native riparian stands dominated by cottonwoods was carried out in Colorado, Idaho, and Utah in late May and June in the early 1980s (Knopf and Olson 1984). The diversity of mammals using the two types of riparian areas differed little, but populations of several mice were considerably higher in Russian olive areas than in cottonwood riparian areas. This was believed to be due to the denser layer of herbaceous plants, primarily exotics, that occurred under Russian olive. On the other hand, forty-five species of land birds occurred in native, cottonwood-dominated riparian areas, compared to thirty-three species in Russian olive stands. The native breeding birds that utilized Russian olive stands were predominantly shrub-nesters, many of which were widely distributed in other lowland habitats. Most species restricted to lowland riparian habitat, particularly hole-nesting species, were absent from Russian olive stands. Almost a third of riparian-breeding bird species are estimated to be negatively affected by the habitat change induced by Russian olive.

Russian olive, when it is a low-abundance component of riparian communities, clearly is not a detrimental species for bird and mammal diversity. The difficulty, however, is that Russian olive tends to replace native riparian vegetation and promote a new riparian community composed largely of exotics. Replacement of riparian cottonwood stands by pure stands of Russian olive will obviously reduce populations of many bird species, especially hole-nesting species.

A number of other exotic trees are joining saltcedars and Russian olive in southwestern riparian areas (Crawford et al. 1996). Siberian elm is an aggressive invader of riparian areas throughout much of the West. Along the Rio Grande in New Mexico, Siberian elm, white mulberry, and Osage orange are becoming important members of the exotic riparian woodland. In southern New Mexico and Arizona, ailanthus has also become naturalized along many rivers and streams. In Phoenix, Arizona, a variety of other horticultural shrubs and trees have invaded drainage channels leading into the Salt River channel (Brock and Farkas 1997).

Along riparian stream channels in arid and semi-arid parts of the southwestern United States, exotic herbaceous plants are especially common. Cottonwood-willow and mesquite woodlands often have a herbaceous understory dominated by exotic Mediterranean grasses and forbs (Stromberg et al. 1997). At the Nature Conservancy's Hassayampa River Preserve in Arizona, a quarter of the 280 herbaceous species found in the riparian zone are exotics (Stromberg and Chew 1997). Bermuda grass, white sweet clover, and yellow sweet clover are dominant perennials on many southwestern stream banks and floodplains, where they displace many native species. Johnsongrass is another coarse exotic grass that is common on southern Arizona streambanks.

The invasion of the riparian zones of larger rivers and streams by exotic trees is, unfortunately, only the most obvious of exotic invasions of riparian habitat. Along smaller streams, exotic herbaceous plants are also becoming serious problems in many places. In the Sacramento Mountains of New Mexico, for example, teasel and musk thistle have become streamside invaders in locations disturbed by human recreational activity. Huenneke (1997) has found that these species are overrunning the habitat of the Mescalero thistle (*Cirsium vinaceum*), a local endemic. Drought and water diversion for human use compound the problem, because the exotics are more resistant to dry soils.

Degradation of Natural Riparian Ecosystems

These riparian invasions show how elimination of flooding and invasion of exotic species can change landscape geomorphology, ecosystem function, and community composition. Being powerful agents of floodplain stabilization, they effectively channelize the invaded rivers. Narrow, deep, swift-flowing river channels bordered by dense brush and woodland replace broad, shallow channels with open riparian vegetation. Decomposition is retarded and woody fuels accumulate. Fire becomes a frequent ecological force, and drier, more saline floodplain soils are promoted. Native riparian species, both woody plants and understory herbs, are replaced by exotics. Animal communities, especially those of birds, become less diverse, and certain groups, such as hole-nesting species, are virtually excluded. Forest-floor invertebrate assemblages are also altered, often with invasion of exotic isopods and insects (Molles et al. 1998). For many riparian plants and animals, these exotic invasions are speeding decline toward extinction.

An essential aspect of southwestern riparian woodlands is their role as stopover sites for migrating birds. Many small passerines make frequent stopovers during spring and fall migration, spending one to several days restoring their body fat reserves. In the southwestern United States, the Colorado River and the Rio Grande form well-vegetated, north-south corridors along which many birds migrate (Yong and Finch 1997a). Stopovers may be made at several points along these river systems. Recent studies of migrating willow flycatchers suggest that the composition of these vegetated corridors is important to this particular species (Yong and Finch 1997b). In the Bosque del Apache Refuge, for example, willow flycatchers were found more commonly in dense willow stands than in mixed cottonwood and Russian olive stands. In addition, the birds netted and examined in willow stands showed greater fat reserves. This suggests that the willow habitat may have more arthropod food than other types of riparian vegetation. Thus, replacement of riparian vegetation rich in willows by saltcedar and Russian olive may reduce the suitability of these riparian corridors for willow flycatchers and other migrants.

Consider the future of the *bosque* forests of central New Mexico. Very little establishment of cottonwoods has occurred since the 1950s, and the oldest trees are now dying without replacement. Without active management to control saltcedar and Russian olive, as well as to promote regeneration of cottonwoods and willows, native *bosque* woodlands will disappear within 50 to 100 years (Howe and Knopf 1991).

Restoration of Western Riparian Ecosystems

Is there hope for restoration of native riparian plant communities? Completely removing exotic riparian trees once they have become abundant is almost impossible. For saltcedar, aboveground cutting or chaining is ineffective in destroying a stand, because the root crowns resprout profusely, producing shoots up to 10 feet high in one growing season. Root plowing, at a depth of 1.0 to 1.5 feet to undercut root crowns, is the most effective mechanical technique. Herbicidal treatments can also be effective but work best when herbicide is applied to aboveground cut stems, or applied subsurface in combination with root plowing. In any of these cases, treatment costs are high, and follow-up procedures are required to prevent regrowth of saltcedar from seed and to restore native plants.

At the Bosque del Apache National Wildlife Refuge, studies initiated after a fire in 1986 have shown that a combination of herbicide treatment, burning, and root plowing could clear dense saltcedar at costs of $300–$525 per acre. Native cottonwoods, willows, and shrubs could then be planted and their natural recruitment encouraged by irrigation. These studies showed that restoration of riparian communities dominated by natives is practical, albeit expensive.

Thus, in the long run, the most effective approach for saltcedar control is likely to be biological control. Several potential biocontrol insects have been identified in field tests conducted in France, where their host is *Tamarix gallica* (Sobhian et al. 1998). These species are able to complete their life cycles on *Tamarix ramosissima*, but not on other saltcedars or plants of other genera and families. These findings are encouraging, but the several problem species of saltcedars and the diversity of climatic conditions under which they exist mean that biological control of these riparian weeds is far from assured.

Russian olive poses an even more complex problem. At present, mechanical and chemical controls are the only approaches than can be used. The popularity of Russian olive as an ornamental tree, and the close relation of this species to the European olive, pose severe limitations to the development of a classical biological control system. In all likelihood, control of Russian olive, as well as that of saltcedars, will require integrated strategies involving mechanical, chemical, and biological controls, coupled with manipulation of stream flows to re-create the conditions required for establishment of native riparian plants (Friedman et al. 1997).

Must these species be controlled? In some places, and in moderate abundance, saltcedars and Russian olives may be beneficial. Where they form monocultures that eliminate native riparian species, however, their impacts are clearly negative. These highly successful exotics are now integral members of riparian communities throughout the West (Taylor and McDaniel 1998). Thus, it is their tendency to completely transform western riparian ecosystems to the detriment of hundreds of native species of plants and animals that must be opposed with the weapons available.

Restoration of riparian ecosystems, including control of invasive riparian trees, is now being viewed in a total landscape perspective (Goodwin et al. 1997). An understanding of the natural hydrology and how it has been modified is critical to restoring the riparian community. Water management techniques that recreate natural hydrological processes can restore basic ecological functions and promote natural recruitment of native species (Molles et al. 1998). New direct control techniques can assist. Recent experiments suggest, for example, that localized aerial application of the herbicide imazapyr may give up to 95 percent kill of saltcedar stems (Duncan 1997). Restored hydrology, improved herbicides, and biological controls may eventually allow exotics to be reduced to moderate, and perhaps even beneficial, components of western riparian ecosystems.

Chapter 12

The Pacific States: Mediterranean Mixing Pot

[The Central Valley was] a large prairie covered with wild oats—which at this season of the year when nothing but the stock [stalk] remains, has much the appearance of common oats.

—Zenas Leonard, *Adventures of Zenas Leonard**

San Diego, California, is largely built on elevated coastal marine terraces, locally called "mesas." Cutting into these mesas are steep-walled canyons that are difficult to carve into building sites. Through these canyons, chaparral and coastal sage scrub extend fingers of native habitat into the urbanized mesa tops. They are surprisingly rich in plant and animal life, and although the dense chaparral shrublands can be a fire hazard in the dry late summer and fall, most people like to preserve them in a natural state. They bring elements of wildness into the city—wild lilac and manzanita, foxes, coyotes, quail, roadrunners, red-tailed hawks—and a host of other native plants and animals frequent the larger canyons.

Slowly, however, native biodiversity in the canyons is fading away. Soulé and his students (e.g., Soulé et al. 1988; Alberts et al. 1993) have found that as the canyons are isolated by highways and other developments, some vertebrates are lost because of the accidents that tend to befall small, isolated populations. Other factors play an even greater role. Native chaparral and coastal sage scrub plants decline because of disturbance and the

*Zenas Leonard (1904). *Adventures of Zenas Leonard*. W. F. Wagner (Ed.). Burrows, Cleveland, OH.

invasion of exotic plants from surrounding urban areas. Hottentot fig, planted for fire protection along the canyon rims, creeps downslope into the canyons, forming dense mats that suffocate other species (Table 12.1). Exotic palms, pampas grass, and sweet fennel invade the stream channels that receive runoff from lawn irrigation and storm drains, creating almost impenetrable jungles. Mediterranean grasses and forbs spring up where paths and sewer lines have disturbed the soil. Exotic plants increase in diversity and abundance over time after the isolation of the canyon by housing developments and roads. In the oldest canyons, as little as 5 to 15 percent of the canyon area still retains native vegetation. As the native vegetation declines, the wildlife of chaparral and coastal sage scrub is replaced by species of the urban perimeter: house mice, black rats, Argentine ants, and birds typical of backyards in southern California. The San Diego canyons are a microcosm of the assault of exotics on Mediterranean habitats of the Pacific Coast.

Table 12.1. Scientific names and regions of origin of exotics of the Pacific coastal region.

PLANTS	
black mustard, *Brassica nigra*	Eurasia
blue gum, *Eucalyptus globulus*	Australia
brome grasses, *Bromus madritensis*, *B. hordaceus*, *B. diandrus*	Eurasia
castor bean, *Ricinus communis*	Europe
crystalline ice plant, *Gasoul crystallinum*	South Africa
European beach grass, *Aimophila arenaria*	Europe
filarees, *Erodium cicutarium*, *E. botrys*, *E. moschatum*	Eurasia
French broom, *Cytisus monspessulanus*	Europe
giant reed, *Arundo donax*	Europe
gorse, *Ulex europaeus*	Europe
Hottentot fig, *Carpobrotus edulis*	South Africa
pampas grass, *Cortaderia jubata*	Argentina
Scotch broom, *Cytisus scoparius*	Europe
sweet fennel, *Foeniculum vulgare*	Europe
tocalote, *Centaurea melitensis*	Europe
wild oats, *Avena fatua*, *A. barbata*	Eurasia
yellow star thistle, *Centaurea solsticialis*	Europe
INSECTS	
Argentine ant, *Iridomyrmex humilis*	Argentina
MAMMALS	
European rabbit, *Oryctolagus cuniculus*	Europe
feral goat, *Capra hircus*	Eurasia
feral hog, *Sus scrofa*	Eurasia

Diversity and Invasibility

Biotic richness of a region is not a guarantee of protection against exotic invasions. The Pacific Coast region has one of the richest floras of any region of North America, both in native species and in exotic species. California has about 4,850 native plant species and 1,045 established exotics (Randall et al. 1998). British Columbia records 2,048 native plant species and 547 exotics (Vitousek et al. 1997). Exotics thus make up 17.7 percent of the total California flora and 21.1 percent of the British Columbia flora. Thus, roughly one out of five species on the Pacific Coast of North America is an alien. Islands off the coast are not immune to invasion. The Channel Islands off southern California have about 353 species of plants, 20 percent of them exotic.

The vulnerability of the Pacific Coast region to exotic invasions seems surprising, in view of the richness of its biota. This vulnerability, in fact, is a general contradiction of the once-held view that "diversity begets stability." Why have so many exotics invaded a region filled with species presumably adjusted closely by evolution to the specific conditions of the region's ecosystems?

One part of the answer lies in the fact that the North American region of Mediterranean climate is one of five such world regions. The other four regions of Mediterranean climate are the Mediterranean Basin of the Old World, central Chile, the tip of South Africa, and south-central and southwestern Australia. These Mediterranean regions have evolved their biotas largely in isolation from one another. Although similar in climate, they have experienced strongly differing patterns of human influence from prehistory to recent history. Collectively, therefore, species adapted to an enormous range of ecological conditions exist in the Mediterranean biota. Plants and animals have been brought to the Pacific Coast from all of these world regions, and several regions have contributed problem exotics.

Some portion of the answer also relates to the fact that the western North American landscape is well suited to annual plants. Annuals are the most numerous problem exotic plants in the Pacific Coast region, and almost all come from another Mediterranean-climate region rich in annuals, the Eurasian Mediterranean Basin. The Eurasian and North American regions of Mediterranean climate share the feature of having the highest percentages of annuals of major regions of the world: 30 percent for the Mediterranean Basin floristic province and 27 percent for the Californian floristic province. The Eurasian Mediterranean region, however, has experienced a long and pervasive history of human tending of livestock, a force of landscape disturbance that favors annual plants. Livestock agriculture was imported full-blown to the Pacific Coast with European settlement in the 1600s to 1800s.

A third part of the answer lies in a clear pattern of facilitation by each set of exotic invaders; each wave of invaders tends to promote future invasions. The dominant species of exotics have changed sequentially over decades and centuries and are still changing. Each group of dominant exotics sets the ecological stage for a new group.

Exotic invasions of the Pacific region raise several basic ecological questions. Some ecosystem types appear, for reasons still not clear, to be much more vulnerable than others to heavy invasion by exotics. Likewise, many exotics appear to have the tenacity to maintain their dominance in the face of intensive restoration efforts. The ecosystems of the Pacific region seem to have been altered in ways that enable exotics to displace a rich assortment of presumably well-adapted natives.

Mediterranean grasslands, woodlands, and related communities that have an extensive herbaceous understory are the ecosystems most heavily invaded by exotics. Coastal strand and dunelands, which share many ecological features with grasslands, are also among the heavily invaded ecosystems. Like other parts of western North America (see Chapter 11), riparian areas are heavily invaded. Forest, chaparral, and desert ecosystems are little invaded, except in sites where an extensive herbaceous ground cover exists. The reasons for this differential sensitivity are unclear.

Furthermore, simple protection of the most heavily invaded ecosystems from disturbance does not lead to recovery of native species. Exotics have evidently switched ecosystem function into a new state, with which the exotics are in stable relation. The degree of transformation is remarkable and is far greater than that of ecosystems in desert-edge and Great Plains regions farther east, but at the same latitude.

History of Exotic Invasions

The history of exotics on the Pacific Coast is a long one. Some exotics had probably arrived in California during the founding of the first missions in the late 1700s (Hendry 1931). Their presence is recorded in the adobe bricks of the first buildings erected by the builders of the California missions. Filaree, or storksbill, a weedy member of the geranium family, was one of the earliest, its seeds being incorporated in bricks of the San Antonio de Padua Mission built in 1771. Filaree, in fact, may have invaded California from Baja California, Mexico, in advance of the founding of the California missions (Mensing and Byrne 1998). Tocalote, a relative of star thistle, and black mustard are other species known to be present in the early 1800s based on remains found in adobe bricks. During the period of Spanish rule, from 1769 to 1824, in which the twenty-one missions from San Diego to Sonoma County, north of San Francisco, were founded, about sixteen annual grasses and broadleaf herbs from the Mediterranean region of Eurasia were introduced with the horses, cattle, and sheep that the Spanish missionaries brought with them (Randall et al. 1998).

The pace of introductions picked up as settlement and commerce increased in California (Randall et al. 1998). During the years that California was governed by Mexico, from 1825 to 1848, about 63 additional exotics appeared. After the Mexican War, early American settlers and gold seekers brought at least 55 other exotics to California. By 1860, at least 134 exotic plants are known to have become established in California. With the arrival of railroads, the growth of population, and the twentieth-century explosion of roads and automobiles, hundreds of additional exotics gained entry. In 1925, 292 exotic species had been identified, and by 1959 the number had risen to 725.

The earliest dominants among exotics were apparently wild oats and black mustard. The widespread abundance of wild oats was mentioned by travelers through the region, such as Zenas Leonard, in the 1830s and 1840s. The displacement of native grassland species by exotics intensified after the mid-1800s, as livestock ranching and dry-land farming expanded with the flood of settlement that accompanied the gold rush. Destruction of the grasslands of the Central Valley was essentially complete by about 1880. In this process, filarees, tocalote, and various annual grasses spread and increased, evidently in several waves of dominance by different combinations of species (Mack 1989).

Impact of Exotics on Pacific Grasslands

The Pacific region's Mediterranean grasslands are one of the most completely transformed ecosystems in the world because of the effects of exotics. In California, almost 23 million acres of the original valley grassland and its neighboring woodlands are now dominated by exotic annual grasses and forbs. This transformation began so early in the period of European settlement that the original composition of the plant community is uncertain. Accounts of rangeland condition in the late 1800s indicate that perennial bunchgrasses were present, but declining, in many locations that are now largely exotic annual grassland (Wester 1981). Analysis of certain crystalline materials derived from plants, known as opal phytoliths, also suggests that some present-day annual grasslands originally had a substantial component of perennial grasses (Bartolome et al. 1986). On the other hand, descriptions from both the Spanish and the American periods suggest that parts of the San Joaquin Valley may have had few perennial grasses. The floor of the valley in places was covered by extensive marshlands, with perennial grassland borders. Away from these marshes, however, the arid slopes may instead have been dominated by native annual broadleaf herbs (Wester 1981). Prairie openings north through Oregon and Washington, still rich in perennial grasses, are now being invaded heavily, in many cases by exotic shrubs.

Why have the dominant bunchgrasses of the California valley grassland been so completely displaced by exotic annuals? Several processes doubtless contributed. Where crop cultivation was undertaken, the native prairie vegetation was destroyed, never to recover. The introduction of livestock grazing greatly increased the impacts of large grazing animals. Prior to European settlement, valley grasslands held only light populations of elk and pronghorn, which exerted only moderate grazing and trampling pressure (Painter 1995). In areas subject to livestock overgrazing, disturbance of the soil surface triggered soil erosion, and the disturbed and eroded surfaces were colonized by exotic annuals. Significantly, where uncultivated prairie subject only to moderate grazing since the late 1800s exists, purple needlegrass and other perennial bunchgrasses still survive, albeit with an intermixture of exotic annuals (Stromberg and Griffin 1996).

Several studies (Bartolome and Gemmill 1981) have shown, however, that simple protection from grazing and cultivation, even with the reintroduction of fire, does not lead to renewed dominance of perennial bunchgrasses such as purple needlegrass (*Nasella pulchra*). Early studies (Biswell 1956), in fact, show that protected stands may decline substantially in purple needlegrass cover. In the San Francisco Bay area, many moderately grazed areas even retain a richer assemblage of native plants than do ungrazed areas (Edwards 1992). In some cases, purple needlegrass cover increases when grazing is done in spring, during the period of maturation of exotic annual grasses. Competition for light and water from annual grasses reduces the growth and survival of purple needlegrass. The annuals germinate immediately after the first autumn rains, utilizing water in the surface soil layers and producing a canopy that shades the foliage of the slower-growing needlegrass. The growth of young clumps of needlegrass is reduced, and its ability to draw water from deep layers is reduced for lack of development of the deep root system characteristic of mature plants.

Other studies of the growth and survival of seedlings of purple needlegrass at Jepson Prairie, near Davis, California, have tried to sort out the influences of burning, grazing by livestock, and competition from exotic annuals. In experimental plots at Jepson Prairie, growth of purple needlegrass seedlings was improved only in plots that were weeded of exotics. Burning and weeding together, however, provided even better growth in the first postfire growing season. Burning also tended to fragment large bunchgrass clumps into smaller units that showed more vigorous vegetative spread. By reducing the heavy mulch that annual grasses produce, grazing may also improve the germination of native species (Randall 1993).

Experimental annual burning of grasslands in the southern Sierra Nevada has shown that only very limited reductions in exotics can be achieved (Parsons and Stohlgren 1989). Both fall and spring burns reduced the dominance of exotic grasses and forbs from about 99 percent to about 80 percent. Annual burns reduced the abundance of exotic annual grasses considerably, but both exotic and native forbs increased in abundance. When annual burning was terminated, however, exotic annual grasses regained their dominance.

Interestingly, many of the dominant exotic annual grasses of the Mediterranean region of California are relatively minor species of severely disturbed or early successional sites in their native Eurasia (Jackson 1985). In California, these exotic annuals, which germinate profusely after the first fall or winter rains, are able to form stable communities that persist indefinitely. The California grassland ecosystem thus appears to have been changed in basic environmental conditions, which now favor the establishment of annual species over native perennial grasses. Very likely, the annuals reduce availability of all basic resources—light, moisture, nutrients—for seedlings of the perennials in early summer. Thus, burning or not burning, grazing or protection from grazing, cannot by themselves restore the original dominance of bunchgrasses.

This raises other interesting questions. How important were annuals in the original valley grassland, and what was their relationship to the establishment and growth of perennial grasses? Estimates of the density of mature clumps of purple needlegrass in the original prairie suggest that it was about 3.3 clumps per square yard, ranging from 0.8 to 5.8 clumps (Dyer and Rice 1997). Such densities leave considerable space for other plants. Good candidates for annual members of native prairie include a variety of herbaceous members of the sunflower family, particularly a group of species known as tarweeds, several legumes, and a variety of plants with bulbs or corms.

Most recently, California grasslands have entered a new phase of exotic dominance. Yellow star thistle is now displacing shallow-rooted annual grasses and forbs over much of northern California (Rice et al. 1997). Star thistle is preadapted to utilizing the deep soil water made available by the displacement of native bunchgrasses by shallow-rooted annuals. Star thistle, itself a winter annual, begins its growth as a rosette of leaves at the ground surface early in the growing season, all the while putting down a deep taproot. In midyear, it rapidly develops a flowering shoot that overtops most other exotic annuals. At maturity, stands of yellow star thistle deplete the deep soil water, removing up to five times the total soil water that was taken out by shallow-rooted species.

Exotic Shrub Invaders

In grasslands, meadows, and old fields from central California northward, exotic shrubs are serious invaders. Scotch broom, introduced to Washington State in 1888 as an ornamental, has become the most invasive shrub of the Pacific Northwest (Parker 1997a). It has now become established in open habitats from central California to southern British Columbia. Its golden yellow legume flowers also make it one of the most apparent exotics during the summer. Scotch broom can grow to a height of 13 feet, and large plants have several thousand flowers. It is an obligatory outcrosser, reproducing only by seed, although seed set appears to be limited by a shortage of efficient pollinators. Enough seeds are produced, however, to enable the shrub to colonize new areas quickly. Its seedpods explode, hurling the seeds from the parent shrub. The seeds of broom also possess eliasomes, oil-rich bodies that attract ants, which carry the seeds to their nests, thus aiding in seed dispersal. Soil disturbance also promotes establishment of Scotch broom (Bossard 1991). French broom has also become naturalized from southern California north to Washington State. French broom is more widespread than Scotch broom in California.

In the relict prairies of the Willamette Valley in Oregon and western Washington, Scotch broom is a particularly serious invader, capable of transforming native prairie into a monospecific shrubland. Scotch broom, in fact, invades undisturbed prairie more rapidly that it does urban fields, even though seed production is more severely limited by a shortage of pollinators in native prairies. One possibility is that in native prairie, seeds fall into a dense layer of moss and reindeer lichens on the ground, where they are protected against bird and rodent predators (Parker 1997b).

Scotch broom has a rich assemblage of specialist herbivores in the British Isles, but only seventeen of these species were found on broom in western North America in a survey done in 1963 (Waloff 1966). In particular, twenty-three species of pod- and seed-feeding insects occur in Britain, but none of these was present in the Pacific Northwest. Five species of true bugs of the family Miridae occur on broom in Britain, but only three were found in western North America. Most of the insects present in western North America appear to be species that are associated with the foliage, stems, or litter of Scotch broom and were probably brought over in vegetative horticultural stock. The deficiency of herbivores is evident in late summer in the North American plants, which lack the "ragged" appearance of those in Britain. Efforts to control Scotch broom biologically have been made. A weevil (*Apion fuscirostre*) with a larval stage that consumes the seeds of the broom has been introduced, which has had only limited success.

Gorse, a spiny legume shrub, was introduced to the Pacific Coast in the late 1800s, again as an ornamental (Boyd 1985). Like the brooms, it has attractive, yellow flowers. Gorse invades moist, open habitats, including pastures, wetland borders, and forest clearcuts. Single plants can grow to a height of 10–12 feet, with a diameter of 30 feet. Areas invaded by gorse can quickly become impenetrable, spiny thickets that exclude all other plants. The shrub flowers profusely and produces seeds that survive in the soil for up to thirty years, making its eradication very difficult.

Other Invasive Plants

Coastal environments have experienced some unique patterns of exotic invasion. Dunelands have suffered heavy invasion of exotics all along the Pacific coast. European beach grass was brought to the San Francisco Bay area in the early 1900s to stabilize coastal dunes (Wiedemann 1984). Since then, it has spread southward to near San Diego and northward to the coasts of Oregon, Washington, and British Columbia. Prior to its appearance, no rhizomatous grass of its sort was present on the Pacific coast, and a distinctive strand vegetation consisting of low, spreading perennials and bunch grasses existed. The topography of coastal duneland was also distinctive, consisting of low mounds capped by spreading mats of these natives. European beach grass spreads by rhizomes to form a dense stand inland from the wave-wash zone. The dense grass stand intercepts windblown sand, creating a tall foredune, and favoring the establishment of tall shrubs. Thus, it alters the geomorphology of the strand environment, and it modifies the course of coastal plant succession (Wiedemann 1984).

Hottentot fig, a succulent perennial brought to California from South Africa as an ornamental and ground cover, is yet another invader of coastal dunes and strands. Because it roots and grows rapidly from stems tucked into raw soil, it has been used in freeway plantings throughout California and southern Oregon. These same features make it a pernicious invader of disturbed areas. Well adapted to sandy soils, it forms dense, prostrate mats that may cover a hundred or more square yards, displacing native strand and dune plants. It also hybridizes with the native sea fig (*Carpobrotus chilensis*). The fruits of Hottentot fig and its hybrids with sea fig are preferred to those of sea fig by native mammals (Vila and D'Antonio 1998). Germination of seeds from these fruits is enhanced by passage through the guts of these animals, promoting the spread of the exotic and hybrid forms.

Another exotic succulent, crystalline ice plant, native to the coast of West Africa, has invaded disturbed soils of coastal areas, including the California Channel Islands (Vivrette and Muller 1977). An annual, it stores salt in its tissues during its growth and releases the accumulated salt when it dies. The resulting salinity of the surface soil inhibits germination of many native plants and other exotics, so that it eventually becomes a single-species dominant over large areas.

The moist soils of riparian borders and floodplains have their own distinctive set of exotics. Coarse grasses such as giant reed and pampas grass are stream-channel invaders. Giant reed can form tall, dense stands that exclude all other wetland plants. Sweet fennel, a tall perennial, invades damp riparian soils and can also form dense stands. On Santa Cruz Island, removal of cattle and sheep in the 1980s allowed sweet fennel to spread rapidly, because of release from grazing pressure. It now covers about 10 percent of the island (Brenton and Klinger 1994). Castor bean, a shrub with large, toxic seeds, is also a common invader of damp habitats.

Woodlands, too, have been invaded, their understories suffering a fate similar to that of grasslands. Exotic trees, primarily Australian species of eucalyptus, have created extensive areas of exotic woodland. Many species of eucalyptus are grown ornamentally in California cities. Blue gums, however, were planted extensively as timber trees and for windbreaks early in the 1900s (Robles and Chapin 1995). More recently, this species has been grown for fuelwood. Often, these plantings were in areas where native woodland trees had been cut. Blue gum has become naturalized and reproduces actively in many of these areas.

The influence of eucalyptus stands on the native biota of the Pacific coastal region has been little studied. Plant species diversity is reduced in eucalyptus groves because of shading and heavy litter accumulation. The chemistry of eucalyptus litter also retards the growth of many native herbaceous plants. The effects on animal diversity are less clear. Allen (1996), surveying animal use of various vegetation types in the San Diego,

California, area found that more nonraptorial bird species, especially foliage-gleaning insectivores, woodpeckers, and hummingbirds, showed more positive associations with stands of eucalyptus than with any other vegetation type. Mammals, reptiles, and amphibians did not show many positive associations with eucalyptus stands and were more frequently associated with riparian plant communities. The eucalyptus stands in the study area were small, however, and thus appeared to add specialized foraging sites for certain birds.

Animal Invaders

Along with exotic plants have come a host of exotic animals that have contributed to disruption of native communities. Many exotic insects have become established in the Pacific coastal region. For example, between 1955 and 1988, about 208 new exotic invertebrates became established in California (Dowell and Gill 1989). Most of these are pests of agriculture and horticulture. Most have come from other parts of continental North America, but increasing numbers are now coming from Asia and the Pacific islands. Hawaii, to a significant degree, is becoming a staging area for invasion of California by exotics.

The Argentine ant is one of the most widespread exotic insects, and the most serious in its ecological impacts. Argentine ants were introduced to New Orleans in about 1891 in shipments of ocean cargo from South America (Suarez et al. 1998). Soon afterward, they appeared near San Bernardino, California. They now occur throughout California. Argentine ants proliferate in moist conditions, such as riparian habitats, irrigated agricultural fields, and watered urban landscaping. Plantings of Hottentot fig and its relatives are centers of infestation in urban areas, as are areas of eucalyptus where moist litter accumulates. As the ant lacks a flying stage, colonies multiply by "budding," the splitting of daughter colonies from a mature parent colony. In urban areas, the spread tends to occur from landscaped and watered areas into native vegetation, with most colonies being concentrated in the 109-yard zone bordering landscaping (Suarez et al. 1998).

Argentine ants have effects that both cascade downward and climb up the trophic ladder. Argentines quickly became a citrus pest in California because they are very effective tenders of aphids that feed on citrus. Argentines displace most native ants in the areas infested, with harvester ants, one of the most important native ants in coastal California, being one of the species more strongly inhibited. At Jasper Ridge, San Mateo County, California, for example, Argentine ants totally eliminated some species and reduced the total abundance of native ants by over 99 percent in the areas they invaded (Human and Gordon 1997). Many groups of ground-living arthropods, especially predators, parasites, and herbivores, were also reduced or eliminated. Isopods and camel crickets, almost completely absent in areas without Argentine ants, increased enormously in numbers. The invertebrate fauna was thus switched from herbivore-dominated to scavenger-dominated.

As native ants decline, specialist ant predators, such as the coast horned lizard (*Phrynosoma coronatum*), disappear (Case et al. 1997). The horned lizards appear to avoid Argentine ants themselves, switching their feeding to beetles as native ants disappear. Ultimately, however, horned lizards disappear in areas of high Argentine ant density.

A variety of exotic mammals have been introduced to the Pacific coastal region. Some fifteen species of mammals have been introduced to California from other continents, from other parts of North America, and even from other parts of California (Lidicker 1991). Some of these species are relatively widespread; others, quite restricted in distribution. Species that are widely distributed include the Virginia opossum, muskrat, red fox, fox and eastern gray squirrels, black and Norway rats, and the house mouse (see Chapters 15 and 16).

An additional seven species of domesticated mammals have become feral in California. Pigs were introduced to California in the late 1700s, with the establishment of the California missions, and feral populations soon appeared (see Chapter 16). Goats were introduced to several of the California Channel Islands, including Santa Catalina and Santa Cruz, in the early 1800s, and translocated to San Clemente Island in about 1875 (Johnson 1975). On San Clemente, an enormous feral population developed, decimating the natural vegetation and threatening many of the island's endemic species (see Chapter 16). European rabbits were introduced to Santa Barbara Island, the smallest of the Channel Islands, and to the Farallon Islands. The native vegetation on both areas was decimated by the rabbit. On Santa Barbara Island, rabbit destruction of the native vegetation was one of the disturbance factors favoring the invasion and spread of crystalline ice plant. Since Santa Barbara Island became part of Channel Islands National Monument, the National Park Service has eliminated its rabbit population.

Problems of Combating Exotic Invaders

The enormous diversity of exotics in the Pacific coastal region makes restoring native communities extremely difficult. This is complicated by the fact that exotic plants have formed mutualistic relationships with a number of native animals. These mutualisms, in fact, may have contributed to the rapid and near complete conquest of the California grassland by exotic annuals.

One native animal, for example, the federally endangered giant kangaroo rat (*Dipodomys ingens*), lives in the arid southwestern portion of the San Joaquin Valley of California. The kangaroo rat constructs a conspicuous nest mound with multiple entrances. The construction and remodeling of the nest mound—burrowing, creation of new entrances, deposition of excavated soil—foraging, and dust bathing of these animals create a wide disturbed zone around the nest site. Within this zone, exotic annuals such as filaree and brome grasses are quick to establish, so that exotics dominate the vicinity of nests. The animals themselves, however, preferentially feed on the large seeds of these exotics. Thus, both the exotics and the kangaroo rats are benefited by their association.

Other such associations exist between pocket gophers of the genus *Thomomys* and a wide variety of exotic annual grasses and forbs. Soil heaps created by pocket gopher tunneling are germination and establishment sites for many grassland plants, both native and exotic. In a northern California coastal prairie, for example, Kotanen (1997) found that pocket gopher heaps covered about 1 percent of the soil surface annually. He also found that such disturbances, although initially colonized by both natives and exotics, became increasingly dominated by exotics as time passed. The dependence of some natives on such establishment sites, however, makes the development of management strategy to reduce exotics very difficult.

In many grasslands, pocket gophers maintain "exotic farms" that consist of 12–24 square yards of soil repeatedly disturbed by deposition of surface heaps and the construction of surface access tunnels that enable animals to reach the surface. At the surface, they graze on plants a short distance from the tunnel opening. Few native plants can survive the intense disturbance and herbivory, but exotics thrive under these circumstances. The heaps of pocket gophers, wherever they occur, represent nuclei of establishment of exotic annuals, making it almost impossible to eradicate exotics without killing their native animal mutualists. In particular, studies have shown that exotic annual grasses with their large seeds germinate and survive better on gopher heaps than does the smaller-seeded purple needlegrass.

The prospect for the Pacific Coast is not a pleasant one. The numbers of exotic plants and animals are growing, and their impacts are escalating. Many of these exotics are also extremely recalcitrant to control efforts. For many exotics, such as the annual grasses, only very limited opportunities appear to exist for application of biological control. In some cases, the impacts of exotics and human disturbance have switched the overall dynamics of invaded ecosystems to new, stable states in which native species have no inherent advantage over exotics. The fact that some native species have become enmeshed in mutualistic interactions with exotics also means that control of exotics must be attempted with caution.

Thus, restoring native ecosystems in the Pacific Coast region requires much more than attacking invasive species one by one. The massive invasion of exotic species is evidence of change in basic ecosystem processes. New ecosystem-based strategies of ecological restoration are needed to protect and restore the native biodiversity of this rich region.

Chapter 13

Hawaiian Islands: Exotics in the Islands of Eden

> Since people first inhabited the Hawaiian Islands, a large proportion of native plants, birds, and other terrestrial animals have been destroyed by introduced species and loss of lowland forests. Although Hawaii comprises less than 0.2% of total land area of the United States, more than 25% of the country's listed endangered species and 72% of recorded historical extinctions are species endemic to Hawaii.
>
> —Wayne C. Gagné, "Conservation Priorities in Hawaiian Natural Systems"*

In July 1990, six European rabbits were seen in a campground in Haleakala National Park on the island of Maui (Loope 1994) (Table 13.1). A survey followed, showing that a feral rabbit population existed in an area of about 60 acres. As it later turned out, this population originated from six rabbits released by a person who had tired of them as pets. Rabbits have been released elsewhere in Hawaii from time to time, but fortunately never in numbers adequate to initiate a wild population.

The discovery of this population in Haleakala National Park led to a campaign to educate the public about the dangers of releasing exotic animals. An eradication program was also begun, and the rabbits were removed by shooting and snaring. More than 100 rabbits were killed, with ten months passing before the last animal was taken. Biologists calculated

*Wayne C. Gagné (1988). "Conservation Priorities in Hawaiian Natural Systems." *BioScience* 38:264–271.

Table 13.1. Scientific names and native regions of the major invasive exotics of the Hawaiian Islands.

TERRESTRIAL PLANTS	
banana poka, *Passiflora mollissima*	South America
broomsedge, *Andropogon virginicus*	North America
bushy beardgrass, *Schizachyrium condensatum*	North America
christmasberry, *Schinus terebinthefolius*	South America
firetree, *Myrica faya*	Canary Islands
Florida blackberry, *Rubus argutus*	North America
fountaingrass, *Pennisetum setaceum*	Africa
gorse, *Ulex europaeus*	Europe
Hilo grass, *Paspalum conjugatum*	Asia
kahili ginger, *Hedychium gardnerianum*	Asia
Kikuyu grass, *Pennisetum clandestinum*	Africa
koa haole, *Leucaena leucocephala*	Central America
Koster's curse, *Clidemia hirta*	Neotropics
lantana, *Lantana camara*	Central America
miconia, *Miconia calvescens*	Neotropics
molasses grass, *Melinus minutiflora*	Africa
red mangrove, *Rhizophora mangle*	Neotropics
strawberry guava, *Psidium cattleianum*	Brazil
tropical ash, *Fraxinus uhdei*	Mexico
velvetgrass, *Holcus lanatus*	Europe
white ginger, *Hedychium coronarium*	Asia
MARINE PLANTS	
red alga, *Acanthophora spicifera*	Western Pacific
red alga, *Hypnea musciformis*	Florida
TERRESTRIAL MOLLUSKS	
giant African snail, *Achatina fulica*	Africa
rosy wolfsnail, *Euglandina rosea*	Southeastern USA
INSECTS	
Argentine ant, *Iridomyrmex humilis*	South America
big-headed ant, *Pheidole megacephala*	Asia
long-legged ant, *Anoplolepis longipes*	Africa
mosquito, *Aedes albopictus*	Asia
mosquito, *Culex quinquefasciatus*	North America
yellowjacket, *Vespula pennsylvanica*	North America
yellowjacket, *Vespula vulgaris*	North America
BIRDS AND MAMMALS	
axis deer, *Axis axis*	Asia
barn owl, *Tyto alba*	North America
black rat, *Rattus rattus*	Europe
cattle egret, *Bubulcus ibis*	North America

European rabbit, *Oryctolagus cuniculus*	Europe
feral cat, *Felis catus*	Europe
feral goat, *Capra hircus*	Europe
feral hog, *Sus scrofa*	Asia, Europe
feral sheep, *Ovis aries*	Europe
Japanese white-eye, *Zosterops japonicus*	Japan
mallard duck, *Anas platyrhynchos*	North America
mouflon, *Ovis musimon*	Europe
Norway rat, *Rattus norvegicus*	Europe
Polynesian rat, *Rattus exulans*	Pacific Islands
small Indian mongoose, *Herpestes auropunctatus*	India
DISEASE ORGANISMS	
avian malaria, *Plasmodium relictum*	Asia
avian pox, virus	Eurasia

that in five years a population like this one would likely grow to about 14 million animals. Once established on that scale, the population would be impossible to eradicate and would cause disastrous damage to the native vegetation of the park and its surroundings.

The Hawaiian Archipelago

The Hawaiian Islands are a showcase of geology and evolution. The archipelago is a set of volcanic islands, atolls, and reefs that spans nearly 1,500 miles of tropical ocean, the young, main islands located at the eastern end of the chain (Carson and Clague 1995). Hawaii, the youngest of the main islands, is probably less than a million years in age, whereas Kauai, the westernmost large island, is 3.8 to 5.6 million years old. The rocks and atolls farther west are eroded remnants of still older islands, so the archipelago as a whole has been in existence for a much longer time, perhaps 70 million years or more. Recent evidence suggests, however, that about 29 million years ago, there was a period when no island existed (Carson and Clague 1995).

The islands possess one of the most highly endemic, fragile, and endangered biotas on earth. At the levels of orders and families, the biota of the Hawaiian Islands is depauperate. Many families of plants are not represented in the native biota of the islands. The insect fauna includes only half of the orders and 15 percent of the families of this class (Howarth and Ramsey 1991). Only a single native land mammal, a bat, occurs in Hawaii. In spite of the extensive speciation that has occurred in some taxa, overall species diversity does not approach that seen in tropical continental areas.

In many groups of organisms, the species tend to have broader habitat ranges than do continental species. Native plants, for example, tend to show a greater elevational range than exotics introduced from continental areas (Kitayama and Mueller-Dombois 1995).

In certain taxa, however, the islands are rich in species, perhaps in part because of the influence of islands, mountains, and other features in isolating populations and promoting speciation of the ancestral forms that reached the archipelago. Almost half of the plants and animals native to the islands are endemic (Eldredge and Miller 1998). Eighty-nine percent of the 1,023 species of native flowering plants are endemic, the result of evolution and speciation by about 270 ancestral colonists (Fosberg 1948). Among invertebrates, 22 to 24 colonist land snails have produced, by speciation, an extraordinary fauna of about 750 species (Cowie et al. 1995). Similarly, speciation of 300 to 400 arthropod colonists has produced about 5,400 descendent species (Howarth and Ramsey 1991; Eldredge and Miller 1998). For land birds, perhaps 20 original colonists gave rise to more than 100 species, many of which are known only as fossils (Loope and Mueller-Dombois 1989). A single ancestral finch radiated to produce 47 species and subspecies of songbirds specialized for diverse feeding niches.

The isolation, low diversity of natural colonists, and complete absence of groups of animals such as grazing or browsing mammals and large terrestrial predators have allowed patterns of evolution that leave the native biota extremely vulnerable to impacts of introduced continental species. Hundreds of plants and animals have already been lost from the islands, and the survivors are threatened by thousands of established exotics, as well as by the continuing appearance of new exotics, such as the European rabbit. Many land bird species disappeared during the Polynesian period, and a large percentage of the survivors is now endangered (Scott et al. 1988).

History of Exotic Introduction

Introductions of exotics began with the Polynesian discovery of the islands in about A.D. 400. Polynesian settlers brought forty to fifty species of plant and animals to the islands, including dogs, pigs, Polynesian rats, several geckos and skinks, and a number of plants (Loope and Mueller-Dombois 1989). Polynesian settlers and their mammalian associates encountered an avifauna lacking experience with terrestrial predators and in many cases were flightless. As a result, many native birds were driven to extinction (Olson and James 1984; Steadman 1995). Some sixty species that became extinct during the Polynesian period have now been identified. Included are flightless species of geese, ibis, and rails, together with flying species of several other families. Again, many of these species were probably easy prey to humans and other mammals because of the previous absence of any such predators.

The Polynesian human population of Hawaii grew to about 200,000 to 250,000. Almost all of the native lowland forests were cleared for farming during the Polynesian period (Athens 1997). Deforestation certainly contributed to the extinction of many birds and probably of many plants and invertebrate animals, as well.

When visited by Captain Cook in 1778, the Hawaiian Islands thus had already sustained major biotic losses due to changes resulting from Polynesian settlement. Rates of environmental disturbance and extinction of species increased following European settlement. Tens of thousands of species of exotic organisms have been brought to the islands, and thousands have become established in the wild. About twenty-five species of new exotic plants and animals appear in Hawaii every year (Randall 1993). Exotics—plants, animals, and microorganisms—span all trophic groups and threaten nothing less than the complete replacement of the Hawaiian terrestrial biota. On Lanai, native ecosystems already are suffering catastrophic collapse because of the various stresses of human disturbance and exotic invasions (Hobdy 1993). Kahoolawe has suffered severe desertification and has lost most of its original biota (Howarth and Ramsey 1991).

Physical disturbance of the landscape and introduction of exotic species by humans have transformed more than 90 percent of the natural environments of Hawaii. Large areas of native highland forests have been logged and replaced by cattle ranches and plantations of pines and eucalyptus. Koa (*Acacia koa*), one of the dominant native trees, has been harvested heavily because of the value of its wood for furniture and cabinetmaking. ʻOhiʻa *Metrosideros polymorpha,* less valuable for its wood, has been harvested for fuel for electric power generation, even after native forests were reduced to less than a quarter of their original extent (Holden 1985).

Exotic Plants

The greatest disturbance of native vegetation, however, has resulted from the introduction of exotic plants and animals. More than 13,000 species of plants have been brought to Hawaii and over 900 have become established in the wild (Eldredge and Miller 1998). Over 100 of these have become serious pests (Smith 1985). Some, such as firetree, have proven to be more invasive in the Hawaiian Islands than elsewhere. Others, including guava and lantana, are serious pests on other Pacific archipelagos, such as the Galápagos. Several categories of alien plants are major problems: trees, shrubs, vines, and grasses. Many of these plants have fruits or seeds that are dispersed by alien birds and mammals. Seeds of exotics now dominate the seed bank, even in native forests, indicating that any disturbance of the vegetation is likely to trigger its germination and growth (Drake 1998).

Among trees, the foremost problem species are firetree, strawberry guava, miconia, and tropical ash. Firetree, native to the Canary Islands, was brought to Hawaii in the late 1800s by Portuguese settlers. During the 1920s and 1930s, the Hawaii Department of Forestry planted firetree extensively as part of an erosion control program. Its fruits are consumed and the seeds dispersed by several native and exotic birds (LaRosa et al. 1985). By the time its invasiveness was recognized, firetree had colonized wet and moist sites, including open volcanic ash areas, on all but one of the main islands. This tree, which has nitrogen-fixing root nodules, forms pure stands that outcompete native plants (Vitousek and Walker 1989). It thus modifies basic nutrient cycling processes of ash-soil ecosystems. Firetree invaded Hawaii Volcanoes National Park in 1961 and has now spread across more than 30,000 acres (Vitousek 1990).

Strawberry guava, a shrub or small tree, was brought to the Hawaiian Islands in 1821 for fruit production. It has become widespread in wet, midelevation forests. Strawberry guava reproduces both clonally and by seed, forming dense thickets that exclude almost all other species. Its fruits are eaten by pigs, and the seeds dispersed in their feces. Huenneke (1997) has found that in forest areas where kahili and white gingers, themselves exotics, occur, the litter of these plants tends to cover seedlings both of native trees and of guava. Guavas, however, are more tolerant of the litter cover, and more of them survive than do native tree seedlings. The displacement of native trees by guava is thus promoted by the guava–ginger interaction.

Miconia, brought to Hawaii in 1960, has spread across more than 11,000 acres on several of the main islands (Stein and Flack 1996). This small tree has proven extremely invasive in Tahiti and is considered a major threat to moist habitats throughout the islands. Tropical ash, introduced for reforestation, has been planted extensively at middle elevations on Hawaii, Oahu, Maui, and Molokai. It has proved invasive in native forest, however, and tends to form dense, monospecific stands.

Problem shrubs include koa haole, Koster's curse, lantana, christmasberry, gorse, and Florida blackberry. Koa haole, a nitrogen-fixing leguminous shrub or small tree, is a widespread fire-tolerant exotic of the dry-to-moist lowland areas of all the main islands. It forms dense stands that crowd out all other plants and stifle efforts to protect and restore remnants of native lowland vegetation in areas such as the Hawaii Volcanoes National Park.

Koster's curse and christmasberry are weedy shrubs that favor moist areas at low and middle elevations. Lantana and gorse, which favor lower elevations, and Florida blackberry, which favors high-elevation forests, are prickly, sprawling bushes that form dense thickets. At least four relatives of Florida blackberry are also problem species.

Banana poka, a South American vine of the passionflower family, was introduced to Hawaii in the early 1900s. Freed from a host of insects that feed on it in South America, it has caused serious damage to native upland forests on Hawaii and Kauai. Its seeds are spread by feral pigs and some alien birds that feed on its fruits. It is a high-climbing vine that invades the crowns of native forest trees and smothers their foliage. About 193 square miles of forest are being attacked by this vine (Mueller-Dombois and Loope 1990). Two other species of passionflower vines also cause similar problems in local areas.

A number of perennial grasses, introduced to Hawaii for cattle fodder and ornamental planting, have become serious problems (Smith 1989). Broomsedge, a perennial introduced from the southeastern United States in the early 1920s, has spread throughout large areas of former wet mountain forest. Although it is a perennial, broomsedge becomes dormant during the temperate zone winter. At this time, broomsedge stands transpire little or no water, so that the soil can become saturated with water for long periods, leading to a serious landslide problem (Mueller-Dombois 1973). During the dormant season, broomsedge also becomes prone to fire and is responsible for a major increase in fire frequency in areas where it occurs (Smith 1985). Velvetgrass, bushy beardgrass, and molasses grass behave in a similar manner (Hughes et al. 1991). Fountain grass, an ornamental that is still sold commercially, is invading parts of the Hawaii Volcanoes National Park on the island of Hawaii. It has also become abundant over the entire western part of Hawaii. Kikuyu grass and molasses grass, introduced by ranchers for forage throughout the main islands, form dense mats that overgrow and kill herbaceous plants and low shrubs. Hilo grass, a species that spreads by rhizomes, invades wet forests, forming dense mats that inhibit understory plants and tree seedlings.

Exotic Vertebrate Animals

At least twenty-one species of mammals have been introduced to the islands (Moulton and Pimm 1986). Two insectivorous bats introduced to control insects in sugar cane fields failed to become established, and individual island populations of several other species have disappeared or have been exterminated. Still, eighteen species of wild or feral mammals exist. The most serious are feral pigs and goats. The pig population is derived from interbreeding of animals introduced by Polynesians with those introduced by Europeans, beginning in the late 1700s. Feral goat populations date from 1778, when Captain Cook introduced them to at least two islands (Coblenz 1978).

Some of these introduced mammals have contributed to the destruction of the native vegetation. Pigs are a major cause of damage to the understory vegetation of native forests. Dispersal of seeds of both banana poka and guava is aided by pigs, which feed heavily on their fruits. Pigs feed selectively on certain native plants, including seedlings of the native canopy trees, reducing their abundance. Rooting by pigs produces germination sites for exotic plants and creates pools in which mosquitos that are vectors of avian diseases breed. Goats are a serious problem in dry-to-moist, more open areas. Goat browsing has seriously reduced a number of native species, as shown by the recovery of these forms in areas fenced to exclude them (Loope and Scowcroft 1985). Eradication of the animals is virtually impossible over large areas of rugged mountain topography. Many resident people also hunt pigs and goats and regard them as a desirable resource. Efforts to control these animals have therefore centered on trying to reduce their numbers to less destructive levels by hunting and by excluding them from specific areas by fencing. Feral sheep and cattle occur in some highland areas of the island of Hawaii, and their control has been necessary to protect habitat of endangered birds.

Several exotic game animals have been introduced (Moulton and Pimm 1986; Stone 1989). The most serious problems have resulted from mouflon sheep, introduced to the islands of Hawaii and Lanai, and axis deer, introduced to Maui, Molokai, and Lanai. The mouflon sheep on Hawaii are controlled for protection of native bird habitats, but both mouflon sheep and axis deer are managed for hunting in other areas.

Rats, mice, mongooses, and feral cats are other problem exotics. Black rats damage many shrubs and trees in the native forests and prey on native invertebrates and birds (see Chapter 16). The mongoose has been a serious predator on birds nesting on the ground or in burrows, especially the endangered Hawaiian black-rumped petrel (*Pterodrama phaeopygia*), Newell's shearwater (*Puffinus puffinus*), and the Hawaiian goose (*Nesochen sandvicensis*). Feral cats are also predators on many seabirds nesting on coastal cliffs.

Some 28 species of exotic passerine birds and doves are now established in the Hawaiian Islands, more than in any other area on earth. Most of them occupy the lowland areas of the islands. At least 130 species of passerine birds and doves have been introduced over the years (Stone 1989), but many of these failed to establish or have become extinct. One exotic raptor, the barn owl, and one ardeid, the cattle egret, have been introduced. Some 15 species of exotic game birds have been established successfully (Stone 1989). Most of these exotic species are confined to cities, ranches and farmland, and lowland areas dominated by exotic plants. The mallard duck has been introduced to Hawaii from mainland North America. Mallards pose a hybridization threat to the Hawaiian duck (*Anas wyvilliana*), as they do to other close relatives of the mallard (see Chapter 15). The prevalence of exotics in lowland areas is illustrated by the fact that even in the 1950s, counts revealed that 91 percent of the small land birds were exotics, with only four species of natives being detected (Moulton and Pimm 1983).

Competition from some of these introductions may be a factor in the decline of native land birds (Mountainspring and Scott 1985). The two clearest instances involve the Japanese white-eye and the native elepaio (*Chasiempis sandwichensis*) and ‘i‘iwi (*Vestiaria coccinea*). The white-eye, one of the most abundant exotics, feeds on insects and nectar in habitats occupied by these two native birds. The introduced barn owl may also compete with the native short-eared owl, or pueo (*Asio flammeus*).

Introduced birds have affected natives in another way. The introduction of avian pox and avian malaria to Hawaii, the latter resulting from the combined introduction of mosquito vectors and exotic birds carrying the malaria parasites, has contributed to the extinction of native birds. These diseases are now one of the major threats to the remaining endemic birds (Scott and Sincock 1985; van Riper et al. 1986; Warner 1968).

Eight species of frogs and toads and 15 species of reptiles, some introduced during the Polynesian period, are present as exotics in the islands (McKeown 1996; F. Kraus, personal communication). Recently, a skink and several geckos have been added to the Hawaiian fauna. These species have displaced earlier Polynesian lizards from many habitats (Case and Bolger 1991).

Exotic Invertebrate Animals

The remarkable fauna of Hawaiian land snails has suffered greater extinction than any other group of organisms. These snails belong to 11 families, most of which have suffered considerable extinction. All but 10 or so of the 300 species of the endemic family Amastridae appear to be extinct (Robert Cowie, personal communication). All 21 species of the genus

Carelia, endemic to Kauai, are believed to be extinct, including the largest species of land snails that occurred in the islands.

Snails of the genus *Achatinella*, one of the best-studied groups, show a spectacular range of colors and patterns, perhaps the most beautiful and varied set of land snails in the world. Many of these snails inhabited trees, feeding on fungi growing on the surfaces of tree leaves and branches. Many were restricted in distribution to single valleys or mountain areas. Many also were slow-growing, long-lived species that required four to six years to reach sexual maturity and produced only one to four offspring per year. Yet their abundance at the time of European discovery was enormous. Descriptions by observers in the 1800s and early 1900s suggest that some of the large species reached densities of 500 to 2,000 per tree (Hadfield 1986). The native Hawaiians collected the snails and strung the shells together to make leis.

Abundance was no guarantee of survival, however. About 80 percent of the 41 species of *Achatinella* known on the island of Oahu have become extinct. Overall, about 50 percent of the known species of land snails—about 500 species in all—are believed to have disappeared since the discovery of the islands by Captain Cook.

Many factors have contributed to snail extinctions, including habitat destruction by both Polynesian and European settlers. The attractive shells of the Hawaiian land snails made them the object of human collectors, who took them by the tens of thousands. Some private collections contained over 100,000 shells of Hawaiian land snails (Hart 1978). Black and Norway rats are also predators on these snails (see Chapter 16).

Most recently, the greatest threat to surviving land snails has been the exotic rosy wolfsnail, a predatory snail introduced deliberately for biological control of the giant African snail, a large herbivorous snail that had become a pest in lowland gardens. The rosy wolfsnail was introduced to the Hawaiian Islands in 1955 by the Hawaii State Department of Agriculture. It has spread into the native mountain forests where the remaining native snails exist and has become the most serious predator on many of these species, including the species of *Achatinella* designated as federally endangered.

The native arthropod fauna has suffered in similar fashion. This fauna included many flightless insects and many forms with highly restricted distribution. Over 2,500 exotic arthropods have been introduced to Hawaii, with more than 400 deliberately (Eldredge and Miller 1998). Amazingly, no species of ant is native to the islands. Many endemic arthropods are now threatened by the introduction of exotic ants, over 40 species of which have been introduced. The most serious of these are the long-legged ant, the big-headed ant, and the Argentine ant. Two species of yellowjacket wasps are also serious predators on native arthropods.

The Argentine ant was first recorded in Honolulu, Oahu, in 1940, and has been carried to the other main islands. The potential effect of this ant is enormous. On Maui, Argentine ants appeared in Haleakala National Park, at elevations above 6,700 feet, in 1967. The species spreads slowly because the queens are flightless, dispersing overland to new-found colonies. By 1985, however, about 1.5 percent of the park was infested by the species. Haleakala Park is home to 235 species of native insects, all endemic to the Hawaiian Islands. Of these, 83 species are found only on Haleakala volcano itself.

Comparisons of the terrestrial invertebrate fauna in areas infested by Argentine ants and areas free of these ants showed that species belonging to many groups of invertebrates were reduced in abundance or were completely displaced in areas with Argentine ants (Cole et al. 1992). These included snails, spiders, springtails, earwigs, true bugs, beetles, flies, bees, and moths. Even some of the largest arthropods, such as wolf spiders, showed reductions in abundance in ant-infested areas. Most of the groups showing reductions in abundance were endemics, but some aliens were also reduced in numbers. Many other aliens, however, increased in abundance, for example, isopods and millipeds that are scavengers on the remains of ant prey or predators on ants themselves.

The most serious effects of Argentine ants, from the standpoint of the ecosystem, are on pollinator insects (Stone and Loope 1987). A ground-nesting bee and a moth with a terrestrial larva were two of the species severely reduced in numbers in ant-infested areas. The adults of these species are important pollinators of plants endemic to Haleakala, including the Haleakala Silversword (*Argyroxiphium sandwichense*, ssp. *macrocephalum*), one of the most remarkable of the endemic plants of Hawaii. Were their pollinators displaced, natural reproduction of the silverswords would likely cease.

Most of the deliberate introductions of insects to Hawaii have been for biological control of agricultural pests, some of which were native moths. Many of the introduced control species have attacked native, nontarget insects. At least twenty-seven species of lepidopterans have become extinct in the Hawaiian Islands, fifteen of these losses being attributed to intentional or unintentional impacts of biological control species (Howarth and Ramsey 1991).

Aquatic invertebrates have not escaped challenge by exotics. Twelve species of snails have become established in freshwater habitats (Cowie 1998). Several North American caddis flies have been introduced to the main islands of Hawaii (Kondratieff et al. 1997). At least one of these caddis flies has become a major food item for endemic fish. Whether or not these introduced snails and insects pose a threat to endemic Hawaiian stream invertebrates, however, is unknown.

Exotics in the Marine Environment

Coastal marine ecosystems are also invaded by exotics. Red mangrove was introduced to Molokai in 1902. It is now established on all of the main islands, invading lagoons, fishponds, and stream estuaries (Smith 1985). Red mangrove forms dense, woody jungles that completely alter the shoreline ecology of invaded areas. Mangrove invasion reduced the suitability of aquatic habitats for four species of endemic waterbirds, most seriously for the Hawaiian stilt (Allen 1998).

At least ten species of marine macroalgae have become established, most coming from Florida or the western Pacific (Russell 1992). At least two species of red algae have restricted the habitat distribution of a related, endemic red alga.

Threatened and Endangered Species

As a result of human disturbance and introduction of exotics, the Hawaiian Islands now contain about 40 percent of the threatened and endangered species in the United States (F. Kraus, personal communication). About 80–90 native plants are known or believed to have become extinct, and 270 were listed as endangered or threatened in 1998 (F. Kraus, personal communication). For animals, 2 mammals, 31 birds, and 41 snails were listed as threatened or endangered in 1985. Over 30 percent of the species in the United States being considered for federal listing as endangered or threatened are from Hawaii. Many extinctions have occurred. On Lanai, for example, about 20 percent of the native plants have disappeared, including several species endemic to that island (Hobdy 1993). Since European discovery, 13 species of Hawaiian land birds have become extinct, and several others have been reduced to less than 100 individuals (Scott et al. 1988). Most recently, in late 1989, the last Kauai o'o' (*Moho braccatus*), a male bird that had responded to taped recordings of its song, could no longer be found. About 30 Hawaiian birds are listed as threatened or endangered, and several of these may now also be extinct.

The potential for invasion by additional exotics is enormous. Brown tree snakes have been discovered at least six times at the Honolulu airport on cargo planes coming from Guam, where this exotic has a disastrous impact on native birds (see Chapter 3). Scent dogs, trained to detect this snake, are used to inspect arriving aircraft. Two piranhas have been found in a lake on Oahu, and many others have been confiscated from private aquaria. This species would certainly have a severe impact on native freshwater invertebrates and fish. Fire ants, which invaded North America by ocean cargo ships, would almost certainly find Hawaii a hospitable habitat, compounding the problems caused by the numerous ants that have already become established.

Protecting the Native Biota

Attacking the problems of exotics in Hawaii obviously requires a multifaceted approach. First and foremost, entry prevention is necessary. This strategy includes legal prohibition against importation, quarantine of living plants or animals before or at entry, inspection of biological materials to detect unwelcome associates, and fumigation of imported materials to kill hitchhiking invaders. Human inspection at ports of entry is being augmented by use of more effective canine or electronic "sniffers" to detect biological materials that might contain unwelcome exotics (Stone and Loope 1987).

Efforts to reduce the impacts of established exotics are also essential. For some established exotics, the rate of spread can be reduced. This approach is being used for the Argentine ant in the Haleakala National Park on Maui and in the Hawaii Volcanoes National Park on Hawaii. Argentine ants have flightless queens and spread by the division of colonies, so that toxic baits can be used to reduce the risk of spread into unoccupied areas.

Protection of pristine areas that have not yet been invaded by exotics is also an effective strategy. Some offshore islets have escaped massive invasion by exotics. In volcanic areas on the island of Hawaii, islands of native vegetation, or kipukas, exist where extensive lava flows have isolated them from the main vegetated landscape. Fencing, monitoring, and weeding such sites may enable these pristine areas to be maintained with relatively small effort.

Local eradication is another strategy. Many exotics can be removed and then excluded by fencing. Goat and pig exclosures have been created in many locations in Hawaii to protect native ecosystems and to allow recovery of native vegetation (Stone and Loope 1987). On Maui, Haleakala National Park is completely fenced against goats, and animals inside the park were removed. In some of these exclosures, plants thought to have been locally extinct have reappeared, and in some cases new species discovered, apparently from seeds lying dormant in the soil. Smaller exclosures have also been constructed for pigs in several locations, and in Hawaii Volcanoes National Park, the 29 miles of pig-proof fence have been constructed (Hone and Stone 1989).

The ultimate example of the exclosure approach is protection of individuals. Exclosures or chemical repellents have been used to protect individuals of critically reduced species. Several exclosures in Hawaii function to protect the last individual plants from destruction by goats or pigs (Stone and Loope 1987). Chemical repellents have also been used to deter rats from climbing into canopies of certain native trees to feed on flowers and fruits, in an effort to increase reproduction of critically threatened species.

Direct control has been attempted for both plant and animal species. In some cases, when exotics were detected early, populations have been eradicated. In most cases, however, exotic species have become so widely established that eradication is impractical. Controlling the populations of some species, such as pigs and bamboos, is complicated by the fact that many Hawaiians regard them as valuable resources. In addition, techniques deemed practical for reducing populations of animals such as pigs may be considered inhumane by animal rights groups.

Efforts may also be made to find means of biological control of the most serious plant and invertebrate invaders. Partial control of lantana and gorse, for example, has been achieved by introduced insect enemies. Controlling problem grasses in this fashion is difficult, however, because the importation of possible enemies is often opposed by the sugarcane industry (because sugarcane, a grass, might also be attacked). Similar objections to biological control exist for gingers, koa haole, tropical ash, and other plants, because of their horticultural or forestry values. Introductions for biological control must be rigorously screened to assure that the risk of impacts to nontarget native species is very low.

Involving the Public

Preserving Hawaii's biota will require the support of the public at large, both residents of the islands and visitors, and the commitment of diverse federal and state agencies to carry out prevention and control programs for exotics. Although the public is aware of the threat of certain species that might affect the islands' tourism image, the cumulative ecological impact of the numerous species that invade annually is not considered serious by most people. Coordination of policies by governmental agencies is also weak. Exotics are the leading environmental problem in Hawaii and are a problem that threatens the state's tourist economy, as well. Conservation organizations must unite to dramatize this problem and must demand concerted action.

Part III

Biotic Perspectives

Chapter 14

Exotic Game and Fish: Addiction to Game and Fish Introduction

> [The exotic game program] has depleted the game funds of 48 states for half a century, and has served as a perfect alibi for postponing the practice of game management.
>
> —Aldo Leopold, "Chukaremia"*

In southern New Mexico, vast areas of desert and dry rangeland support very low populations of ungulates, such as mule deer, pronghorn, and desert bighorn sheep. Some scientists (e.g., Martin 1970) have suggested that this condition is the result of a massive extinction of large animals in North America because of the arrival of Paleo-Indian hunters from Asia about 12,000 years ago. According to this theory, numerous "empty niches" might exist in the American Southwest, perhaps enough to support 10–20 million animal units (one animal unit equals 1,000 pounds of cow) of animals consuming forage not utilized by cattle (Martin 1970). The New Mexico Department of Game and Fish, serving the interests of hunters, was eager to fill those empty niches. Under a cooperative federal–state program, gemsbok, the southern African form of the oryx, were introduced to White Sands Missile Range between 1969 and 1977 (Table 14.1). The objective was to establish a trophy hunting program that would benefit sport hunters and earn the department additional revenue. The gemsbok was selected because of its physiological adaptation to desert cli-

*Aldo Leopold (1938). "Chukaremia." *Outdoor America* 3:3.

Table 14.1. Exotic game, fish, and fur-bearing species that have been introduced into North America and Hawaii.

Fish	
bitterling, *Rhodeus sericeus*	Europe
blackchin tilapia, *Sarotherodon melanotheron*	Africa
blue tilapia, *Oreochromis aureus*	Africa
brown trout, *Salmo trutta*	Europe
common carp, *Cyprinus carpio*	Europe
ide, *Leuciscus idus*	Europe
Mozambique tilapia, *Oreochromis mossambicus*	Africa
oscar, *Astronotus ocellatus*	South America
peacock bass, *Cichla ocellaris*	South America
rudd, *Scardinius erythrophthalamus*	Europe
tench, *Tinca tinca*	Europe
Birds	
black francolin, *Francolinus francolinus*	Asia
chestnut-bellied sandgrouse, *Pterocles exustus*	Old World
chukar partridge, *Alectoris chukar*	Asia
common peafowl, *Pavo cristatus*	Asia
emu, *Dromaius novaehollandiae*	Australia
Erckel's francolin, *Francolinus erckelii*	Africa
gray francolin, *Francolinus pondicerianus*	Asia
gray partridge, *Perdix perdix*	Europe
Himalayan snowcock, *Tetraogallus himalayensis*	Asia
Japanese quail, *Coturnix japonica*	Asia
kalij pheasant, *Lophura leucomelana*	Asia
red-billed francolin, *Francolinus adspersus*	Africa
red junglefowl, *Gallus gallus*	Asia
ring-necked pheasant, *Phasianus colchicus*	Asia
ostrich, *Struthio camelus*	Africa
Mammals	
axis deer, *Axis axis*	Asia
Barbary sheep, or aoudad, *Ammotragus lervia*	Africa
blackbuck antelope, *Antilope cervicapra*	Asia
eland, *Tragelaphus oryx*	Africa
fallow deer, *Dama dama*	Eurasia
gemsbok, *Oryx gazella*	Africa
Himalayan tahr, *Hemitragus jemlahicus*	Asia
mouflon sheep, *Ovis musimon*	Europe
nilgai antelope, *Boselaphus tragocamelus*	Asia
nutria, *Myocastor coypus*	South America
Persian wild goat, *Capra aegagrus*	Asia
red deer, *Cervus elaphus*	Europe
reindeer, *Rangifer tarandus*	Eurasia
sambar deer, *Cervus unicolor*	Asia
Siberian ibex, *Capra ibex*	Asia
sika deer, *Cervus nippon*	Japan
wild boar, *Sus scrofa*	Eurasia

mate. Little consideration was given to the possible impact of this species on the desert ecosystem or on the native wildlife. It was apparently assumed that the gemsbok would stay where they were put.

The original 93 gemsbok have grown to a herd of 2,700, and they have spread over an area about 100 miles in diameter. In spite of $885,000 spent to build a 68-mile game fence, their range now includes White Sands National Monument, where 100–200 animals have overbrowsed the native vegetation. The gemsbok also range into land administered by other state and federal agencies, and into private land. Although they are hunted, no strong barriers exist to their dispersal, and the population continues to grow and spread. On White Sands Missile Range, gemsbok coexist with 300–350 native pronghorn (*Antilocapra americana*). Until the late 1990s, they coexisted with about 1,800 feral horses. Studies of food habits of these animals suggest that horses and gemsbok are food competitors, because both feed heavily on grasses (Smith et al. 1998). Most feral horses have been removed in recent years, with the last scheduled to be rounded up in 1999. Pronghorn occupy only a portion of the White Sands Range. Their diet is mainly shrub browse and broadleaf herbaceous plants. Nevertheless, gemsbok consume many plant species utilized by pronghorn, and expansion of gemsbok populations into range more suitable for pronghorn could lead to food competition. Eliminating gemsbok from the national monument by shooting is opposed by a New Mexico animal protection group, and capturing and relocating the animals is very expensive. The question that must be faced is this: Has the value of this species as a game animal outweighed its potential impacts on native ecosystems and species?

History of Game and Fish Introduction

Hunting, fishing, and trapping have been traditional activities in North America since the first European settlement. Hunters and fishers have also been a strong political force, the result being that federal and state agencies were created to serve their interests. Although wildlife management as a science emerged in the 1930s under the influence of Aldo Leopold, who regarded wildlife as members of entire ecosystems, management practice largely remained focused on individual game species and short-term results. The concept that more species mean better hunting and fishing pervaded wildlife management agencies, and introduction of exotic game and fish was considered a useful practice. Even during and after the decade of the 1960s, when endangered species became an environmental concern and strong endangered species legislation was enacted, game and fish agencies continued to introduce exotics with little concern about their impacts. The legacy of these actions is a hodgepodge of exotic game and fish that compound the difficulties of protecting and managing native plants and animals, both game and nongame.

The history of introduction of exotic game, furbearers, and fish to North America goes back to the 1700s. Prior to the 1940s, however, introductions were carried out in a virtually unplanned manner. Exotic ungulates, game birds, and sport fish were released in many locations in the hope that they would become established and add new species to the sportsmen's menu. Translocations of native game species were made freely to new locations within North America. State wildlife agencies and private sportsman's groups carried out these introductions and translocations with almost no restriction. This chapter examines introductions of exotic game and fish species brought to North America from elsewhere. Chapter 15 considers translocations of North American natives to other North American locations.

In 1948, the U.S. Fish and Wildlife Service organized the State–Federal Cooperative Foreign Game Investigations Program. This program compiled life history and ecological information on potential exotic game animals and furnished advice on the advisability of introductions in particular situations, but it was basically designed to encourage the establishment of exotic game animals in habitats that seemed to lack good populations of native species (Bump 1968). As noted previously, the gemsbok was one of the large game animals promoted by this program. Many exotic game birds, including several African and Asian francolins and Asian pheasants, were studied, and information about them was published so that state game agencies could identify species that might flourish in their regions. Between 1960 and 1970, the program led to the introduction of at least nineteen species of exotics to the southern United States and Hawaii (Bland and Temple 1993). The federal program was terminated in 1970, and in 1977 President Carter issued an executive order effectively canceling the use of federal funds and resources for introduction of exotics.

Exotic Ungulate Game Animals

Exotic large ungulates have been introduced to all states and provinces of the United States and Canada. The largest populations and greatest variety of exotic species, however, are found in the southern and southwestern United States. Most of these exotics are foreign to North America. Some, however, involve translocations of North American species to locations outside their original range (see Chapter 15). Many of the early introductions were made by state or provincial game and fish agencies, but more recent introductions have been made largely by private groups. In most cases, the objective of the introductions has been to create additional sport-hunting opportunities, including commercial hunting and fishing on private lands. Game farming, the rearing of animals for commercial production of meat, hides, and other products, however, is a growing factor. Game farming now involves native species such as elk (*Cervus canadensis*) and intercontinental exotics, such as African ostriches and zebras.

Early introductions of exotic ungulates were often made into unconfined areas, with the objective of establishing free-ranging wild populations. Sika deer, for example, were introduced to James Island, Maryland, in Chesapeake Bay, in 1916 (Flyger 1960). The 4 or 5 original animals increased to about 270 in 1957. In 1958, at least 161 animals died from malnutrition. Sika deer from James Island also colonized nearby areas on the Eastern Shore of Maryland and were translocated to Assateague and Chincoteague Islands, as well as to other mainland areas of Maryland.

Unconfined populations of both sika and axis deer can displace native white-tailed deer (*Odocoileus virginianus*). Both utilize a wider range of habitats and food plants and tolerate overbrowsed habitat better than white-tailed deer (Demarais et al. 1998). In some places, Sika deer have increased in numbers at the apparent expense of native white-tailed deer (Keiper 1985). During the 1970s and early 1980s, for example, the percentage of sika deer in the overall deer harvest in eastern Maryland increased greatly, with more than 80 percent of the kill being sika deer and less than 20 percent whitetails on Assateague Island (Feldhamer et al. 1978; Keiper et al. 1984). Sika and axis deer have also been shown to displace white-tailed deer in experimental enclosures in Texas (Demarais et al. 1998).

A somewhat different objective was involved in the introduction of reindeer to North America from Eurasia. Beginning in 1891, reindeer, the Old World semidomesticated relative of the caribou, were introduced to several mainland and island areas in Alaska and Canada in attempts to create a herding economy for native peoples of the Arctic (Klein 1980). Reindeer numbers increased to about 600,000 in the 1930s but have declined since then to less than 25,000. The largest herds were in western, southwestern, and northern coastal areas of mainland Alaska. Unsuccessful attempts were made to establish reindeer herding in several locations in Canada. Reindeer were also introduced to several Alaskan islands, including Nunivak, St. Matthew, the Pribilofs, Kodiak, and two islands in the Aleutians. On St. Matthew, 29 animals introduced in 1944 grew to an estimated 6,000 in 1963, a number that greatly exceeded the carrying capacity of the vegetation. In the winter of 1963–1964, all but 50 animals, all females, starved (Klein 1968). On Nunivak Island, the herd grew to 20,000 animals, and then crashed to 5,000 because of degradation of the tundra range. Reindeer still survive on several Alaskan islands, where their numbers are controlled by hunting.

Interest in reindeer herding has increased in recent years, largely because of a growing market for reindeer antlers in Asia, where the antlers are used in medicines. Unfortunately, reindeer herding creates several problems in regions where wild caribou populations occur (Klein 1980). Reindeer tend to desert domestic herds and join wild caribou herds. Reindeer herding tends to be concentrated on the same ranges throughout the year, whereas arctic caribou herds are highly migratory. Reindeer, as a result, exploit tundra food plants, especially ground lichens, more intensively than do caribou. Thus, where winter ranges of reindeer and caribou coincide, competition for food may impact caribou populations. Diseases may be passed from reindeer to caribou, as well. When reindeer were introduced to West Greenland, for example, warble flies (*Oedemagena tarandi*) and nasal bot flies (*Cephenemyia trompe*) were introduced to caribou populations that had previously been free of these parasites.

More recently, private introductions of exotic large game have usually been made to fenced game ranches, most of which are managed for commercial hunting. When large exotic animals are released in considerable numbers and at several or many locations, however, some animals almost always escape confinement and establish wild populations that come into contact with native wildlife.

Game ranching is more popular in Texas than in any other state. As of 1994, at least 195,000 animals of 71 species of exotic ungulates were living on Texas ranches (Demarais et al. 1998). These included more than 10,000 animals each of axis, sika, and fallow deer, blackbuck and nilgai antelope, and Barbary sheep (Teer 1991). Recent additions to the ranks of exotic animals are the African ostrich and the Australian emu, which are ranched for their meat, oil, leather, and eggs. Of the more than 195,000 animals, about 77,000 were free-ranging animals that were not confined by game fences. The numbers and diversity of these exotics in 1994 had increased almost 14-fold over their totals in 1963. Although exotic game animals are found in more than half the counties in Texas, most are concentrated in the south-Texas plains and Edwards Plateau ecological regions (Mungall and Sheffield 1994).

The popularity of game ranching in Texas dates to the period from

1930 to 1950, when native game species had been depleted, hunting seasons and limits were being imposed, and trophy hunting was becoming popular (Mungall and Sheffield 1994). Nilgai antelope were introduced to the King Ranch in southernmost Texas in the 1920s, and several species of exotic deer and African antelopes to a large ranch on the Edwards Plateau in the 1930s. Between the 1930s and the early 1960s, over 175 additional ranches acquired exotic ungulates, and commercial trophy hunting became popular.

As the populations of exotic game increased on Texas ranches, a number of species escaped confinement and established free-ranging populations. These include axis deer, blackbuck antelope, nilgai antelope, Barbary sheep, fallow deer, sika deer, and mouflon sheep. In addition to these species, with numbers totaling nearly 75,000 animals, the population of wild boar and pigs in Texas is estimated at about 122,000. Several thousand hybrids of red deer and elk and of ibex and domestic goats are also free ranging. These numbers may be placed in perspective against estimates of 4 million white-tailed deer, 200,000 mule deer, 200,000 javelina, and 15,000 pronghorn for the state. When the market for emu products collapsed in 1994, emus were set free by some ranchers, but, as yet, no wild breeding populations appear to have become established.

Many other states also have free-ranging populations of exotic game. California, for example, has localized populations of fallow, sambar, and axis deer, as well as Barbary sheep and Himalayan tahr. In New Mexico, the Department of Game and Fish has introduced four ungulates—Barbary sheep, gemsbok, Persian wild goat, and Siberian ibex—to unconfined wildlands. Barbary sheep were released into the rugged Canadian River Canyon in northeastern New Mexico in 1950 (Laycock 1966). Barbary sheep in New Mexico are now estimated to number about 4,000 animals.

Eurasian wild boar have been introduced to a number of locations in the United States (Mayer and Brisbin 1991). In some places these animals have maintained their genetic characteristics, but in many places they have interbred with feral domestic hogs. Several wild boar introductions have failed to establish viable populations. In 1908, however, Eurasian wild boar were introduced to a poorly fenced enclosure in the Great Smoky Mountains. Many of these animals escaped and initiated a wild population that has spread through mountain areas in North Carolina and Tennessee. Both private individuals and state game departments have made subsequent releases in mountain areas of North Carolina and Tennessee with the goal of establishing boar as a game animal. Animals from the Great Smoky Mountains have also been translocated to a wildlife management area in western Tennessee, to locations in the coast ranges of central California, and to locations in Florida, West Virginia, and South Carolina. In turn, animals from the initial California introduction have been translocated to other locations in central and northern California. In Texas, Eurasian wild boar have also been introduced to the central coast and to the Edwards Plateau. Still other wild boar introductions have been made to Gulf Coastal states, where populations of feral domestic hogs have existed for a long time, and to New Hampshire.

In the Great Smoky Mountains National Park, the effects of rooting by Eurasian wild boar can be serious for forest-floor plants and animals. Population densities of wild boar in the park in 1979 ranged from about thirteen to sixty animals per square mile (Mayer and Brisbin 1991). Rooting was especially destructive for herbaceous vegetation of the forest floor, reducing cover by 90 percent or more (Bratton 1975). Rooting also thinned the litter layer and increased leaching losses of several mineral nutrients. Red-backed voles (*Clethrionomys gapperi*) and short-tailed shrews (*Blarina brevicauda*) were nearly eliminated in areas of severe rooting (Singer et al. 1984). Boar are also predators on many forest-floor vertebrates and on-ground nests of birds, including game birds such as the wild turkey.

Exotic ungulates pose a variety of dangers to native wildlife. Foremost among these are competition, the potential for introduction of diseases and parasites to native species, and hybridization with native species. The diet of Barbary sheep, for example, is very similar to that of desert bighorn sheep, a species that has declined greatly in the American Southwest. Simpson et al. (1978) found that over 75 percent of important desert bighorn foods were also consumed by Barbary sheep. Barbary sheep also have a higher reproductive potential and greater dispersal tendency than do bighorns. Furthermore, Barbary sheep carry bluetongue, a viral disease, that can be fatal to desert bighorns. Thus, invasion of bighorn habitat by Barbary sheep is likely to lead to displacement of bighorns, if they are present, or prevent their reintroduction (Robinson and Bolen 1989).

Barbary sheep are also dietary competitors with mule deer (*Odocoileus hemionus*), the most important native large game animal in western North America. In Palo Duro Canyon, Texas, where both species were introduced, the overlap in diet is about 74 percent (Krysl et al. 1980).

Hybridization with exotic game animals poses a danger to some North American natives. Sika and red deer hybridize freely in England, Scotland, and Ireland (Abernethy 1994), and both can interbreed with North American elk (*Cervus canadensis*). The translocation of native game species to other areas within North America has also resulted in hybridization with local forms (see Chapter 15).

Exotic Furbearing Animals

Exotic furbearing animals have also contributed to disruption of North American ecosystems. The nutria, a large rodent native to southern South America, was introduced to Louisiana in 1938 for fur production under captive conditions in breeding farms. Escapes from a Louisiana breeding farm damaged by a hurricane founded a wild population in coastal marshlands in 1940. These animals, augmented by other escapes and releases, created a population that grew to an estimated 20 million animals by the late 1950s (Hesse et al. 1997). In the 1960s and 1970s, fur trappers harvested millions of nutrias annually. The declining popularity of fur clothing in the 1980s and 1990s, however, led to a greatly reduced harvest and to correspondingly greater populations of nutrias in the wild (Hesse et al. 1997). Nutrias have spread along the Gulf coast and have also been introduced at several localities in Florida, Maryland, California, and even Ohio.

Nutrias are herbivorous, and their intensive feeding has had significant impacts on coastal salt-marsh vegetation. In the Baratraria Basin of Louisiana, for example, nutrias feed preferentially on bulrushes (*Scirpus* spp.) as opposed to cordgrass (*Spartina* spp.). This has apparently caused a shift in dominance from bulrush to cordgrass since the 1930s, as well as increasing the conversion of marsh to open water, now occurring at a rate of about 2 percent per year. Where nutrias have grazed heavily on maidencane (*Panicum hemitomon*), a dominant grass of floating marsh vegetation, lawnlike stands of spikerush (*Eleocharis* sp.) have replaced maidencane (Hesse et al. 1997). In Atchafalaya Bay, nutria and waterfowl grazing in combination reduces the biomass of palatable marsh plants much more than does waterfowl grazing alone (Evers et al. 1998), so that nutrias are clearly competing with wintering waterfowl.

Nutrias are also serious predators on seedlings of bald cypress (*Taxodium distichum*) in Louisiana (Blair and Langlinais 1960; Conner and Toliver 1990). Nutria impacts, together with other disruptions of cypress swamp ecology, threaten to eliminate this vegetation type from much of coastal Louisiana.

Exotic Game Birds

Five exotic game birds have been successfully introduced to continental North America: ring-necked pheasant, gray partridge, chukar partridge, black francolin, and Himalayan snowcock. The first three are widespread, the last two very localized. Black francolins occur in parts of Louisiana and Florida; the Himalayan snowcock, in high mountains of Nevada.

Ring-necked pheasants were first brought to North America about 1730 from Europe, where they had been introduced from Asia in Roman times. Early releases in eastern North America failed or survived for only a few years. In 1881, however, they were introduced directly from China into the Willamette Valley, Oregon. Within ten years, numbers had increased to the point that the species was a popular game bird. From Oregon, pheasants were translocated to neighboring states and to the upper Midwest. They now occur across much of southern Canada and the United States. Ring-necked pheasants occur both in farmland, especially in irrigated areas, and in native grassland. The largest populations are found in the northern plains states and southern prairie provinces.

The ring-necked pheasant is usually regarded as a beneficial game species, but it may have a detrimental competitive effect on native prairie chickens. Both Sharp (1957) and Vance and Westemeier (1979) found that pheasants tended to dominate prairie chickens in behavioral encounters and to displace them from booming grounds. Pheasants often laid eggs in prairie chicken nests, reducing the reproductive success of the latter. The spread of the ring-necked pheasant has probably contributed to the decline of the greater prairie chicken (*Tympanuchus cupido*) in the northern and eastern parts of its range (Sharp 1957).

The gray partridge was released in many locations in North America, beginning in the 1700s. All introductions prior to 1900 failed, however. Extensive introductions in the early to middle 1900s have led to its establishment in many locations across southern Canada, from Nova Scotia to British Columbia, and in the northern United States. Gray partridges are largely restricted to farmland and do not appear to have had detrimental effects on either crops or native birds.

Chukar partridge have been released in at least forty-two states and in six provinces of Canada. The first successful establishment, however, was in Nevada in 1935. They have become well established in mountainous, arid, and semiarid parts of the intermontane West from southern British Columbia to northwestern New Mexico. Chukars apparently do well on rangeland that has been degraded and invaded by cheatgrass (*Bromus tectorum*), itself an exotic. The spread of cheatgrass through the intermontane region, in fact, may be the factor that allowed chukars to become established. Some of this habitat was degraded from sagebrush and native grass to cheatgrass, and was originally occupied by sage grouse (*Centrocercus urophasianus*), a native species.

One remarkable introduction under the Foreign Game Investigations Program was of the Himalayan snowcock, a large relative of the chukar partridge, to the Ruby Mountains, northeastern Nevada (Bland and

Temple 1993). This large grouse lives in treeless alpine habitats in Pakistan, and it is unsuited to all but a few high-mountain locations in Nevada. Nevertheless, the Nevada Department of Wildlife created a captive breeding population of this species and released over 2,000 birds in several mountain ranges in northern Nevada between 1963 and 1979, at a cost of over $750,000. This very expensive effort resulted in the establishment of one small population of a species that is very difficult to hunt. No consideration was given to the possible effects of this bird on the vegetation of this unique alpine environment, much of which is now a federal wilderness area. Nor has the species been monitored effectively to determine its population and impact on the alpine environment.

Several species of native game birds have been translocated to new locations in North America (see Chapter 15). The mallard duck, native to mainland North America, has also been introduced to Hawaii (see Chapter 13).

More exotic game birds, seventy-five species in all, have been released in Hawaii than to any other place on earth. Of these, only fourteen species, still the world's record for exotic game bird introductions, are currently established (Long 1981). The island of Hawaii alone has twelve established species. Exotic game birds in the Hawaiian Islands include four francolins, three quail, two pheasants, the chukar partridge, wild turkey, peafowl, red jungle fowl, and chestnut-bellied sandgrouse. The red jungle fowl occurs in the wild only in the montane forests of Kauai; elsewhere feral domestic birds are occasional. Most of these species appear to have little impact on native ecosystems. The diet of the ring-necked pheasant overlaps considerably with that of the nene goose (*Branta sanvicensis*), but it is uncertain whether or not competition from pheasants might eventually retard recovery efforts for the nene (Cole et al. 1995). Chukar partridges and ring-necked pheasants, which occur together in many locations, feed on fruits of many native woody plants and may provide a useful service in their dispersal (Cole et al. 1995). However, they may also be a reservoir of exotic avian diseases and prey for exotic predators. On the island of Hawaii, the kalij pheasant, a forest bird, feeds heavily on the fruits of the banana poka (see Chapter 13) and is probably an important disperser of this problem exotic. Erckel's francolin, wild turkey, and peafowl may also disperse the seeds of banana poka (Scott et al. 1986).

Exotic Sport Fish

A number of exotic game fish have been introduced to North America from other continents. The common carp was introduced to most parts of

North America between 1830 and 1890 (see Chapter 7). Tench, ide, bitterling, and rudd are other European game fish that were introduced to eastern North America in the 1800s and early 1900s, but they have now disappeared or survive only in a few locations (Courtenay et al. 1984; Radonski et al. 1984).

Perhaps the best-known exotic game fish is the brown trout. Introduced from Europe in 1883 by the U.S. Fish Commission, brown trout have been stocked into almost all streams and lakes suitable for trout in North America (Radonski et al. 1984). From the sportsman's point of view, brown trout have been a success, in that they do well in many waters that are not preferred by native trout. Where endemic trout occur, however, the brown trout may be a detrimental competitor. In the Kern River of California, for example, the endemic golden trout (*Onchorhynchus aquabonita*) has been threatened by introduced browns, and nearly $1 million has been spent to eradicate the browns (OTA 1993).

Several exotic game fish have been introduced to Florida, most of them African or South American cichlids. These include the blue, Mozambique, and blackchin tilapias, the oscar, and the peacock bass. The blue tilapia has become a predominant species in many waters and is probably reducing spawning success by various native fish. The peacock bass, introduced to Florida in 1986, has reduced populations of bass and other native game fish. Peacock bass, which reach a length of 20 inches and a weight of over 4 pounds, are predatory fish that proved to be a keystone predator when introduced to Gatun Lake, Panama (Zaret and Paine 1973). There, several fish were eliminated, zooplankton and mosquito populations exploded, and fish-eating bird populations were reduced.

Ecological Risks of Exotic Game and Fish

Although many of the exotic game and fish that have been introduced have caused only minor ecological disruptions, several have had negative impacts on native species, and a few have had serious consequences at the ecosystem level. Many exotic game and fish are similar in their ecological relationships to native species and thus act as "substitutes." In rare cases, as with the ring-necked pheasant and chukar in Hawaii, introduced species may benefit native plants by replacing native birds lost by human-caused extinction. More often, their impacts tend to be competition between an introduced exotic and a similar native, and in some cases the transmission of a disease to native species. Because they are similar to native species, Aldo Leopold's assessment is telling. Their introduction has often been at the expense of sound management of native ecosystems and native game and fish.

Introduced species such as the wild boar, nutria, and peacock bass, however, are unique in their ecological impacts and may cause major disruption of community or ecosystem structure and function. Introductions of game animals to Hawaii and other oceanic islands are especially risky in this regard, because of the fact that the biotas and ecosystems of such islands have evolved in the absence of nonflying mammals. As Magnuson (1976) stated, managing with exotics is a "game of chance."

Consequently, little justification exists for introduction of exotic game and fish. In 1966, this realization led the U.S. Department of Interior to specify a restricted set of conditions for introducing exotics on federal lands. In 1976, the Wildlife Society, a scientific organization devoted to wildlife management, adopted a policy of opposition to introduction of exotics unless there is a clear need, evidence that no deleterious impacts will result, and assurance that the species can be contained in the area to which it is introduced. President Carter's 1977 executive order effectively terminated the introduction of exotics on federal lands and discouraged state and private groups from introducing exotics.

Regulation of Game and Fish Introduction

Two recent economic trends make the impetus to import exotic animals by the private sector very strong. One is the trend toward privatization of hunting and fishing, with its emphasis on trophy animals and trendy exotics. The second is the trend toward farming of exotic game species for commercial harvest of meat, skins, and other products. Regardless of the assurances offered by private operators, the successful propagation of large numbers of almost any game or fish species means that feral or naturalized wild populations will become established by the escape of individuals.

Part of the desire for introduction of exotic game and fish reflects an educational shortcoming. The members of hunting and fishing organizations, and sometimes even the personnel of governmental game and fish agencies, are too often unaware of the problems that exotics can cause. Management efforts to improve habitats and resources for native game and fish are often abandoned for the "quick fix" of a "better adapted" exotic. For good ecological reasons, these introductions rarely live up to their expectations. An improved understanding of wildlife ecology can thus help minimize the careless introduction of exotic game and fish.

Nevertheless, a clear need exists for tighter regulation of the importation of exotic wildlife for sport hunting and game ranching. In some locations, such as Hawaii, importing wildlife for such purposes is too risky to be permitted. Elsewhere, one reasonable proposal is that an importer must post a bond for exotic animals that are brought to North America. The bond would serve to provide funds for the eradication of exotics that escape confinement and establish wild populations. Even with such a requirement, a system of monitoring to detect incipient wild populations is also needed. As several exotic game and furbearing animals such as Eurasian wild boar and nutrias have shown, populations of such species can quickly become so firmly established that eradication is next to impossible.

The issues raised by exotic game and fish are echoed by those of native North American species that have become introduced to regions of North America outside their original ranges and by feral domestic animals and other species that live in close association with humans.

Chapter 15

Homegrown Exotics: Natives Out of Place

> In contemplating the cowbird and its hosts, we tend vaguely to assume that all of these species have come down through the ages together and that, therefore, the parasitic relationship must be tolerable to the hosts. . . . I believe this assumption may be hasty. Rather, I believe, the opening of the forests by civilized man has allowed the cowbird to penetrate into new regions where it has access to host species that have had little or no ancestral experience through which to develop effective defenses against it. Here the cowbird is exploiting the new hosts with exceptional benefit to itself and with extraordinary potential for damage to them.
>
> —Harold Mayfield, "The Brown-Headed Cowbird, with Old and New Hosts"*

The least Bell's vireo was originally one of the most abundant birds in California, ranging from the Mexican border to the upper Sacramento Valley. This subspecies of the Bell's vireo is an obligate riparian nester, and it is no surprise that the species declined with the destruction of perhaps 95 percent of the riparian habitat in the state. What was surprising was its disappearance from much of the remaining riparian habitat. By 1978, it was almost extinct, an estimated ninety pairs surviving in a few river valleys in southern California and northern Baja California, Mexico. The culprit in the decline of the species in its remaining riparian habitat was the

*Harold Mayfield (1965). "The Brown-Headed Cowbird, with Old and New Hosts." *Living Bird* 4:13–28.

brown-headed cowbird, a bird native to the North American Great Plains (Table 15.1).The cowbird is a nest parasite, laying its eggs in the nests of other birds, an adaptation that permitted it to follow a nomadic life in association with migratory herds of bison.

Table 15.1. Scientific names and native regions of invasive homegrown exotics.

PLANTS	
black locust, *Robinia pseudoacacia*	eastern North America
honey locust, *Gleditsia triacanthos*	midwestern United States
Osage orange, *Maclura pomifera*	Texas, Oklahoma, Arkansas
Washington fan palm, *Washingtonia filifera*	California, Arizona
FRESHWATER INVERTEBRATES	
crayfish, *Procambarus clarkii*	eastern United States
opossum shrimp, *Mysis relicta*	northern North America
rusty crayfish, *Orconectes rusticus*	Ohio drainage
FISH	
brook trout, *Salvelinus fontenalis*	eastern North America
channel catfish, *Ictalurus punctatus*	eastern North America
chinook salmon, *Onchorhynchus tshawytscha*	Pacific Coast
coho salmon, *Onchorhynchus kisutch*	Pacific Coast
cutthroat trout, *Onchorhynchus clarki*	western North America
flathead catfish, *Pylodictis olivaris*	Mississippi Valley
golden trout, *Onchorhynchus aguabonita*	western North America
kokanee salmon, *Onchorhynchus nerka*	Pacific Coast
lake trout, *Salvelinus namayaish*	northern North America
largemouth bass, *Micropterus salmoides*	eastern North America
mosquitofish, *Gambusia affinis*	eastern North America
northern pike, *Esox lucius*	eastern North America
pink salmon, *Onchorhynchus gorbuscha*	Pacific Coast
rainbow trout/steelhead, *Onchorhynchus mykiss*	western North America
smallmouth bass, *Micropterus dolomieu*	eastern North America
AMPHIBIANS AND REPTILES	
bullfrog, *Rana catesbiana*	eastern United States
Rio Grande leopard frog, *Rana berlandieri*	Texas
BIRDS	
barred owl, *Strix varia*	eastern North America
bobwhite quail, *Colinus virginianus*	eastern United States
brown-headed cowbird, *Molothrus ater*	central North America
California quail, *Callipepla californica*	California
Canada goose, *Branta canadensis*	northern North America
Gambel's quail, *Callipepla gambellii*	western United States
house finch, *Carpodacus mexicanus*	western North America

mallard duck, *Anas platyrhynchos*	mainland North America
mountain quail, *Oreortyx pictus*	western United States
plain chachalaca, *Ortalis vetula*	Texas and Mexico
ruffed grouse, *Bonasa umbellus*	North America
scaled quail, *Callipepla squamata*	western United States
white-tailed ptarmigan, *Lagopus leucurus*	western North America
wild turkey, *Meleagris gallopavo*	western United States
willow ptarmigan, *Lagopus lagopus*	arctic North America
MAMMALS	
arctic fox, *Alopex lagopus*	northern North America
bison, *Bison bison*	central North America
coyote, *Canis latrans*	western North America
eastern cottontail, *Sylvilagus floridanus*	eastern North America
elk, *Cervus Canadensis*	western North America
mountain goat, *Oreamnos americanus*	western North America
raccoon, *Procyon lotor*	North America
red fox, *Vulpes fulva*	North America
Virginia opossum, *Didelphis virginiana*	eastern United States

The cowbird arrived in California about 1890 and found a variety of new hosts that were not adapted to counter cowbird nest parasitism, including the least Bell's vireo. In the 1980s, somewhere between 50 and 100 percent of all least Bell's vireo nests were parasitized, and its reproductive success was abysmal. The subspecies was listed as federally endangered in 1986. In 1983 trapping was begun to remove cowbirds, and by 1992 trapping was being conducted in all riparian areas where the vireo survived (Griffith and Griffith 1999). In some areas, nests were also monitored to detect and remove cowbird eggs. In the San Luis Rey Valley in San Diego County, for example, trapping alone was inadequate to suppress cowbird parasitism. Monitoring all nests in the drainage, however, increased fledging of young by up to 44 percent over the success that was predicted for cowbird trapping without nest monitoring (Kus 1999). The result of trapping and monitoring efforts has been a rapid increase in numbers of least Bell's vireos in southern California. Now, perhaps 1,500 pairs inhabit the river valleys in southern California. To maintain this recovery, and to encourage the vireo to recolonize its lost range in California, however, requires a continuing, labor-intensive program of cowbird management.

The cowbird example shows that problem exotics originate not only outside of North America, but also from locations within North America. "Homegrown" exotics include mammals, birds, reptiles, amphibians, fish, invertebrates, and plants. Homegrown exotics occur, in part, because humans have transported native species across barriers, such as the Great Plains, the western mountains, or major watershed divides. At least fifty-one species of native terrestrial vertebrates and fifty-seven species of native fish have been introduced to areas outside their native range (OTA 1993). Transformation of the North American landscape has also created habitats that allow formerly restricted species to expand their range or follow habitat corridors into once isolated parts of the continent. Many natives, such as the brown-headed cowbird and the coyote, have expanded their range far into new regions.

This chapter considers only native terrestrial and freshwater species that have invaded or been introduced to other locations in mainland North America. As noted in other chapters, marine and estuarine species have been translocated from Atlantic to Pacific coasts, and vice versa (see Chapter 5). Many mainland North American species have also been introduced to Hawaii (see Chapter 13).

Homegrown Exotic Plants

Only a few native North American plants have become homegrown exotics. The most significant of these are several trees that have been planted for ornamental reasons. Osage orange, native to parts of Texas, Oklahoma, and Arkansas; black locust, native to eastern North America; and honey locust, native to the Midwest, have been planted widely and have become established in the wild in many locations. Along southwestern rivers, these species contribute to the displacement of native riparian trees (see Chapter 11). The Washington fan palm, native to desert springs in California and Arizona, has also been planted widely as an ornamental and is a common invasive tree in riparian areas of coastal California.

Homegrown Exotic Invertebrates

Likewise, only a few native invertebrates have become problem exotics in other parts of North America. In the northern Midwest, however, a remarkable sequence of sequential displacement of a native crayfish by homegrown exotic congeners is occurring. The native crayfish of northern Wisconsin lakes is *Orconectes virilis*. Two other members of this same genus have been introduced to the region, probably in bait buckets of fishermen. *Orconectes propinquus* apparently became established in the 1950s and *O. rusticus* in the 1960s (Lodge 1993). *Rusticus* is native to streams and lakes from Illinois, Michigan, and Ohio south to Missouri and Alabama. All three species are generally similar in their ecology. However, *propin-*

quus tends to displace *virilis,* and *rusticus* to displace both congeners. In addition to displacing its congeners, *rusticus* causes major changes in the littoral community of lakes and the benthic community of streams (Charlebois and Lamberti 1996; Houghton et al. 1998). It has displaced the native *virilis* in many lakes and streams. It suffers lower predation by fish, allowing it to become more abundant than *virilis.* Its feeding reduces the abundance of other benthic invertebrates. It is also strongly herbivorous, and decimates submerged rooted vegetation, altering the entire ecology of shallow-water zones. *Orconectes rusticus* is also a threat to several *Orconectes* species with restricted ranges in Kentucky and Tennessee.

Some native insects are being deliberately transported from one part of North America to another. A recent fad is to release butterflies, rather than throw rice, at outdoor weddings. Butterfly breeders in the United States rear the larvae of butterflies such as monarchs and swallowtails and ship chrysalises to buyers throughout the country. Most of these butterflies are unlikely to survive, but some might interbreed with local individuals of their own or closely related species. These hybrids might produce offspring that are not well adapted to local environments. In some cases, releases might introduce species in new geographical areas, where they might competitively displace local endemics. Insect diseases or parasites might also be spread by such releases. This practice, especially if expanded, might ultimately play havoc with butterfly faunas in many areas.

Homegrown Exotic Fish

Numerous native fish have been stocked into freshwaters throughout North America. Many natural waters are now dominated by introduced species, in numbers of individuals if not numbers of species (Chandler 1997). Fish introductions have restructured these freshwater ecosystems and have resulted in the extirpation of many populations of fish, amphibians, and invertebrates.

Large predatory fish, in particular, have been stocked into lakes and streams with little consideration to their potential impacts on native aquatic animals, especially amphibians and other fish. In Ontario, ponds stocked with predatory fish, mostly North American natives, showed reduced diversities of amphibians, especially species of small body size and those that produced small egg clutches (Hecnar and M'Closkey 1997). In New England, lakes with predatory fish showed a much reduced diversity of native minnow species (Whittier et al. 1997). Many of these lakes had been stocked with smallmouth bass, largemouth bass, and northern pike.

Throughout the West, trout of various species have been introduced into high-elevation lakes that were originally fishless. In the California Sierra Nevada, at least five species of trout have been introduced to alpine lakes. These include North American natives such as the golden trout, rainbow trout, brook trout, cutthroat trout, and the exotic brown trout. In this region, the endemic mountain yellow-legged frog (*Rana muscosa*) and the Yosemite toad (*Bufo canorus*) have disappeared from many lakes. For the mountain yellow-legged frog, strong evidence indicates that predation by these salmonids has led to the disappearance (Bradford et al. 1993). Mountain yellow-legged frogs require about three years as tadpoles, before transforming into adults. Thus, they require lakes deep enough not to freeze solid in winter. Such lakes can also support trout, and if trout are present, the tadpoles are at high risk from predation. The 60-lakes Basin is one of the remaining strongholds of the mountain yellow-leg. Here there are many lakes with frogs only, as well as some lakes with trout only (Vredenberg and Matthews 1997). Both trout and tadpoles can be found in very few lakes. It is doubtful, however, that these lakes can produce enough mature frogs to maintain a viable population, and their presence probably depends on continued immigration from trout-free lakes.

Similarly, in high-elevation lakes in North Cascades National Park, densities of larval long-toed salamanders (*Amblystoma macrodactylum*) were reduced in lakes that were stocked with rainbow and cutthroat trout (Tyler et al. 1998). The reduction was greatest in nitrogen-rich lakes with dense populations of crustacean zooplankton foods—lakes in which salamander larvae should have been abundant.

The Great Lakes have also been the object of deliberate biotic tinkering with exotics. Several salmon have been introduced to replace fisheries damaged or destroyed by overfishing and sea lamprey predation (Mills et al. 1993b). Rainbow trout and steelhead, the anadromus form of the rainbow trout, have been introduced to the Great Lakes beginning in the late 1800s. Steelhead have established fall and spring spawning runs in streams flowing into Lake Superior. Kokanee salmon were introduced to Lakes Huron and Ontario and have established a spawning population in Lake Huron. Pink salmon were introduced to Lake Superior in 1956, apparently by releases of fish into the Current River on the north shore of the lake. Pink salmon now spawn successfully and have spread to the other Great Lakes. Coho salmon, stocked into Lake Michigan in 1966, have established a spawning population in that lake. Chinook salmon have been stocked into several of the Great Lakes at various times from the late 1800s to the present. This species also spawns successfully in tributaries of Lake Michigan. Coho salmon, chinook salmon, and rainbow trout are still stocked in the Great Lakes.

In Lake Tahoe, lake trout were introduced to replace the fishery of native cutthroat trout, which collapsed in the 1930s (Morgan et al. 1978). Kokanee salmon were accidentally released into the lake in 1944. Once the kokanee became established, efforts were made to manage it to create an important sport fishery, and stocking was carried out from 1950 onward. In an effort to improve food supplies for lake trout, opossum shrimp were introduced from 1963 to 1965. Kokanee and opossum shrimp apparently caused the extinction of at least three species of water fleas in the lake during the 1970s. Unfortunately, opossum shrimp and the native copepods that now dominate the zooplankton do not provide an adequate food supply for kokanee, which have declined. Lake trout feed on opossum shrimp, but this addition to their diet has not led to increase in lake trout abundance. Tinkering with the biotic structure of Lake Tahoe has effectively precluded the restoration of the cutthroat trout population. Lake trout have also been introduced to Yellowstone Lake, Wyoming, creating a serious threat to cuthroat trout there.

Kokanee salmon and opossum shrimp have been introduced to many other lakes, with varying results. In Flathead Lake, Montana, their introduction led to changes up and down the food chain, and in the terrestrial as well as the aquatic biota (see Chapter 18).

The channel catfish and flathead catfish, native to the Mississippi drainage of central North America, have been introduced throughout most of the rest of temperate North America. The channel catfish has also been introduced to Puerto Rico and Hawaii (Ruesink et al. 1995). Both catfish can be major predators on smaller fish, including the razorback sucker (*Xyrauchen texanus*) and Colorado pikeminnow (*Ptychocheilus lucius*), both federally endangered, in the Salt River, Arizona. They may also interact with native species in unusual ways. In the Colorado River, predation on channel catfish by Colorado pikeminnow is apparently leading to significant mortality of pikeminnows (Osmundson et al. 1997). Many dead pikeminnows have been found with channel catfish lodged in their throats, the catfish's spines piercing the pharynx or esophagus.

Northern pike are another game fish, native to eastern North America, that have been introduced outside their range. In California, for example, pike were illegally introduced to Lake Davis, in Plumas County, creating a population that threatened to spread to other streams and lakes (see Chapter 21).

Homegrown Exotic Amphibians

Several native amphibians have been introduced to new locations, the most serious being the bullfrog. Bullfrogs were absent from North America west of the Rocky Mountains. They have been introduced to most parts of the West, however, and from these introduction points have spread widely on their own. In California, for example, overharvesting of the native red-legged frog (*Rana aurora*) for food led to the introduction of bullfrogs in 1896 (Jennings and Hayes 1985). Bullfrogs, as well as other North American ranid frogs, have been introduced to Vancouver Island, British Columbia, where red-legged frogs are native (Green 1978).

Bullfrogs are responsible for the decline and local disappearance of many wetland amphibians and reptiles in the western United States. Endemic ranid frogs have been locally extirpated in several locations. In some places, these local endemics are being replaced by the Rio Grande leopard frog, another homegrown exotic species that coexists with the bullfrog in Texas (Rosen and Schwalbe 1995).

At San Bernardino National Wildlife Refuge in southern Arizona, for example, lowland leopard frogs (*Rana yavapaiensis*), Mexican garter snakes (*Thamnophis eques*), and Sonoran mud turtles (*Kinosternon sonoriense*) were reduced in abundance in ponds and marshes with bullfrogs (Schwalbe and Rosen 1988). Both garter snakes and mud turtles were represented only by large adults in areas having bullfrogs, whereas juveniles were common where bullfrogs were absent. Analysis of stomach contents of bullfrogs also revealed that they were feeding on two endangered native fishes: the Yaqui chub (*Gila purpurea*) and the Yaqui topminnow (*Poeciliopsis sonoriensis*). These observations led refuge personnel to undertake a bullfrog reduction program, which initially resulted in increased reproductive success of the leopard frog and the garter snake. In spite of this, the lowland leopard frog was apparently extirpated on the refuge a few years later (Rosen and Schwalbe 1995).

On the upper Eel River in California, bullfrogs, moving upstream, began to appear in the late 1980s. Experimental studies of the interaction of bullfrogs and the native yellow-legged frog (*Rana boylii*) have revealed that feeding on benthic algae by bullfrogs leads to almost a 50 percent reduction in survivorship of yellow-legged frog tadpoles (Kupferberg 1997). Some reduction was also noted in the survival of Pacific tree frog (*Hyla regilla*) tadpoles, as well.

North American Birds

Only a few nongame birds have been introduced to new regions of North America by direct human action. The house finch became established on Long Island, New York, as a result of the release of caged birds by a bird dealer who was handling them illegally (Elliott and Arbib 1953). This population grew exponentially and has now spread over much of the eastern United States and Canada.

Among North American game birds, the bobwhite, Gambel's, California, scaled, and mountain quail have been introduced to various locations outside their original ranges. Willow ptarmigan, native to much of arctic and subarctic North America, were introduced to Nova Scotia, and white-tailed ptarmigan, native to alpine areas of Alaska, western Canada, and the Rocky Mountains of the United States, were introduced to the California Sierra Nevada. Ruffed grouse, widespread in North America, were introduced to Newfoundland. The plain chachalaca, which occurs in the Rio Grande Valley of Texas, has been introduced to some Georgia sea islands.

Wild turkeys have been introduced to many localities outside their original range, including several western states and Canadian provinces. The wild turkey was originally widely distributed from the eastern and central United States and southernmost Canada south into Mexico. The species was extirpated by hunting in many areas. Because of its popularity as a game bird, it has been reintroduced to some of these localities. It has also been introduced to many locations in western states and Canadian provinces.

Several native nongame birds have spread to new regions because of the transformation of the North American environment. The brown-headed cowbird was originally restricted to the Great Plains of central North America (Mayfield 1965). It was commonly called the "buffalo bird," because it followed the herds of bison, feeding on insects stirred up by the animals. Thus, its eggs were spread among hosts over the wide range of bison migration, with host species being at risk for only part of the breeding season. In developed North America, the cowbird has expanded its range enormously by switching from bison to livestock, particularly cattle and horses. In the late 1700s and early 1800s, cowbirds and other native grassland birds moved eastward into the cleared farmlands of the Midwest and East. In the 1900s, cowbirds spread west through the mountain and Pacific states and provinces. The range of the cowbird now includes all of subarctic Canada, the United States, and northern Mexico. Cowbirds forage in association with livestock of all sorts and parasitize more than 200 species of songbirds (Robinson 1997). Now, they are also more sedentary, and they put their hosts at risk throughout the breeding season (Chace and Cruz 1998).

Many songbird species parasitized by cowbirds within the cowbird's original range recognize and reject cowbird eggs or abandon parasitized nests. In contrast to the least Bell's vireo of California, for example, Bell's vireos from the borders of the Great Plains rear only about 6 percent of the cowbird eggs laid in their nests (Mayfield 1965). Outside their original range, cowbirds are having a serious impact on many naive hosts. In the 1960s, Mayfield (1965) found that 55 percent of the nests of the Kirtland's warbler (*Dendroica kirtlandii*), at the time a declining species with a very restricted breeding range in the Lower Peninsula of Michigan, were parasitized by cowbirds. The warblers were rearing 41 percent of the cowbird eggs in these nests, at the expense of few or no warblers. Cowbird control has helped the Kirtland's warbler to increase considerably in numbers in recent years.

Nest parasitism by cowbirds is a particularly serious problem in several midwestern states, where forests have largely been reduced to woodlots in an agricultural landscape (Robinson 1997). Although cowbirds forage in open country, they search wooded areas for nests to parasitize. Forest thrushes, such as the wood thrush (*Hylocichla mustelina*) and veery (*Catharus fuscescens*), are very heavily parasitized, with 75–95 percent of all nests usually receiving one or more cowbird eggs. Regional populations of these species are apparently population "sinks," that is, being maintained only by dispersal of birds from more heavily forested regions where a reproductive excess is realized. Forest-inhabiting warblers, vireos, tanagers, flycatchers, and orioles are also very heavily parasitized.

Other native North American species have expanded their ranges, leading to possible impacts on other native species. The barred owl, for example, was native to the forested region of eastern North America. With the growth of riparian forests along rivers of the Great Plains, because of control of spring floods, and with the planting of trees in shelterbelts, barred owls have crossed the Great Plains and colonized the northern Rocky Mountains and the Pacific Northwest. Barred owls have now spread south as far as central California (Dark et al. 1998), and they show no signs of stopping. Some of this spread may result from changes in structure of western forests, due to fire exclusion, that have made the forest habitat optimal for barred owls (Wright and Hayward 1998). Unfortunately, barred owls are interbreeding with the spotted owl (*Strix occidentalis*), one of the western forest species of greatest conservation concern.

Two native waterfowl, the mallard duck and the Canada goose, have become established well outside their original ranges. Wild mallards, often augmented by escaped semidomestic birds, have spread into the southeastern United States and eastern Canada. The spread of the mallard duck has impacted several of its close relatives. In eastern Canada, competition and hybridization between mallards and the American black duck (*Anas rubripes*) have been hypothesized to contribute to the decline of the latter (Johnsgard 1967), but evidence for these influences as being the cause of decline is weak (Mcauley et al. 1998). In the southeastern and southwestern states, however, the mallard hybridizes with the mottled duck (*Anas fulvigula*) and the Mexican duck (*Anas diazi*). The latter, once recognized as a distinct species, has hybridized with the mallard so extensively that it has been merged with the mallard.

Canada geese originally bred from the upper Midwest north into the Arctic. Feral birds have now established nonmigratory breeding populations in many locations along the Atlantic coast and in inland areas south of their original breeding range.

North American Mammals

Several native North American mammals have been introduced to new areas, often islands, to create game populations for hunting. Eastern cottontail rabbits were introduced to western Oregon in 1939 and 1941, the objective being to establish a new game animal (Verts and Carraway 1980). Their population has spread through much of the Willamette Valley, and some hybridization has occurred with the native brush rabbit (*Sylvilagus bachmani*).

Native ungulates have been a favorite group for translocations within North America. Elk were introduced to Santa Rosa Island, another of the California Channel Islands, and a herd of about 1,000 animals now occurs there. Black-tailed deer have been introduced to the Queen Charlotte Islands, British Columbia, as well as to Kodiak Island, Alaska, and other northern islands. Bison, almost driven to extinction in North America in the 1800s, survived in several small populations. Under protection they have increased greatly in numbers and have been introduced to many places, in some cases well outside their original range. On Santa Catalina Island, off southern California, for example, a herd of 500 has resulted from 14 animals introduced in 1924 for a western film production.

The mountain goat provides an extreme example of issues arising for homegrown exotics. Endemic to North America, it ranges naturally from southern Alaska and the Yukon Territory through the mountains of

Canada to Washington, Idaho, and Montana. Mountain goats have been translocated to locations in several states, including the Black Hills of South Dakota.

Mountain goats were introduced to the Olympic Mountains of Washington in the late 1920s (Scheffer 1993; Houston et al. 1994). About 11 or 12 individuals were apparently released on several occasions between 1925 and 1929. By the late 1930s, the population had grown and spread through the high elevations of the Olympics. By 1983, it had reached about 1,175 individuals (Houston et al. 1994).

This population explosion raised concern that mountain goats might be a threat to alpine ecology. Because mountain goats are foraging generalists, their impact on alpine meadows is potentially extensive. Studies have revealed significant effects as a result of grazing, trampling, and wallowing. Trampling reduces the cover of mosses and lichens in meadows. Grazing reduces the cover of native bunchgrasses and increases the abundance of forbs that invade disturbed areas. Mountain goats also create wallows where weedy plants tend to establish. Given the impacts of feral domestic goats in many other locations, national park biologists were justifiably concerned about the possible impacts of mountain goats on endemic plants. Olympic National Park has eight endemic vascular plants, seven of which occur in mountain goat habitat. Four other species that are endemic to the Olympics and to Vancouver Island or the Queen Charlotte Islands are also eaten by goats.

Goat removal in Olympic National Park was undertaken in the 1980s. About 509 animals were captured, translocated to other locations, or killed. In 1990, only about 389 animals remained (Houston et al. 1994). In 1995, after extensive study, the National Park Service concluded that all goats should be removed from the park. Because of the rugged terrain, shooting was proposed as the removal method. This proposal was contested by animal rights organizations, which contended that the goat was not a proven exotic, and that damage to vegetation was exaggerated. Schultz (1994), however, has reviewed the historical evidence relating to mountain goat distribution in the region and showed conclusively that these early reports are either incorrect or refer to other animals.

Several small carnivores or omnivores have been introduced to new localities, usually to create a furbearer population for trapping. Often, these localities were formerly predator-free islands or other isolated locations. Examples of such introductions include raccoons, red foxes, and arctic foxes to islands off the Pacific coast of North America and red foxes to the Olympic Mountains.

Raccoons were introduced to the Queen Charlotte Islands, British

Columbia, in the 1940s to provide an additional furbearing species. In the 1980s, coincident with the observation that seabird colonies had disappeared on several small islands, raccoons were found to have colonized the same islands. Raccoons have reached most islands less than 1,312 feet offshore, 29 percent of islands between 1,312 and 2,625 feet offshore, and one island 3,116 feet offshore (Hartman and Eastman 1999). In 1991, on a 119-acre island used for nesting by ancient murrelets (*Synthliboramphus antiquus*), four raccoons were estimated to have killed 488 murrelets and consumed or destroyed 188 eggs and chicks (Hartman et al. 1997). About 74 percent of the world population of ancient murrelets breeds in the Queen Charlotte Islands, so that raccoon predation is a serious threat to the species. Red foxes, which invaded Shaiak Island, Alaska, had a similar impact on several nesting seabirds (Peterson 1982).

Arctic foxes were introduced to various islands in the Aleutian chain and off the coast of the Alaska peninsula. In some cases, they have subsequently been removed in order to improve seabird breeding success. Dramatic recovery of pigeon guillemot (*Cepphus columba*) populations was noted after several of these removals. On the outer Shumagin Islands, off the Alaska peninsula, for example, arctic foxes were introduced in the late 1800s as part of a fur-farming enterprize. Arctic ground squirrels (*Spermophilus undulatus*) were also introduced as food for the foxes. In 1994 and 1995, foxes were removed from two of the islands (Byrd et al. 1997). The response of breeding seabird populations is now being monitored. Black oystercatchers (*Haematopus bachmani*) and pigeon guillemots, two species reduced in numbers by the effects of the *Exxon Valdez* oil spill, were of particular concern. Both species are expected to increase substantially over a period of several years.

With the widespread reduction of wolves and deforestation of large areas of eastern and far-western North America, the coyote has greatly expanded its range. In the twentieth century, the coyote invaded the once continuously forested region of western Washington and adjacent areas of the Pacific Northwest. It has also spread far eastward through the Great Lakes region.

The Virginia opossum is another native omnivore that has expanded its range greatly, because of both landscape change and translocation by humans. Originally, it was limited largely to Mexico and the eastern United States, north to Pennsylvania. With European settlement, the opossum has spread north to southern Ontario and southern New England and west through much of the Great Plains. Virginia opossums were also introduced to several locations in the Pacific states and now occur from southern British Columbia to northern Baja California.

Dealing with Homegrown Exotics

The effects of homegrown exotics are thus varied, but two stand out. First, some have proven to be serious predators or, as in the case of the brown-headed cowbird, parasites on other native species. Second, the expanded ranges of many have brought them into contact with close relatives with whom they compete or hybridize. For these two reasons, homegrown exotics are a serious threat to many localized North American endemics. Their status as North American natives should give them no preferential treatment, as they are often as destructive as intercontinental exotics.

Unfortunately, the fact that homegrown exotics are congeners of other native species makes their control very difficult; techniques that reduce their populations may affect their close relatives. This difficulty means that efforts must be concentrated in preventing spread of homegrown exotics into regions where they may impact close relatives.

Homegrown exotics also raise serious issues of control, because many are native species that many people value, aesthetically or for sport hunting or fishing. In some cases, they also raise a basic issue of the role of parks and other protected areas in preserving species native to a general region. For example, should Olympic National Park be responsible for preservation of the mountain goat, which occurs in the general region but is not native to the park itself? As the human population of North America grows, this sort of question will become more frequent. Climatic warming may also force parks and preserves to broaden their responsibilities for protection of species native to areas made unsuitable to them by climatic change. Chapters 20 and 21, which consider aspects of management and policy related to exotics, return to some of these issues.

Chapter 16

Human Domesticates and Associates: Our Best Friends and Closest Associates

> Overgrazed conditions existed on all areas ranged over by burros. In many places herbage growth was cropped to the roots and some species of shrubbery were totally destroyed. Soil erosion was greater in burro infested areas. . . .
>
> —J. P. Brooks, in "Feral Asses on Public Lands"*

San Clemente Island lies 49 miles off the coast of southern California. Although not volcanic, the island has never been connected to the California mainland, and its biota has thus colonized over water. It is the richest of the California Channel Islands in endemic plants, with fourteen species and varieties known only from the island itself, and thirty-four additional species and varieties endemic to San Clemente and one or more of the other Channel Islands (Oberbauer 1989). Four subspecies of birds are known only from San Clemente, and six other Channel Island endemics occur there (Johnson 1975). Endemic forms of two mammals and two lizards are also found there. The island is also a microcosm of introduced human domesticates and associates. Over recorded history, it has been home to free-ranging populations of cattle, sheep, horses, and swine and to feral populations of domestic goats, house cats, dogs,

*In S. W. Carothers, M. E. Stitt, and R. R. Johnson (1976). "Feral Asses on Public Lands: An Analysis of Biotic Impact, Legal Consideration and Management Alternatives." *North American Wildlife Conference* 41:396–405.

and house mice (Table 16.1). Mule deer were also brought to the island. Several of the larger exotic species no longer survive.

No exotic has done more damage than feral domestic goats. Goats may have been introduced to the island in 1875 (Johnson 1975) or even earlier. In the absence of ungulate predators, the goat population grew to perhaps 10,000 animals and decimated the vegetation everywhere except on the inaccessible walls of canyons. In deep canyons, remnants of woodland hung on, but when goats were present the youngest individual trees were over seventy years in age. Between 1972 and 1989, the United States Navy, which administers the island, carried on a goat removal program, and about 29,000 animals were either killed or captured and taken to the mainland. Final removal was accomplished with the use of Judas goats, radio-collared animals that were released to join remaining groups of feral animals. Between June 1989 and September 1991, this technique allowed sharpshooters to locate and kill all remaining goats. The goat-free island is now experiencing slow recovery of native vegetation.

Goats are just one of many human domesticates and associates that can invade natural ecosystems. These animals include domesticated household and farm animals, undomesticated animals that have evolved a close

Table 16.1. Scientific names and native regions of invasive domesticates and other animals associated with humans.

BIRDS	
cattle egret, *Bubuculus ibis*	Africa
domestic pigeon (rock dove), *Columba livia*	Europe
English sparrow, *Passer domesticus*	Europe
European starling, *Sturnus vulgaris*	Europe
house finch, *Carpodacus mexicanus*	North America
monk parakeet, *Myiopsitta monarchus*	South America
mute swan, *Cygnus olor*	Eurasia
MAMMALS	
black rat, *Rattus rattus*	Eurasia
burro, *Equus asinus*	Africa
cattle, *Bos taurus*	Eurasia
dogs, *Canis familiaris*	Eurasia
domestic cat, *Felis catus*	Egypt
goat, *Capra hircus*	Eurasia
horse, *Equus caballus*	Eurasia
house mouse, *Mus musculus*	Europe
Norway rat, *Rattus norvegicus*	Eurasia
sheep, *Ovis aries*	Eurasia
swine, *Sus scrofa*	Eurasia

association with humans, and, increasingly, wild animals that are captured and held as pets. Many of these animals share the goat's capability to disrupt native North American ecosystems and their native plants and animals. Most of these close human associates are not native to North America.

Domesticated animals often establish free-ranging and feral populations. Free-ranging animals are those that retain a close connection with the human environments; they forage, but do not breed, in wild areas. Feral animals are those that have established breeding populations in the wild and have become independent of humans. Free-ranging and feral domesticates include household pets and livestock animals. Some are predators, some herbivores.

Cats and Dogs

Among the most frequently encountered feral or free-ranging domestic animals are some of our best-loved species: cats and dogs. In the United States, the total domestic cat population is probably in excess of 100 million animals (Coleman et al. 1996). The number of dogs is usually estimated to exceed the number of cats by 10 percent or so (Denney 1974). A recent survey conducted by the National Biological Service of nonnative species of greatest concern in 937 public and private parks and management areas found that cats were a problem in 180 areas and dogs in 123 areas (Drost and Fellers 1995).

Cats were domesticated in Egypt about 4,000 years ago. They quickly became popular and had spread throughout the ancient world, including Britain and continental Europe, by about 2300 B.P. (Coleman et al. 1996). Cats were carried to all parts of the world during the period of European exploration and colonization. In many places, they quickly established feral populations, often with severe impacts on native faunas.

In North America, cats can be household pets that range outdoors near their homes, free-ranging farm animals that are never allowed indoors, or completely feral animals that avoid human habitations. The impact of cat predation on small mammal and bird populations is substantial. Domestic cats, excluding completely feral animals, feed about 70 percent on small mammals, 20 percent on birds, and 10 percent on lizards and other small animals (Fitzgerald 1988). Household and farm cats alone probably kill over a billion birds annually in North America (Stallcup 1991). Several billion small mammals, many of them exotics themselves, doubtless fall prey to cats.

In several special situations, cat predation can be extremely serious. On

San Clemente Island, off southern California, for example, feral cats are a major predator on nestlings and young of the endangered San Clemente loggerhead shrike (*Lanius ludovicianus mearnsi*) (Scott and Morrison 1990). On this island, eradication of cats was judged to be infeasible because of the great difficulty in observing and capturing them. Poisoned baits could not be used because of the danger they would pose to the endemic island fox (*Urocyon littoralis*).

Like cats, dogs range from pets that are allowed to run free to stray animals and feral animals. Strays are unowned animals that still live in close association with humans. Truly feral animals, dogs that are completely independent of the human environment, are probably rare in North America or Hawaii. Packs of feral dogs have been reported in some areas of the Southeast, including Arkansas and Alabama. Free-ranging dogs are widespread, however, and often kill livestock and wildlife. Predation on wildlife by free-running and feral dogs can be substantial but is usually not limiting to their populations. Denney (1974) estimated that about 20,000 deer were killed annually by dogs in the thirty-two states that responded to a survey. In New England states, about 3 percent of deer mortality was attributed to killing by dogs. Small numbers of various other game animals and birds were also reported to be killed by dogs. Free-ranging dogs are significant predators on gopher tortoises in some parts of the southeastern United States (Causey and Cude 1978).

Feral Livestock

Feral livestock animals can cause major ecological impacts. Feral hogs occur in many parts of the southern and western United States and Hawaii (Wood and Barrett 1979; Mayer and Brisbin 1991). In many locations these animals have interbred with the Eurasian wild boar, which was introduced as a game animal (see Chapter 14). Feral hogs are common in many southeastern forests from Virginia to Texas and Oklahoma (Wood and Lynn 1977). They are also found in isolated locations in New Mexico and Arizona (Mayer and Brisbin 1991). Since the early 1980s, their populations in the southern states have grown considerably, and they have spread, or been introduced, to more northern states in a band from Ohio and West Virginia to Colorado (Gipson et al. 1998). Unauthorized releases by hog-hunting groups are thought to be responsible for much of this spread. Many of the southeastern populations date from early European settlement, however, when hogs were allowed to forage freely in wooded areas (See Chapter 4). Domestic hogs were released in Florida as early as 1539 by Hernando de Soto, and in California as early as 1769 by founders of the Spanish missions. In Hawaii, Polynesians introduced pigs about A.D. 1000, and Captain Cook brought European animals to the islands in 1778. Many regions of North America and Hawaii thus have lived with feral domestic pigs for several hundred years. Many states, including Hawaii, now treat feral hogs as game animals, and annual harvests run into the tens of thousands.

In the United States, feral hogs, wild boar, and hybrids of boar and feral domestic hogs are estimated to number between half a million and 2 million animals (Mayer and Brisbin 1991). Thus, they are the most numerous of North America's exotic ungulates. Feral hogs disturb native vegetation by their feeding, rooting, and dispersal of exotic plants. In the southeastern states, where they feed heavily on mast, they are serious competitors for native mast-feeding wildlife, including white-tailed deer, wild turkey, gray squirrels, and black bear. An adult hog can consume 1,300 pounds of mast in half a year. Hogs are also predators on small vertebrates and ground-nesting birds and reptiles.

In California, feral hogs occur widely in the coast ranges, in foothills of the Sierra Nevada, and on several of the California Channel Islands. They are now considered a game animal, and 15,000–20,000 are killed annually by hunters. On Santa Cruz Island, they may have become established in the mid-1800s, but they were certainly present by the 1920s (Van Vuren 1984). Rooting by these animals, especially in canyon bottoms during fall and winter, causes serious disturbance of the island's vegetation, which is rich in endemic plants.

The coastal prairies of northern California have also experienced disturbance by feral hogs, which have probably been present there for over 100 years. At the University of California's Northern California Coast Range Preserve in Mendocino County, feral hogs root up between 7 and 8 percent of the ground surface annually (Kotanen 1995). The disturbed areas are colonized by both native and exotic annuals, but the existing level of disturbance does not seem to be leading to progressive reduction in native components of the vegetation. Some native species might have disappeared in the early years following hog introduction.

Horses and burros have also established feral populations. Feral horses occur in several western states and on barrier islands of the mid-Atlantic coast. Feral burros are found only in the Southwest. Feral horses and burros pose difficult problems in western parks and public lands. On the one hand, they are clearly exotics. On the other, they are part of the folklore of western exploration and settlement. Where their numbers are uncontrolled, however, they can have severe negative impacts on native species. A census in 1993 revealed populations of about 39,000 horses and 7,500 burros on Bureau of Land Management (BLM) lands in the western states (Pogacnik 1995). Altogether, about 43 million acres of BLM lands have feral horse and burro populations. Most feral horses occur on BLM lands, but large numbers of burros also occur on national park lands. Most feral horses are in Nevada and Wyoming, and most feral burros in Arizona and California.

In Death Valley National Monument, for example, feral burros have been held responsible for the decline of the desert bighorn sheep. Originally, about 5,000 bighorns lived in Death Valley. By the early 1970s, only about 600 were left. Feral burros were estimated to number 1,500 animals, with their population showing a growth rate of about 18 percent per year (Norment and Douglas 1977). Thus, nearly three bighorns were lost for every additional burro. Burros and bighorns have very similar diets, but burros can browse woody plants more heavily because of adaptations of their digestive system for processing woodier materials (Seegmiller and Ohmart 1981). Also, burros are highly aggressive, and they displace bighorn sheep from desert springs.

Feral burros have also created problems of overbrowsing in Grand Canyon National Park and other western areas. This is shown clearly by the change in vegetational characteristics and small mammal communities in areas accessible to burros, compared to inaccessible areas (Carothers et al. 1976). Accessible areas have less than a fourth the total plant cover, fewer plant species, and about a fourth the total small mammals as inaccessible areas. Some mammal species common in inaccessible areas are absent in accessible sites, and vice versa.

Overgrazing by burros has long been a problem in the Grand Canyon. Prior to 1969, however, burro populations were reduced periodically by hunting. In 1971, under the Wild Horse and Burro Act (Federal Law 92-195), the killing of feral burros and horses on most federal lands was made a felony. As a result, their populations increased from about 17,000 in 1971 to 64,000 in 1985. Overgrazing is still a serious problem in many areas.

Recent policy on federal lands has been to reduce feral burro and horse populations by capturing them and placing them for adoption. From 1973 through 1997, over 165,000 horses have been removed from federal lands in the western states. To adopt a burro or horse, one must pay a fee of $125 and agree to conditions relating to ownership and care. Over 150,000 animals have been adopted, but this program has not reduced feral animal populations, which have a growth rate of about 24 percent annually. Willingness to adopt is becoming exhausted, and large numbers of burros and horses are being maintained in captivity at great expense. In

1997, for example, 10,443 horses and burros were captured, 1,751 more than were adopted. About $2.6 million in federal funds were used in the maintenance of about 6,285 animals.

Feral cattle and sheep populations occur on some of the California Channel Islands and in Hawaii. Cattle were introduced to all of the major islands of Hawaii, and feral populations developed in many places. Feral animals still occur on Maui and Hawaii (Stone 1985). Feral sheep also existed on several islands and are still present on Hawaii. Cattle and sheep, both tended and feral, have contributed to severe overgrazing of the native Hawaiian vegetation and to the probable extinction of some plants.

Feral goats have probably done more ecological damage to island ecosystems than any other ungulate. In Hawaii, feral goat populations occurred on all major islands, although they have now been eradicated on Lanai and Niihau. Goats favor dry, open habitats, although in some places they also invade wet forest, especially during dry weather. Goat browsing has severely impacted woody plants in many low and midelevation areas and have contributed to the endangerment of a number of endemic plants.

In California, goats were introduced to Santa Catalina Island, as well as to San Clemente Island. The population on Santa Catalina dates from 1827, so that goat browsing has impacted the vegetation even longer than on San Clemente (Schoenherr 1992). About forty-eight species of plants are believed to have been lost from the island, primarily as a result of damage to the vegetation by goats.

Free-Ranging Livestock

Free-ranging livestock animals can also be serious ecological problems. The grazing of livestock, primarily cattle and sheep, on public lands is a long-standing and contentious issue. Evidence is overwhelming that uncontrolled livestock grazing from the late 1700s through the 1800s caused massive alteration of vegetation and landscapes in western North America (see Chapters 10, 11, and 12). Direct vegetational impacts include the transformation of desert grasslands to desert scrub, widespread invasion of grasslands by junipers and other shrubs, destruction of dominant bunchgrasses, and reduction of native riparian vegetation. Indirectly, grazing in forest areas reduced the ground cover of herbaceous plants that compete with tree seedlings, reducing ground-level fuel loads that support surface fires. These changes have contributed to the development of dense forest stands that are much more vulnerable to stand-replacing crown fires (Belsky and Blumenthal 1996).

Overgrazed western landscapes also experienced heavy erosion and arroyo-cutting, and downstream floodplains received increased sand and silt deposition. Livestock allowed to graze unrestricted in areas with streams spend a greatly disproportionate fraction of time in the riparian zone (Fitch and Adams 1998). The result is channel and bank disruption, altered seasonal patterns of flow, increased water temperature, increased nutrient and silt loads, degradation of riparian vegetation, and decrease in fish and riparian wildlife (Ohmart 1996).

Even though protection of biodiversity on public lands has become recognized as a major national priority, livestock grazing continues to be one of the most intensive uses of these lands. As of 1988, about 7 million livestock animals were grazing on 268 million acres of BLM and Forest Service lands, including some wilderness areas, in the western United States (Fleischner 1994). Some grazing is also permitted on some national wildlife refuges and national park lands. From the standpoint of range quality, more than half of this land is rated as fair to poor in condition, but grazing leases are still arranged at cheap rates and high stocking levels.

Ecologically, grazing influences the composition of plant and animal communities and the overall structure and function of the ecosystem (Fleischner 1994). Most changes in plant and animal species composition involve reduction in species diversity. Riparian vegetation in western rangelands and forests is among the communities most severely impacted by livestock grazing. Cattle, in particular, browse and trample streamside vegetation that is used by a variety of breeding birds, such as the federally endangered willow flycatcher (*Empidonax traillii*). In both riparian and upland situations, furthermore, livestock impacts are promoting the invasion of a host of other exotics.

The impact of livestock grazing is shown dramatically by the few exclosure studies that have looked at effects of grazing on the native biota. Dobkin et al. (1998), for example, examined bird communities in a livestock exclosure on a national wildlife refuge in southeastern Oregon. The exclosure had been maintained in a wet-meadow habitat since 1958. In 1990, however, livestock grazing was terminated throughout the refuge. Meadow vegetation bordering a stream channel inside the exclosure was dominated by sedges and grasses; that outside the exclosure, by cold-desert shrubs. Ten wetland bird species occurred within the exclosures but not outside, and several others were more abundant within the exclosure than outside. The areas outside the exclosures showed a low density of semidesert upland birds. Creation of the exclosure had allowed the water table associated with the stream channel to rise and extend outward, restoring the moisture required by wet-meadow plants. Although in four years after the complete exclusion of livestock, some improvement was beginning to occur outside the exclosure, the restoration of wet-meadow hydrology and vegetation was clearly a process requiring decades.

Even in desert areas, seasonal livestock grazing impacts the native biota. In the Mojave Desert of California, cattle are trucked in during late winter and spring, when new shrub and herbaceous plant foliage is most abundant. During these seasons, cattle and tortoises may consume the same plants, but how often cattle seriously deplete the food supply of tortoises is unclear (Oldemeyer 1994). Normal tortoise populations require about 0.47 ounces per square yard of green herbaceous material to realize their reproductive potential (C. Richard Tracy, personal communication). Forage biomass may fall below this level frequently in some places, but only uncommonly in others. Tracy recommends that cattle be removed from desert range when their grazing has reduced herbaceous forage to the level required by tortoises.

Rats and Mice

A special category of exotics consists of species that possess a close ecological association with humans as household, urban, or agricultural commensals. These are Old World species that evolved adjustments to living in close association with humans in cities and farms. They include house mice, Norway rats, and black rats. Polynesian rats are other animals that have adapted to life in close association with humans. Commensal mice and rats are serious pests of agricultural crops and stored foods, as well as being nuisance pests in houses and other buildings. They are vectors of a number of human and livestock diseases. In continental areas, however, they do not seem to be serious problems in natural ecosystems.

The house mouse, as a human associate, probably originated in the region of India and Pakistan. The house mouse occurs as a wild animal from Sweden and the Mediterranean region of Europe eastward through most of temperate Asia. Its commensal subspecies *domesticus* has been carried worldwide by humans, however, and is abundant in and near human habitations. Populations have also reentered the wild in many places. In some exotic areas, house mouse populations may erupt to astounding levels. In North America, for example, densities in excess of 81,940 per acre were recorded in the southern San Joaquin Valley, California.

Impacts of the house mouse are most serious on islands, some of which were free of rodents, others of which hold populations of species vulnerable to competition. In the Hawaiian Islands, house mice occur on all major islands and many small islands and atolls (Stone 1985). They range upward to nearly 13,123 feet in elevation and are abundant in many areas of natural vegetation. House mice sometimes erupt in large numbers (Stone 1985). They may feed on seeds of native plants and are supporting prey for exotic predators, such as cats and mongooses, that prey on native animals (Stone 1989). Specific serious impacts have not been documented in Hawaii, however. On barrier islands along the Atlantic coast of Florida, in contrast, one subspecies of the beach mouse (*Peromyscus polionotus*) has apparently been driven to extinction by competition from the house mouse and predation by cats (Humphrey and Barbour 1981). Two other barrier island subspecies have been reduced in abundance.

Norway, black, and Polynesian rats are three of seven members of the genus *Rattus* that have become commensal with humans. Norway and black rats have become worldwide in distribution, and the Polynesian rat has become widespread in Southeast Asia and the Pacific islands. The Norway rat is more common in temperate regions, especially in urban areas, and the black rat more common in the tropics. Black rats are more arboreal than Norway rats, and they are often called "roof rats."

On oceanic islands, exotic rats may also exert serious effects. Black and Polynesian rats occur on all major islands in Hawaii and range from sea level to high in the mountains (Stone 1985). In general, however, black rats tend to be more abundant at higher elevations, and Polynesian rats more common in the lowlands. Black rats damage the flowers, fruit, and bark of many shrubs and trees in the native forests. They are also predators on native invertebrates and birds. Although the Polynesian rat seems not to have been a serious predator on snails, black and Norway rats are major predators on snails.

The black rat led to the extinction of the Laysan rail on Midway Island. Laysan rails had been translocated there from Laysan Island, where their habitat had been decimated by the grazing of introduced rabbits. Black rats reached Midway from ships that brought military equipment and personnel to the island during World War II.

Birds of Cities and Farms

A number of farm and city birds have been introduced to North America and Hawaii by organizations or individuals dedicated to such introduction (Temple 1992). Several of these are urban birds that evolved a close association with humans in the Old World. In North America, they quickly filled niches in urban environments for which native North

American birds were not adapted. The domestic pigeon was the first. It is derived from the rock dove, formerly a widespread bird in mountainous portions of Eurasia. The association of this bird with humans probably goes back to the origin of the first cities, which provided artificial nesting sites for these cliff-inhabiting birds. Domesticated birds have established feral populations worldwide, including urban areas throughout North America. The house sparrow was deliberately introduced to New York City by the Brooklyn Institute in 1852 or 1853 and spread rapidly through urban environments. By 1900, it had become one of the most abundant birds in North America. The European starling was also introduced to New York City, with the release of 60 birds in 1890 and 40 more in 1891. These releases were made by Eugene Schieffelin, whose interest was the introduction to North America of birds mentioned in the works of Shakespeare.

The European starling is the most serious exotic bird in North America. In fifty years, Schieffelin's releases of starlings had grown exponentially to 120 million birds (Davis 1950). In eastern North America, where starlings breed primarily in tree cavities, they are serious competitors for some woodpeckers, particularly red-bellied woodpeckers (*Melanerpes carolinus*) and northern flickers (*Colaptes auratus*) (Ingold 1994), and some secondary cavity-nesters, such as the great crested flycatcher (*Myiarchus crinitus*).

The mute swan, native to Eurasia, was brought to North America in the 1800s as a bird to grace parks and private estates. Feral birds were noted as early as 1919, and by the late 1920s a substantial wild population existed in the lower Hudson Valley and along the south shore of Long Island, New York. A large wild population, over 10,000 birds, now exists from Massachusetts to Virginia. About 3,300 feral birds now occur in Michigan and Wisconsin. Still another population exists in the Puget Sound region of Washington and southwestern British Columbia. These populations are growing rapidly, with a recent estimate for the Atlantic coastal population being 5.6 percent per year.

Along the Atlantic coast, mute swans are now creating serious problems for many native waterbirds. Consuming 8 to 9 pounds of aquatic plants per day, swans are depleting plant foods used by tundra swans that winter in and near Chesapeake Bay. They are highly territorial birds, aggressively defending areas up to 25 acres in size. Around Chesapeake Bay, their territorial and feeding activities are displacing colonies of waterbirds such as black skimmers (*Rhynchops niger*), least (*Sterna antillarum*), common (*S. hirundo*), Forster's (*S. forsteri*), and royal terns (*S. maxima*)(Williams 1997). In the Midwest, their populations are displacing those of the common loon (*Gavia immer*), a bird declining as a result of a variety of anthropogenic effects.

Although the detrimental impacts of mute swans are clearly recognized by conservation and wildlife groups, action to control or reduce mute swan populations is often thwarted by public sentiment. Mute swans are large "charismatic exotics," and ecologically naive humans can easily be motivated to protect them. Animal rights organizations have risen to the defense of swans in many instances, opposing efforts to kill or capture adults or to reduce their breeding success by shaking the eggs to kill the embryos (Williams 1997).

The cattle egret is a bird that is adapted to foraging in association with herds of wild ungulates, but it has now become an associate of livestock animals throughout most of its range. Native to Africa and Asia, it apparently colonized northeastern South America in the late 1870s, crossing the South Atlantic from West Africa. This crossing had probably occurred more than once, but in the 1870s the birds found herds of livestock and a climate similar to that of Africa. The colonizing population grew exponentially and reached Florida in the early 1940s (Bock and Lepthien 1976). Cattle egrets have now colonized most of North America, including southern Canada.

Exotic Pets

The panorama of exotic animals associated with humans is now becoming very diverse. Exotic pets are in fashion, and the trade in vertebrates, and even some invertebrates, has become enormous. Most of these animals come from outside North America, but the reptile trade also exploits many North American natives. The urban population of tortoises in Los Angeles, for example, includes not only the tortoise of the Sonoran Desert, but also the three additional congeneric species of Texas, Mexico, and Florida. Many desert lizards and snakes are also taken from nature as pets.

Exotic fish are in high demand for personal aquaria and, increasingly, for aquaria in restaurants, hotels, and other businesses. Breeding facilities for exotic fish have been the source of introductions to the wild in several locations (see Chapters 7 and 9).

The international wild bird trade, much of it illegal, involves the importation or smuggling of about 850,000 birds into the United States annually (Bolze 1992). The United States is the world's largest importer of these animals. About half of the legal imports now are parrots. In addition, about 150,000 parrots are smuggled annually into the United States across the Mexican border.

The wild bird trade is largely responsible for the fact that nine or more species of parrots now have wild breeding populations in the United States. The monk parakeet is the most widely distributed, having been found in seventy-six locations in fifteen states (Van Bael and Pruett-Jones 1996). Wild birds of this species were first noted in New York City in 1967. Releases and escapes of the species initiated populations in other areas. For several years, the U.S. Fish and Wildlife Service carried on a control program, but this ended in 1975, by which time the species was widely distributed. The nationwide population of this species now appears to be in the early stages of exponential growth. Monk parakeets feed on a wide variety of seeds, fruits, and buds and are able to survive in parts of the northern United States that have severe winters (Hyman and Pruett-Jones 1995). The species could potentially become a serious agricultural pest.

The house finch, a bird native to western North America, has also been sold as a cage bird. Birds released on Long Island, New York, gave rise to a population that has spread throughout much of the East (Elliot and Arbib 1953) (see Chapter 15).

Policy Issues

Several difficult policy issues relate to the management of domesticated animals and other close human associates. Foremost is the issue of domestic animal grazing on public lands in western states and provinces. Livestock grazing on U.S. National Forest and Bureau of Land Management lands has been destructive to wildlife habitat, especially in riparian situations. Ranchers in many areas have been awarded cheap grazing leases for decades, and they assume a continuing right to such leases. Increases in lease fees and restrictions imposed to protect endangered species are opposed by ranching interests, some with considerable political strength.

A second issue is the management of feral populations of animals deemed by many to have cultural and historical significance. The wild horse and burro adoption program has proved to be expensive and difficult to administer. The whereabouts of more than 32,000 animals that were adopted is unknown. Many animals that were adopted were later sold for slaughter, a violation of program conditions. The number of animals that have been captured but not adopted is increasing. Obviously, the conditions of the Wild Horse and Burro Act must be modified, or new approaches should be taken to control equid populations on public lands. In the latter vein, the BLM is supporting research into immunocontraception techniques that can be used to reduce the growth rate of feral horse and burro populations (Pogacnik 1995).

A third set of issues involves animal rights and treatment. The treatment of feral domestic animals and escaped pets involves many conflicting issues of ecology and ethics (Temple 1992). Because the species in question have special economic or aesthetic values, efforts to eradicate or control them often generate controversy and legal opposition. Efforts to eradicate feral goats from San Clemente Island by shooting, for example, were opposed and legally delayed by an animal rights organization. Religious beliefs sometimes oppose conservation concerns. For example, some Buddhist groups practice the liberation of captive animals such as goldfish and turtles. Liberation is regarded as a compassionate act that will bring good karma (West 1997). Conservation biologists, on the other hand, see the possibility of such releases leading to competition or disease in populations of endangered native species.

Managing exotics that are our close associates is thus largely a "people" problem. Dealing effectively with these species requires recognition of their ecological, economic, and social benefits and detriments. Conservation biology can contribute by clarifying the ecological impacts that these species have in the natural ecosystems that they invade. Education, however, is essential to enable the public to evaluate the benefits and detriments of these species in an unbiased manner.

Part IV

Theoretical Perspectives

Chapter 17

Exotics and Community Structure: Biodiversity Bombs

> Details of the ecology of species, and the communities into which they are being introduced are bound to be extremely important in determining whether a particular invasion will succeed. Yet, considering introductions in total, it is clear that there are some simple theoretical expectations and there are some equally simple general patterns.
>
> —Stuart L. Pimm, "Theories of Predicting Success and Impact of Introduced Species"*

> However, until our predictive ability improves in a more general arena, it is best to assume that the Frankenstein Effect is the one firm rule: new invasions are likely to have unexpected consequences.
>
> —Peter Moyle and Theo Light, "Biological Invasions of Fresh Water"*

Vernal pools are one of California's most distinctive ecosystems. Vernal pools are shallow, temporary freshwater ponds that form during winter and spring in grasslands of the Central Valley and the southern coastal

*S. L. Pimm (1989). "Theories of Predicting Success and Impact of Introduced Species." Pp. 351–367 in J. A. Drake, H. A. Mooney, F. di Castri, R. H. Groves, F. J. Kruger, M. Rejmánek, and M. Williamson (Eds.), *Biological Invasions: A Global Perspective*. John Wiley and Sons, New York.

*P. B. Moyle and T. Light (1996). "Biological Invasions of Fresh Water: Empirical Rules and Assembly Theory." *Biological Conservation* 78:149–161.

region. This once extensive microecosystem apparently has existed for a long period of recent geological time, evidenced by the fact that it has accumulated a rich flora and fauna of temporary pond organisms. Zedler (1987) lists forty-seven species of annuals and perennials that, in California, are almost entirely restricted to vernal pools. Many of these plants, and some animals, are endemic to the California vernal pools.

California vernal pools occur mostly in regions of the original valley grassland. These grasslands have been invaded by numerous Old World exotic plants, mostly annuals (see Chapter 12), which now dominate almost all areas still in grassland. Yet, despite the introduction of more than a hundred exotic wetland plants to North America, only a handful of exotic plants has entered the vernal pool flora, and these are usually of low abundance. If one examines a transect, only a few feet in length, that passes across the border of a pool basin, the transition from vegetation dominated by exotics to one nearly free of exotics can be seen. Quite obviously, either the vernal pool habitat or its plant community is resistant to invasion of exotics, or else the exotics that have reached California grasslands are mostly incapable of invading this ecosystem.

Two goals of the emerging field of invasion ecology are to predict whether a certain exotic species will fail or succeed in invading a particular community and to predict the impact of species that become established. These goals can be translated into several specific questions that are examined in this chapter: First, are there specific characteristics of exotic species that make them effective invaders of native communities? Second, do specific characteristics of native communities influence the probability of invasion attempts being successful? Third, do particular characteristics of successful invaders determine their impact on native communities? Fourth, does the structure of communities influence the degree of impact that successful invaders will have? Last, what determines the time frame within which the community impacts of successful invaders appear?

Characteristics of Invaders

Let us begin with the first question: Do successful invaders have predictable characteristics? As discussed earlier (see Chapter 2), predictability has proved possible in a very general, statistical sense (Lodge 1993). The "tens rule" suggests that about 10 percent of invasion attempts are successful, and about 10 percent of the successes lead to deleterious impacts. The tens rule, however, does not offer any help in predicting the invasiveness of a particular species.

Trying to identify the special biological characteristics that make species good invaders has taxed the ingenuity of many scientists. The first efforts were simply to seek generalizations about the characteristics of successful exotics. In one of the first attempts to characterize invasive plants, Baker (1965) proposed a set of characteristics that, altogether, would create an "ideal" weed. This hypothetical species was suggested to be self-compatible, flower early, produce abundant seed, disperse its seed widely, germinate under varied conditions, grow rapidly, spread vegetatively, and be a strong competitor. No single species combines all of these characteristics, but many invasive plants combine several of them. On the other hand, species that possess several or many of these characteristics do not necessarily become weedy. Thus, plants that possess several or many of these characteristics might well be considered "high-risk" species from the standpoint of weediness or invasiveness (Noble 1989).

Several other very general correlations about the characteristics of invasive plants have been offered. Daehler (1998) examined taxonomic patterns and biological characteristics of invasive plants at a global level. He found, first of all, that certain plant families contained significantly more invaders of natural areas than expected on the basis of their species richness. These families included the grasses, legumes, and the family to which waterweeds (*Elodea* spp.) belong. Considering biological characteristics, he also found that families rich in species that are pollinated abiotically, fix nitrogen, or exhibit a viny growth form contain more invasive species of natural areas than expected. Families of woody plants with clonal growth and plants of aquatic or semiaquatic habitats were also overrepresented in invaders of natural areas.

Several investigators have also found that invasiveness in plants is correlated with low nuclear DNA content, which is, in turn, correlated with small seed size (Rejmánek 1996). Low nuclear DNA content is associated with faster cell division and thus more rapid plant growth. In several plant genera, invasiveness is inversely correlated with nuclear DNA content. The occurrence of weediness in many polyploids is something of an exception to this pattern, however.

Attempts to quantify these biological patterns more rigorously have met with only marginal success. Perrins et al. (1992), for example, asked a group of sixty-five scientists to score forty-nine annual plants or a range of families on a 5-point scale of weediness. They then looked for correlations of the consensus score for weediness with some thirty-nine characteristics of varied nature. Although multivariate analysis predicted weediness better than expected, the degree of accuracy was not high, and the effort was not judged a success.

Several geographical patterns of invasiveness by plants have also been suggested. For example, in North America, invasive exotics tend to belong to genera that are not native to North America (Rejmánek 1996). For example, of 112 species of European grasses established in California, over 61 percent belong to genera not native to California. However, only 39 percent of grasses in Europe belong to genera not shared with California.

A second pattern is that invasive exotics also tend to be more widespread in their native region than noninvasive species belonging to the same genera. In an analysis of 165 European plants that have invaded eastern Canada, together with paired noninvasive European congeners, Goodwin et al. (1999) found that 70 percent of invasives could be predicted from information on how widespread their distribution was in Europe.

Invasive plants in North America also tend to be species that have proven invasive in other regions. About 235 species of nonindigenous woody plants have become established in the wild in North America. Of these, about 76 have been judged to be invasive enough to merit "pest" status. Analysis of the traits of these invasive species, compared to a sample of noninvasive exotics, revealed a strong tendency for species that were invasive in other world locations to be invasive in North America (Reichard and Hamilton 1997). This suggests that invasiveness is determined much more by the characteristics of the species than by those of the general region. Woody species that were capable of vegetative reproduction and those that bore fruit for much of the year also tended to be invasive.

Unfortunately, these general patterns offer only very weak predictions about the likelihood that a particular plant species will prove invasive in a particular region. More quantitative approaches focused on specific groups of plants have given somewhat improved predictive power. Rejmánek and Richardson (1996) have attempted to develop a strong predictive relationship for the genus *Pinus.* This genus contains about 100 species worldwide, with a number being extremely invasive in areas of the Southern Hemisphere, where pines are not native. On the basis of an analysis of life history characteristics of 24 species, both invasives and noninvasives, Rejmánek found that small seed size, rapid growth to maturity, and short interval between large seed crops were features associated with invasive species. These could be combined in a mathematical relation that correctly predicted invasiveness of 34 additional members of the genus. In addition, this relationship, when extended to trees and shrubs of angiosperm plants, correctly predicted invasiveness of 95 percent of a sample of 40 species belonging to 40 genera. Clearly, there are limits to the extension of a relationship derived for a particular group of plants, such as pines.

Reichard and Hamilton (1997) applied a multivariate analysis to the characteristics of 149 species of invasive and noninvasive woody plants for which they had compiled information on fourteen variables. They then tested the resulting relationship on 58 species. Of 34 invasives, 33 were correctly identified; of 24 noninvasives, however, 7 were incorrectly classified as invasive. On the basis of this analysis, they prepared a decision tree that considered the characteristics of a species and accepted or denied its admission to North America, or they recommended further analysis and monitoring. Tested against 291 species, only 2 percent of invasives were accepted, with none of these being species considered to be serious pests. Again, in a sense erring on the side of caution, 54 percent of noninvaders were denied admission or were judged to deserve further study before admission. Thus, it appears that improved prediction of invasiveness is possible when many life history features of species are considered.

Less success has been achieved with predicting invasiveness of animals. Brown (1989) suggested several invasiveness "rules" for vertebrates. The first was that successful invaders were natives of continents or of extensive, nonisolated continental habitats. This rule rests on the idea that such species would thus be invading less diverse biotas in which the chance of competitors or predators opposing successful establishment was low. A second rule was that similarity in the physical environments of both source and introduction areas enhanced the chances of success of an introduced species. A third rule was that an invading species was more likely to be successful if native species with similar niches were absent. Ehrlich (1989) presented a comparable set of characteristics for invasive vertebrates. He suggested that successful invaders tended, among other things, to be abundant and widespread in their native region, to be vagile and gregarious, to have a broad diet and habitat tolerance, to possess much genetic variability, to have a short generation time, to be larger than close relatives, and to be associated with humans to some degree.

Unfortunately, for both plants and animals, characteristics that at first examination appear to be significant often turn out to be confounded with many other potential causal relationships. For example, the preponderance of European species among North American invaders suggests that centuries of human disturbance of the European environment have created an evolutionary class of "weedy" species, highly adapted for colonization of new areas. But this preponderance may reflect nothing more than the magnitude of commerce from Europe to North America and the great opportunities for dispersal of weedy species along this route (Simberloff 1989).

Similarly, occupation of a region with similar climate and vegetation to North American localities seems a likely characteristic for predicting the likelihood of successful invasion (Simberloff 1989). As logical as this seems, data on successful versus unsuccessful colonization by species from similar versus different source areas are almost completely lacking. The best evidence on this point comes not from problem species, but from biological control species. With these forms, success of establishment only appears more likely for forms coming from elsewhere on the same continent than for those coming from other continents.

Invasibility of Communities

Let us turn to the second major question about exotic invasions: Do specific characteristics of native communities influence the probability of their being invaded? Why are exotic species seemingly able to invade communities of native species with a long history of evolutionary adaptation to their particular environment and to each other? This is also a difficult question to address because of the many confounding variables and the lack of rigorous experimental tests.

The question of community invasibility is related to an issue that has divided ecologists in recent years: Are communities tightly structured and their composition regulated homeostatically? Or are they simple accumulations of species that tolerate basic conditions of the physical environment, have found food resources that they can exploit, and are able to avoid enemies well enough to survive?

One view—regulated or equilibrial—is that the member species of mature communities are tightly constrained by patterns of competition for resources and by food-chain interactions with higher and lower trophic levels. The species that occur together, under this view, are those most efficient at dealing with competitive and food-chain pressures. They have been tested by generations of ecological interaction under prevailing environmental conditions, and they have a high degree of mutual evolutionary adaptation. The result is that under prevailing conditions of climate or the physical environment, an equilibrium structure and composition exists. Deviations from such a condition, due to variation in physical conditions or to disturbance, result in gradual recovery toward the equilibrium condition.

If this pattern is true, one should be able to define community assembly rules, that is, rules that predict what species can invade a community at different stages of its development. Several suggested generalizations about invasions emerge from the community assembly theory (Moyle and Light 1996b). First, communities with high species diversity and many strong species interactions should resist invasion. Pimm (1989) has presented a model of community assembly showing that the difficulty of invasion increases as the total number of accumulated invasion attempts increases, so that the number of species present eventually reaches an asymptote. Second, established species should be able to resist invasion by superior competitors, if the latter is not abundant. Third, the composition of the final community depends on the order in which particular species are added, so that different mature communities may develop from the same pool of potential invaders, if the invasion order differs.

The second view—nonequilibrial or open—is that communities are simply assemblages of species that, at the moment, are simply able to grow and reproduce. Competitive and trophic relations are resultant and variable rather than strongly regulatory. Furthermore, constancy of prevailing conditions is itself an illusion, and the various species are imperfectly tracking their changing physical and biotic environments. The makeup of the community is always changing at some rate in response to changing physical conditions and disturbance (Lodge 1993). Under this pattern, assembly rules could not predict differences in species that can invade at different stages of community development.

Under the equilibrial view, a mature community is almost "closed" to newcomers, and opposition to the invasion of an exotic species is strong. Some unusual circumstance must therefore exist to enable such an invasion. Disturbance, the appearance of a species freed from its natural enemies, or some genetic change that enables a species to overcome the community's resistance to invasion must occur. Under the nonequilibrial view, the community is open to invasion by any species that finds favorable conditions for growth and reproduction. The simple crossing of dispersal barriers may be all that is necessary, although disturbance, escape from enemies, and changing environmental conditions may assist the invasion process.

The equilibrial view, furthermore, suggests that change in community composition is "unnatural," a view that is, in effect, espoused by many conservation organizations that strive to preserve existing communities of native species. The nonequilibrial view is that comings and goings of species are "natural" and that some degree of change must be accepted, especially in a world of climatic change, even welcomed.

Testing these alternatives, or determining where reality sits along the spectrum between these extremes, is difficult. First of all, all communities appear to be invasible to some degree, so that invasibility is a matter of degree. Rejmánek (1989), for example, compiled a list of sixty plant species, representing forty families, that have been reported to invade undisturbed natural communities. Determining the relative contributions of abiotic conditions, internal community structure and organization, and external influences on invasibility is difficult in complex, uncontrollable natural ecosystems. Furthermore, the number of attempted invasions, as well as successful invasions, is rarely known. When one observes many or few successful invasions, therefore, it is difficult to rule out differences in the number of attempted invasions. The available data are mostly observational, the results of a mix of natural and human influences. Controlled experiments on invasion are virtually unavailable.

Some communities, especially aquatic, appear to be nonequilibrial or open, so loosely structured that many ecological resources are available without effective competition. Moyle and Light (1996b), for example, argued that streams and estuaries in California are largely open to invaders. The primary factor that seems to determine success in establishment of exotics in these ecosystems is adaptation to conditions of the physical environment. In general, this appears to be true of most lake and stream ecosystems, for which successful invasions seem to occur without relation to richness of the native biota.

Other communities, especially terrestrial, appear to be somewhat regulated and lie somewhere on the scale between nonequilibrial and equilibrial. Following one suggested assembly "rule," simple communities, with few species, seem to be more susceptible to invasion than complex communities. This rule is supported by much observational evidence. Brown (1989), for example, noted that there is a virtual lack of cases in which island vertebrates have successfully invaded continental areas, southern Florida being somewhat of an exception. Ehrlich (1989) also supported this view, on the basis of the success of introduced mammals in locations such as New Zealand, which lacked native terrestrial mammals, and the few successes in New Guinea, which has a rich mammal fauna. Some thirty-three species of mammals have successfully colonized New Zealand. Some groups of organisms do not support this "rule." For biological control insects, Simberloff (1989) examined the success of introductions to islands versus to continental areas. Although more successful introductions have been made of continental species to islands than of island species to continents, the probability of success of any individual introduction attempt was not different for these cases.

What factors enable exotics to invade communities that seem to have a substantial degree of regulation? One frequent suggestion about invasibility is that disturbed habitats, particularly those associated with humans, are more easily invaded than undisturbed areas (Brown 1989). In aquatic ecosystems, intermediate degree of disturbance appears to enable almost any species adapted to the physical conditions and basic resources to invade (Moyle and Light 1996b). As discussed in previous chapters, fire, livestock grazing, hydrological change, and physical disruption or transformation of habitats open the door to terrestrial species from regions where these types of disturbance have been long standing. Exotics themselves often create disruptions of native ecosystems that favor invasions of still other exotics.

The importance of disturbance can often be seen in local situations where disturbed and undisturbed habitats are in close proximity. In New Zealand and the Hawaiian Islands, for example, introduced birds are largely absent in areas of undisturbed native forest, although they abound in exotic-dominated vegetation, agricultural land, and native forest disturbed by livestock and feral domestic animals (Ehrlich 1989).

Support for the importance of disturbance also comes from studies of biotic succession. For terrestrial plant communities, the frequency of exotics is greatest in early stages of secondary succession, that is, in communities that develop following disturbance (Rejmánek 1989). This suggests that an organization develops during succession that reduces the ease of invasion of exotics. The difficulty, however, is that the processes of natural and human-assisted dispersal likely bring more early-successional species than late-successional species to any given area. Needed: an experiment that introduces propagules of early- to late-successional exotics to a range of early- to late-successional communities of nonexotics.

Also, for terrestrial communities, the greatest frequencies of exotics seem to appear in communities of intermediate moisture status, neither those of arid habitats nor those of wet habitats (Rejmánek 1989). This suggests that aridity may reduce the success of exotic establishment, on the one hand, and severe competition, their establishment in wet habitats, on the other. As for succession, however, the diversity of exotic propagules that reach habitats along this moisture gradient is unknown.

Another relationship that may permit invasions of equilibrial communities is the fact that many exotics escape their own enemies—parasites, predators, competitors—and are unchallenged by potential enemies in the community invaded. For example, for introductions of the house sparrow and the European starling (Table 17.1) to North America, the evidence of reduced parasite load is strong (Table 17.2). North American house sparrows, for example, have only thirty-seven species of mites, fleas, and lice, compared to the sixty species recorded in Europe. European starlings in North America have only twenty-two species of parasitic worms, only 31 percent as many as have been recorded in European starlings in Europe (Table 17.3). In both cases, however, the North American populations have some species that do not appear in the European populations, so that some new parasites have been acquired in the New World. Thus, counteradaptation is, in fact, gradually occurring to these new species on the North American scene. Many other cases exist of reduced disease, parasite, and predator loads on successful exotics. In California, for example, the bullfrog has only two of the parasites that it carries in the eastern United States (Lafferty and Page 1997).

Still other possible reasons for invasion of exotics into equilibrial communities include changes in climate that cause both adaptation patterns of community members to become poorly integrated and ecological resources to become available. This possibility is considered serious under the scenario of rapid global climate change. In addition, exotics may possess or acquire genetic characteristics that enable them to overcome community resistance to invasion. For example, species in the Old World may have been exposed to selection for adaptation to Mediterranean climates for a long period of geological history and might therefore be able to invade regions where such climates have been of more recent origin.

Table 17.1. Scientific names and native regions of exotics that illustrate relationships to community structure.

PLANTS	
Brazilian peppertree, *Schinus terebinthefolius*	South America
cut-leaved teasel, *Dipsacus laciniatus*	Europe
European cordgrass, *Spartina anglica*	Europe
fennel, *Foeniculum vulgare*	Europe
small cordgrass, *Spartina maritima*	Europe
smooth cordgrass, *Spartina alterniflora*	Eastern North America
AMPHIBIANS	
bullfrog, *Rana catesbiana*	Eastern North America
BIRDS	
European starling, *Sturnus vulgaris*	Europe
house sparrow, *Passer domesticus*	Europe

Table 17.2. External parasites recorded in house sparrows in North America and Europe.

Parasite	*Europe*		*North America*		*New in North America*	
Group	*Genera*	*Species*	*Genera*	*Species*	*Genera*	*Species*
Mites	8	35	5	24	2	8
Lice	8	18	4	9	1	4
Fleas	1	7	1	4	1	2
TOTAL	17	60	10	37	3	14

Data from Brown and Wilson (1975).

Table 17.3. Parasitic worms recorded in European starlings in North America and Europe. (Data from Hair and Forrester 1970)

	Europe		*North America*		*New in North America*	
Parasite Group	*Genera*	*Species*	*Genera*	*Species*	*Genera*	*Species*
Flukes	7	26	4	4	1	2
Tapeworms	9	12	4	5	1	2
Nematodes	14	26	6	10	3	4
Acanthocephalans	4	6	2	3		
TOTAL	44	70	16	22	5	8

Data from Hair and Forrester (1970).

Impacts of Successful Exotics

Now to the third major question: Do particular characteristics of successful invaders determine their impact on native communities? Most successful invaders—the tens rule suggests about 90 percent—do not create major disruption of the invaded ecosystems. Others can have a keystone role, greatly modifying much of the structure and dynamics of the invaded system. Although there is probably no clear dichotomy between keystone exotics and exotics with negligible impacts (Hurlbert 1997), the impacts of some exotics can be profound. In some cases, deliberate introductions that were expected to be beneficial have proven to be disastrous, a pattern that Moyle has termed the Frankenstein Effect (Moyle and Light 1996b).

This book has presented many examples of how keystone exotics can almost completely transform community structure, especially in southern Florida (Chapter 9), western riparian areas (Chapter 11), and Hawaii (Chapter 13). New food chains, made up of exotic species, have become established in many ecosystems, particularly in aquatic systems such as

West Coast estuaries (Chapter 5) and the Great Lakes (Chapter 6). In some of these situations, new community types, composed largely of exotics, have developed.

The ecological distinctiveness of an exotic, relative to the species that exist in a community, appears to be an important factor in the impact expressed (Cox 1997). Exotic plants that are different from native species in life-form and exotic animals that are different from natives in feeding ecology often exert keystone impacts. Animals of high trophic level, by their presence or absence, often define the intensity of animal impacts. Mammalian herbivores introduced to islands in the absence of their predators, for example, are likely to create keystone impacts. In aquatic ecosystems, on the other hand, piscivorous fish are most likely to exert keystone effects, whereas omnivorous and detritivorous species were unlikely to have major impacts (Moyle and Light 1996b).

Community Structure and Impacts of Exotics

The fourth basic question concerns whether the structure of communities influences the degree of impact that successful invaders will have. This question has not been examined in detail. Moyle and Light (1996b), however, concluded that communities of aquatic environments with either extremely high or extremely low variability in abiotic conditions are most likely to experience severe impacts by introduced exotics. For streams with high variability of flow and temperature, this may be equivalent to the existence of a nonequilibrial community. At the other extreme, tropical and subtropical lakes seem to be very vulnerable to exotic impacts, the extreme perhaps being Lake Victoria in East Africa, where an introduced piscivorous fish has caused the extinction of hundreds of native fish species. Communities of oligotrophic lakes, such as Lake Tahoe (see Chapter 15), may also be sensitive to disruption by introduced exotics (Li and Moyle 1981).

Time Lags in Impacts of Exotics

Let us consider the final major question: What determines the time frame within which the community impacts of successful invaders appear? Although many established exotics appear to have negligible impacts, this is no guarantee that they will not become serious problems in the future, because of time lags in population increase and geographical spread. Many, if not most, exotic species show a long period following initial establishment during which their populations remain small and their impact minimal. Following this lag period, they may explode in abundance and impact. For plants, lag times may range from 20 or so years up to more than 300 years (Wade 1997). The causes of such time lags are poorly understood, but these seem to be ecological in some cases, genetic in others (Crooks and Soulé 1999).

Ecological time lags in some cases involve a demographic growth period—simply the time that it takes for an established exotic to build up populations large enough to have significant dispersal potential. Cut-leaved teasel, for example, was introduced to North America in the 1800s by early immigrants from Europe (Solecki 1993). It remained a localized species in northern New York State until the mid-1900s. In the past few decades, however, it has spread rapidly through the upper midwestern states, where it forms dense colonies that crowd out most other herbaceous plants. In Florida, the Brazilian peppertree showed a similar lag (Ewel 1986). Introduced in the late 1800s, it did not begin to spread aggressively until the 1950s.

In other cases, an ecological change can create conditions favorable to the rapid growth and spread of a species. On Santa Cruz Island, off the California coast, removal of 36,000 sheep and 1,500 cattle, themselves exotics, allowed an exotic weed, fennel, to increase explosively in abundance. It now infests about 10 percent of the island (Brenton and Klinger 1994).

The possibility that genetic change may enable a long-established exotic to become a widespread and destructive species has been suggested by many evolutionary biologists, but very few possible examples can be recognized (Crooks and Soulé 1999). The clearest example comes from Europe and involves the North American smooth cordgrass and its European congener, small cordgrass (Williamson 1996). Smooth cordgrass was introduced to Europe in 1803, or possibly earlier, and became established in a number of estuaries on both sides of the English Channel. A sterile hybrid of these two species was recognized in 1870 in locations where smooth and small cordgrasses came together. In 1892, however, a vigorous, fertile cordgrass with doubled chromosome number appeared, now known as common cordgrass. Common cordgrass has spread and been planted widely in western Europe, and it has a wider intertidal and geographical range than the European native small cordgrass. As noted in Chapter 5, common and smooth cordgrasses have been introduced to the North American West Coast, where additional hybridization events have occurred.

Invasion Theory

None of these basic questions about the relationships of exotics and the communities they invade has yet been answered with any degree of satisfaction. Many intriguing patterns can be seen in particular groups of organisms, and in particular types of ecosystems, but robust generalizations are few. The answers to these questions, together with those to questions that are examined in Chapters 18, 19, and 20, are basic to the development of sound ecological strategies for managing exotics.

The urgent need for theory that addresses these questions is rapidly giving rise to a new field of applied ecology: invasion ecology. Theory that explicitly predicts the risk of invasion for particular plant species is just now beginning to emerge (Rejmánek 1998). To understand not only which species tend to be invasive, but also how they will interact with the ecosystems they encounter, however, will require a theory that spans several scales of organization: those of the organism, the population, the community, and the ecosystem. To apply this understanding to management of invasive exotics, ecological theory and practice must be integrated to a higher degree than exists in the current field of pest management.

Chapter 18

Exotics and Ecosystem Impacts: Changing the Way Nature Works

> Invasions that alter ecosystem processes are important to ecological theory because such effects are less well characterized than are population or community effects of invasion, and they represent a clear example of species control over ecosystem processes. In addition, invasions that alter ecosystems represent a particularly significant threat to native populations and communities: they don't merely compete with or consume native species, they change the rules of the game by altering environmental conditions or resource availability.
>
> —Carla D'Antonio and Peter Vitousek, "Biological Invasions by Exotic Grasses, the Grass/Fire Cycle, and Global Change"*

Asian cogon grass was first introduced accidentally to the United States near Mobile, Alabama, in 1912, and later deliberately planted in Mississippi as a forage grass (Table 18.1). It has now spread into Florida, where it occurs in twenty-seven of the state's sixty-seven counties. A tall, fast-growing, perennial grass, its abundant rhizomes penetrate to a depth of as much as 4 feet. In a mature stand, the rhizomes alone may have a mass of 3 tons per acre. In Florida, cogon initially invaded open, disturbed soils

*Carla D'Antonio and Peter Vitousek (1992). "Biological Invasions by Exotic Grasses, the Grass/Fire Cycle, and Global Change." *Annual Review of Ecology and Systematics* 23:63–87.

Table 18.1. Scientific names and native regions of invasive exotics with ecosystem impacts.

PLANTS	
annual ryegrass, *Lolium multiflorum*	Europe
blue gum, *Eucalyptus globulus*	Australia
broomsedge, *Andropogon virginicus*	eastern North America
buffel grass, *Pennisetum ciliaris*	Africa
bushy beardgrass, *Schizachyrium condensatum*	eastern North America
cheatgrass, *Bromus tectorum*	Eurasia
Chinese tallow, *Sapium sebiferum*	eastern Africa
cogongrass, *Imperata cylindrica*	southeast Asia
crystalline ice plant, *Gasoul crystallinum*	South Africa
European beach grass, *Aimophila arenaria*	Europe
firetree, *Myrica faya*	Canary Islands
hydrilla, *Hydrilla crassipes*	Sri Lanka
molasses grass, *Melinis minutiflora*	Africa
Molucca albizia, *Albizia falcataria*	Asia
saltcedar, *Tamarix chinensis*	Eurasia
saltcedar, *Tamarix ramosissima*	Eurasia
water hyacinth, *Eichhornia crassipes*	South America
water lettuce, *Pistia stratiotes*	South America
MARINE INVERTEBRATES	
Asian clam, *Potamocorbula amurensis*	Asia
Australian isopod, *Sphaeroma quoyanum*	Australia
Japanese mussel, *Musculista senhousia*	Asia
FRESHWATER INVERTEBRATES	
opossum shrimp, *Mysis relicta*	North America
quagga mussel, *Dreissena bugensis*	Russia
zebra mussel, *Dreissena polymorpha*	Russia
FISH	
Alewife, *Alosa pseudoharengus*	Atlantic Ocean
rainbow smelt, *Osmerus mordax*	Atlantic Coast
sea lamprey, *Petromyzon marinus*	Atlantic Ocean
MAMMALS	
feral hog, *Sus scrofa*	Asia, Europe

along roads, in pastures, and on mined phosphate lands. Now, however, it is invading pine woodlands and other natural ecosystems.

The sandhill ecosystem of central Florida is one of the sensitive natural environments now being invaded by cogon. The sandhill vegetation consists of an open canopy of long-leaf pine (*Pinus palustris*) and an understory of palmetto, shrubs, and perennial bunchgrasses. It is a fire-dependent ecosystem and tends to be invaded by hardwood trees in the absence of fire. Cogon, however, is not just another perennial grass similar in characteristics to the native grasses. Rather, it is a rhizomatous grass, distinct in morphology and physiology, that modifies basic sandhill ecosystem properties (Lippencott 1997). Cogon produces a greater fuel load, so that fires burn higher and hotter, killing more plants, including long-leaf pine seedlings, than fires fueled by native grasses. Its deep, protected rhizomes enable cogon stands to recover rapidly from such fires. The dense leaf canopy and large rhizome mass of cogon lead to depletion of soil water. This, together with the thick litter mat that develops under cogon, adversely affects seedling recruitment of native plants. As stands of cogon reach 60 percent cover, most native understory species are displaced, and plant diversity is reduced. Many burrowing animals, such as scarab beetles, are also reduced in abundance. The invasion of this new type of plant, a rhizomatous grass, has thus altered fire intensity, soil moisture regime, native plant reproduction, and overall vegetation structure. The long-term result of these changes is predicted to be the transformation of a diverse pine woodland into a monospecific stand of cogon grass.

Ecosystem Impacts

As seen throughout this book, a single keystone exotic species can often have far-reaching impacts on an entire biotic community. Many of these impacts also involve collective ecosystem properties, physical as well as biotic. Vitousek (1990) has suggested that exotics are able to exert ecosystem effects in three basic ways. First, exotics may differ from natives in the basic ways that they modify physical and chemical conditions of the ecosystem, such as solar radiation, nutrient availability, and salinity. Typically, this is a result of the manner in which they acquire and utilize resources. Second, exotics may act in ways that influence the composition of major trophic groups, such as producers, herbivores, and carnivores. This can result from either "top-down" influences of exotic species at high levels in the food web or "bottom-up" effects at the plant-producer level. Third, exotics may alter the frequency or intensity of some type of disturbance, such as fire frequency or intensity, flooding influences, or pest or pathogen outbreaks. These are perhaps the most long-lasting kinds of impacts of exotics, but at the same time, some of the least obvious.

Specific ecosystem impacts include alteration of the physical landscape, hydrology, soil moisture and salinity regimes, nutrient cycling pattern, primary productivity, food web structure and energy flow pattern, decomposition rate, frequency and intensity of fires or floods, structure and dynamics of the vegetation, and properties such as stability and resilience. These impacts may, in fact, shift ecosystem structure and function to a considerably altered, but stable, state, from which recovery becomes nearly impossible. This chapter examines the nature of these sorts of ecosystem impacts by exotics in detail.

Physical Structure and Geomorphology of Ecosystems

In both aquatic and terrestrial ecosystems, the basic structure of the landscape, or its geomorphology, can be altered by invasive exotics. The influence of "exotic ecosystem engineers" in California estuaries illustrates how important impacts on the physical environment can be in aquatic environments (see Chapter 5). The burrowing activities of Australian isopods are causing banks of tidal channels to collapse, leading to the widening of channels and the loss of vegetated salt marsh (Crooks 1997). Outbreaks of the Japanese mussel, on the other hand, stabilize mudflat surfaces and create mussel-bed communities where eelgrasses once grew and shorebirds probed for burrowing organisms (Crooks 1996, 1998).

The interface of freshwater and terrestrial systems is particularly vulnerable to landscape impacts of exotics. Along the Green River in Canyonlands National Park, for example, the invasion of saltcedar has caused a basic change in the form of the river channel (Graf 1978). Spring floods tended to scour away stands of the native willows (*Salix exigua*) that had colonized banks and sandbars, thus maintaining a broad, shallow channel. Tamarisk, on the other hand, is able to withstand the annual floods and stabilize islands and sandbars. As a result, channel width has become reduced by an average of 27 percent, and in some places by more than 50 percent. Transient habitats of open, shifting sand have become densely vegetated terraces and islands. Islands themselves have broadened, lengthened, and often become fused with the banks, eliminating narrow channels and backwaters. Flood flows through the narrowed channel now tend to overflow the banks, rather than flushing bars away and carrying the flow in a deepened channel. Farther downstream on the Colorado River, below Glen Canyon Dam, the primary effect of reduced flooding and saltcedar invasion has been to stabilize bars and create a new band of dense shrubs in the zone formerly scoured by spring floods.

Similar changes have occurred on the Brazos River in Texas following saltcedar invasion. Dense stands of saltcedar narrowed the river channel by 64 percent between 1941 and 1979 (Blackburn et al. 1982). Impeded water flow has resulted in increased sedimentation, the material deposited over the thirty-eight-year period averaging about 10 feet in depth. Because of the narrower active channel, flooding now inundates a much greater area of the floodplain. A flood flow in 1971, for example, crested at a level 8 feet higher than a flow of equal volume in 1941.

In terrestrial ecosystems, landscape structure can also be altered by exotics. Along the coast of Oregon, for example, European beach grass was planted extensively in the 1930s and 1950s for dune stabilization (Wiedemann 1984). Prior to establishment of this grass, the zone immediately behind the beach consisted of sand hummocks capped by native perennial herbs, shrubs, and rushes. These hummocks were up to 16 feet high. European beach grass has displaced many of these natives along the Pacific coast from central California to British Columbia. European beach grass spreads by rhizomes and forms dense clumps up to 3 feet in height. It traps blowing sand very effectively and creates a steep foredune immediately behind the beach, eliminating the broader zone of hummocks that existed before. Its effectiveness in trapping sand might even reduce the sand supply to active inland dunes, changing their structure and dynamics.

Ecosystem Hydrology and Salinity

The hydrology of aquatic and terrestrial ecosystems is frequently affected by exotics. Saltcedar invasion of desert watercourses leads to increased transpiration of water, relative to that by native cottonwoods and willows, and can cause cessation of surface water flows and deepening of the groundwater table (Randall 1993). As the water table drops, many native plants and animals are eliminated. At Eagle Borax Springs in Death Valley, for example, saltcedar spread through the large marsh area during the 1950s (Neill 1983). By the late 1960s, surface water had disappeared, native marsh plants were being replaced by saltcedar, and even the native mesquite trees bordering the marsh were in decline. A ten-year program of saltcedar eradication was undertaken at that point, with the result that surface water returned and the native vegetation recovered. At the Nature Conservancy's Dos Palmas Preserve in the Coachella Valley, California, saltcedars threatened the existence of permanent stream pools occupied by a desert pupfish (*Cyprinodon macularius*), a federally designated endangered species. Saltcedar has also been eliminated from this preserve.

In terrestrial ecosystems, exotics can influence hydrology in quite different fashion. In Hawaii, for example, broomsedge, a perennial grass native to the southeastern United States, has invaded disturbed slopes in wet mountain forest areas on the island of Hawaii. During the Northern

Hemisphere winter, however, this grass becomes dormant and transpires very little water. As a result, the soil on broomsedge slopes becomes saturated for long periods of time, and landslides frequently occur (Mueller-Dombois 1973).

Several exotics exert strong influence on the vegetation by increasing soil salinity. The saltcedars are one such group of plants, and their impacts on riparian vegetation are due partly to salination of the surface soil (see Chapter 11). In coastal California, crystalline ice plant also increases soil salinity in coastal areas and on the Channel Islands (Vivrette and Muller 1977). An annual, the plant is named because its surface is covered with tiny, clear, bubble-like structures that give the plant an "icy" look. This ice plant accumulates salt during its annual growth phase and releases it into the soil when it dies. Crystalline ice plant becomes established in areas of disturbance, but by salinating the surface soil, it inhibits the germination and growth of other plants. Thus, crystalline ice plant spreads as a monospecific stand.

Ecosystem Fertility and Productivity

Decomposition, soil nutrient cycling, and primary production are ecosystem processes that many, if not most, exotics influence to some degree. Along the Texas Gulf Coast, for example, the Chinese tallow tree has had a striking effect on these processes. As noted in Chapter 9, this species has displaced coastal prairie in many locations. Chinese tallow is deciduous and produces a leaf fall of about 1.7 tons per acre, an amount comparable to that of native deciduous trees in the southern United States (Cameron and Spencer 1989). The leaves of Chinese tallow are low in lignin and decay very rapidly. Rapid decomposition is perhaps aided by the action of high populations of detritus-feeding isopods that develop in the tallow litter. Concentrations of phosphorus, nitrate nitrogen, potassium, zinc, manganese, and iron are higher under tallow stands than in coastal prairie soils. Chinese tallow thus seems to act as a nutrient "pump," acquiring mineral nutrients from deep soil layers and depositing them in the surface soil where leaf decay occurs. Net primary production of tallow stands is double that of coastal prairie (Harcombe et al. 1993). Clearly, greatly altered ecosystem function resulted from the invasion of Chinese tallow and probably functions to maintain this forest ecosystem in what was originally a grassland habitat.

Another classic example of the influence of an exotic on soil nutrient relationships is the effect of firetree on nitrogen levels in young nitrogen-deficient volcanic soils on the island of Hawaii. Firetree invaded Hawaii

Volcanoes National Park in 1961 and spread over more than 30,000 acres by 1985. Firetree is a small early-successional tree that can colonize volcanic ash deposits and lava flows less than fifteen years old. Plant growth in sites like these is strongly limited by shortage of nitrogen, as revealed by fertilization experiments.

Unlike native early-successional plants in Hawaii, however, firetree has root nodules with nitrogen-fixing actinomycete symbionts. In areas thickly colonized by firetree, symbiotic nitrogen fixation is about 16 pounds per acre. In areas without firetree, the total nitrogen income is, at best, about 4.9 pounds per acre, so that firetree more than quadruples the nitrogen input to the ecosystem (Vitousek 1990). A fraction of this nitrogen enters the soil through decomposition of firetree litter, thus becoming available to other members of the plant community. In all likelihood, the increased availability of nitrogen created by firetree will favor the invasion of other exotic trees, which tend to be more successful on fertile sites than on infertile sites in Hawaii. Strawberry guava (*Psidium cattleianum*), for example, is a species that has been shown to grow rapidly in soil taken from beneath firetrees.

Another fast-growing tropical tree widely planted in Hawaii is Molucca albizia. This leguminous tree harbors symbiotic nitrogen-fixing bacteria. Thirteen-year-old stands of Molucca albizia on abandoned sugarcane land showed several to many times the available nitrogen levels that prevailed in eucalyptus stands of similar age (Garcia-Monteil and Binkley 1998). Thus, like firetree, this leguminous tree is likely to encourage late-successional dominance of nonnative species.

In California, the introduced blue gum, an Australian eucalyptus, also modifies basic soil ecosystem processes. The deep-rooted blue gums are able to tap deep soil water and nutrients and thus grow throughout much of the year in areas of Mediterranean climate. Consequently, their annual litter production is more than twice that of annual grassland in similar soil (Robles and Chapin 1995). Eucalyptus litter is dense and slow to decompose, and the accumulated litter is up to nine times that found in annual grassland. Nutrients such as nitrogen remain tied up in the accumulated litter. This accumulated, dense litter also modifies the influence of fire. Fires in eucalyptus stands are likely to be less frequent, but more intense, than grassland fires.

Aspects of soil ecology can also be influenced by exotic animals. The rooting of feral hogs, for example, produces major change in the litter and surface soil of forest ecosystems. In Great Smoky Mountains National Park, areas heavily rooted by hogs had a thinner litter layer, a more mixed organic and mineral surface soil, and much more bare soil than areas not used by hogs (Singer et al. 1984). Calcium, magnesium, and phosphorus were much reduced in the surface soil of rooted areas. Levels of soil nitrate were higher in areas of rooted soil, indicating more rapid release of nitrogen from litter components. In stream outflow, nitrate was more than doubled and total nitrogen increased by more than four times. Deep seepage losses of nitrate were also markedly increased. In areas used heavily by hogs in Hawaiian montane forests, similar effects were noted by Vitousek (1986). The litter layer was thinner and release of mineral nitrogen by decomposition faster in areas rooted by hogs, compared to litter in protected exclosures.

Trophic Structure and Energy Flow

Exotics can influence the energy flow pattern of ecosystems in both a top-down and a bottom-up fashion. Top-down influence refers to the effect that species at the high levels in the food chain can have on lower levels. Top-down influences are also said to "cascade" down the food web. Bottom-up influence is the effect that species can induce in basic nutrient and primary production processes that influence food chains upward from their producer base. These effects "climb up" the food web.

Top-down effects are clearly illustrated by the invasion of the sea lamprey into the Great Lakes (see Chapter 6). With the decimation of the native salmonids, the alewife and rainbow smelt filled the niches of native zooplankton feeders, and in the absence of larger predatory fish, they became enormously abundant. These fish, themselves exotics, altered populations of zooplankton, depleting those of larger species and promoting increased abundance of medium-size species (Wells 1970). The change in composition of zooplankton was correlated with the further decline of native plankton-feeding fish, presumably because of competition for food with the alewife and rainbow smelt (Stewart et al. 1981). Thus, the addition of a new top predator, the lamprey, cascaded down through the carnivore and planktivore levels to impact the composition of the zooplankton.

The fact that exotics can alter bottom-up influences on food-chain dynamics in ecosystems is illustrated clearly by exotic mollusks that have invaded aquatic ecosystems. In San Francisco Bay, for example, the Asian clam, an efficient filter feeder on phytoplankton, has reduced the summer phytoplankton biomass in parts of the bay to less than a tenth, and annual primary production to less than a fifth, of their former levels (Cloern and Alpine 1991) (see Chapter 5).

In lake ecosystems, exotic zebra and quagga mussels have caused even more profound change in production dynamics. Zebra and quagga mussels are filter feeders, and one of their most profound effects is the removal of planktonic material, including phytoplankton and particulate organic matter, from the water. In western Lake Erie, where an average density of 41,830 per square yard prevails, the mussel population filters the entire water volume of this part of the lake every ten days. In this portion of Lake Erie, filtration removes about 7.0 million short tons of phytoplankton, 5.5 million short tons of which are consumed by the mussels, and the rest is deposited in the bottom sediments as pseudofeces (material entrapped by the mussels' filtration system, but not ingested). This filtration removal was calculated to remove about 26 percent of the total primary production of western Lake Erie (Madenjian 1995). Impacts of zebra and quagga mussels on phytoplankton in the deeper central and eastern basins of Lake Erie, however, are minimal (Makarewicz et al. 1999).

Feeding by zebra and quagga mussels also influences nutrient cycling patterns in lakes. In experimental microcosms, mussels promote the transformation of nitrogen and phosphorus from particulate organic matter into soluble forms of these elements (James et al. 1997). In effect, they are speeding the regeneration of critical nutrients to plant producers. These experimental results are similar to the changes seen in western Lake Erie following invasion of zebra and quagga mussels (Holland et al. 1995).

The secondary effects of zebra and quagga mussels on the rest of the lake biota are striking. The most general effect has been to increase the importance of the benthic sector of the food web (Ricciardi et al. 1997). Mussel filtration clears the water of particulate material, depositing much of it in the bottom sediments. Populations of deposit-feeding invertebrates and their predators thus increase. Increased light penetration favors the growth of benthic algae and submerged vascular plants, and thus of the food chains that are based on these plants. As benthic producers increase their activity, they remove nutrients from the water, and phytoplankton production in the water column is decreased. Energy flow through open-water food chains leading to zooplankton and pelagic fish is thus diminished. The mussels themselves represent a major new benthic food resource. In addition, their feces and pseudofeces, deposited in the sediments, represent an increased organic food supply for other benthic invertebrates. These organisms are, in turn, exploited by benthic fish and ultimately by large piscivorous fish. In short, a major shift from pelagic to benthic energy flow occurs in lakes invaded by zebra and quagga mussels.

Similar effects occur in large rivers. In the Hudson River, for example, the invasion of zebra and quagga mussels has greatly increased the importance of the benthic food web and has encouraged the growth of macrophyte beds in shallow waters (Strayer et al. 1999).

In quite a different way, invasive aquatic plants such as water hyacinth, water lettuce, and hydrilla transform the pattern of energy flow of the lakes and rivers they blanket. These species can switch the balance of grazing and detritus food chains of the subsurface ecosystem. Because they are surface mat-formers, they profoundly change the physical and biotic conditions below the surface. They effectively eliminate open-water, plankton-based food webs. They block out light and reduce the oxygen supply to deeper waters, suffocating benthic plants and animals. The subsurface zone of the lake or stream becomes essentially a detrital ecosystem.

Some exotics have influences that are both top-down and bottom-up. Chapter 12, for example, noted that Argentine ants, as predators, greatly modify the ground invertebrate fauna on which they prey. In turn, they influence the abundance of higher-level species, such as horned lizards, that also feed on ants and other invertebrates.

A striking example of combined top-down and bottom-up influences was provided by the introduction of the opossum shrimp into the Flathead Lake drainage in Montana, with the objective of increasing the food supply for kokanee salmon, itself an exotic in the drainage (Spencer et al. 1991). Opossum shrimp were introduced into the drainage between 1968 and 1975 and appeared in Flathead Lake in 1981. Opossum shrimp, however, spend the day in deep waters and migrate to the surface to feed at night. Kokanee forage during the day in surface waters. The result was that the opossum shrimp decimated populations of smaller zooplankton on which the kokanee were dependent, and the kokanee population crashed. As a result of the kokanee decline, the wintering population of bald eagles at the lake declined as birds were forced to move elsewhere because of lack of fish. Zooplankton impacts similar to those in Flatland Lake have been observed in many of the more than 100 lakes in the northwestern United States and western Canada into which opossum shrimp have been introduced (Spencer et al. 1999).

The extreme influence of exotics on ecosystem function occurs in arid and semiarid regions, including many oceanic islands, where exotic herbivores promote the desertification of the terrestrial ecosystem. Browsing and grazing by goats on San Clemente Island, off the California coast, virtually stripped the island of herbaceous vegetation in the 1960s and probably caused the extinction of several species (see Chapter 12). Livestock grazing throughout the arid and semiarid West during the 1800s and early 1900s led to massive erosion that has altered both the eroded areas and the valley areas where deposition occurred for centuries to come (see Chapter 11).

Alteration of Disturbance Regimes in Ecosystems

Alteration of the disturbance regime in ecosystems may well be the most serious general impact of exotic species (Mack and D'Antonio 1998). This type of impact is perhaps best illustrated by the effects of exotic plants on fire frequency and intensity in terrestrial ecosystems. As noted earlier, exotic annual grasses such as cheatgrass have introduced fire into the cold deserts of the Columbia Plateau and Great Basin. Likewise, exotic perennial grasses such as buffel grass have done the same in the hot deserts of southern Arizona. In both cases, fire has been introduced to desert communities composed largely of fire-intolerant plants (see Chapter 10). The invasion of native riparian woodlands along western rivers has been promoted by saltcedar's flammability and its rapid recovery from fire (see Chapter 11).

Through this influence on fire, exotic grasses can alter the composition even of fire-adapted forests and shrublands. As shown in the introductory example, dense stands of cogon grass in Florida burn hotter and higher, reducing the survival of young longleaf pines. These changes all favor increase in abundance of cogon itself. Thus, the long-term fate of invaded areas is conversion into a monospecific cogon grassland. In California, one of the common practices following wildfire has been to seed burned areas with exotic annual ryegrass. In San Diego County, California, for example, a wildfire burned nearly 5,000 acres on Otay Mountain in August, 1979 (Zedler et al. 1983). The burn was seeded with annual ryegrass, which germinated and grew during the winter and spring, producing a dense stand. The ryegrass died during the summer, creating a highly flammable ground layer. An arson fire in July 1980 reburned much of the 1979 fire area. The effect of two fires in less than a year was to virtually eliminate obligate seed-reproducing shrubs, such as coast blue lilac (*Ceanothus oliganthus*), by killing their seedlings. Some shrubs that normally recover by resprouting from the subsurface root crown, including mission manzanita (*Xylococcus bicolor*), were also greatly reduced. The result of short-interval fires like this, promoted by exotic annual grasses, has been the shift of chaparral vegetation to a less woody coastal sage scrub vegetation in many areas.

In Hawaii, broomsedge, bushy beardgrass, and molasses grass have invaded native seasonal woodlands in the Hawaii Volcanoes National Park

(Hughes et al. 1991). The original ʻoʻhia (*Metrosideros polymorpha*) woodlands had a sparse herbaceous understory that burned only occasionally and in small areas. Bushy beardgrass and broomsedge, invading these woodlands, form a much more continuous cover and grow upward into the crowns of shrubs. The invaded woodlands burn more frequently and over larger areas. Burning also favors the invasion of molasses grass, an even more flammable species. The change in frequency and the extent of fires create a positive feedback system that gradually converts native woodland into exotic grassland.

Several of the exotics discussed here influence several ecosystem processes. As noted at the chapter outset, cogon grass in the Florida sandhill ecosystem provides a good example (Lippencott 1997). Cogon outcompetes other herbaceous species by shading them and reducing their germination by forming a thick litter layer. It depletes soil moisture, increases fire intensity, reduces longleaf seedling survival, and reduces populations of burrowing invertebrates. The multiple impacts caused by exotic animals ranging from feral hogs to zebra mussels were also noted earlier in this chapter.

Alternative Stable States

Species with such multiple impacts are, in effect, "superkeystone species" whose effects can be nearly permanent. Superkeystone exotics, which usually differ markedly in form and function from native species, can reorganize the ecosystem so that the change is irreversible. In other words, they shift the system to a new, stable, structural and functional state. Ecologists now realize that in many regions ecosystems exist in one of several alternative stable states (Laycock 1991). In a particular state, such as grassland, a slight disturbance or short-term stress may modify the structure of the system, but recovery is possible. Greater disturbance or stress, however, may push the ecosystem across a threshold, so that the ecosystem moves to a different state, such as shrubland. This new state persists, even if the disturbance or stress is removed.

In some regions of North America, alternate stable states composed of native species can be recognized, an example being desert grassland and desert scrub in parts of southern Arizona. In other regions, invading species have created new stable states, dominated by exotics. In California, as noted earlier, the native valley grassland has been displaced by Mediterranean annual grassland (see Chapter 12). Similarly, in the northern intermontane region, sagebrush and perennial bunchgrasses have been replaced by cheatgrass and other exotic annuals and perennials from the

Old World (see Chapter 10). In these cases, the entire ecosystem has shifted to a new set of dynamics: altered soil moisture and nutrient regime, changed patterns of production and decomposition, new biotic interactions, modified fire cycle, and an exotic-dominated community adjusted to these factors.

Alternate stable states induced by exotic invaders are also exemplified by the Chinese tallow forests that have replaced coastal prairie in Texas, Australian paperbark stands that have displaced marshland and cypress swamps in the Everglades region of Florida, and the saltcedar and Russian olive jungles that are replacing cottonwood and willow riparian woodlands in the West.

Alternate stable states can also be induced in aquatic ecosystems by the introduction of exotics. Undisturbed north temperate lakes tend to have oligotrophic characteristics. They have abundant populations of piscivorous fish that limit populations of planktivorous fish. As a result, these lakes have abundant zooplankton that, in turn, graze phytoplankton heavily. Thus, algal blooms are rare, water transparency high, and rooted macrophytes abundant. A variety of factors can alter this system (Ludwig et al. 1997), one being any factor that decimates the piscivorous fish that control the system in top-down fashion. As described earlier, the sea lamprey invasion of the Great Lakes did just this. Loss of piscivorous fish allows planktivorous fish to increase in abundance, reducing zooplankton and their grazing on phytoplankton. Blooms of algae result, water clarity declines, and submerged macrophytes are shaded out. These are characteristics of eutrophic lakes and constitute a state that is stable as long as piscivore populations remain low. Thus, the exotic sea lamprey probably contributed significantly to the eutrophication of the lower Great Lakes during the mid-1900s.

Significance of Ecosystem Impacts

Although ecosystem effects of exotics are probably the most profound and long lasting, ecologists have barely begun to understand the extent of such impacts. In Florida, for example, Gordon (1998) found that twelve to twenty of the thirty-one most invasive plants had potential ecosystem impacts. Thus, the impacts discussed in this chapter are simply the most obvious. In fact, all exotics probably have some ecosystem impact. Thus, the many exotics that are considered to be without detrimental impact—90 percent of established exotics, according to the tens rule—should not be dismissed as totally harmless. In all probability, we have simply not been able to assess their real impacts.

The transformation of ecosystems to alternate states poses enormous difficulties for management and restoration of invaded ecosystems. Keystone exotics test the resilience of natural systems, their ability to recover from disturbance or stress. In some cases, even if the exotic culprit can be controlled, many of the changed ecosystem conditions may persist for decades or perhaps indefinitely. Where basic aspects of the physical environment have been altered, this is likely to be true. In other cases, removal or reduction of the exotic may allow rapid recovery of the ecosystem to near its original condition. Where an exotic has exerted strong top-down or bottom-up control over biotic structure, this may be the response.

Keystone exotics can have still more profound impacts. As shown in Chapter 19, alteration of basic ecosystem conditions also changes the rules of the evolutionary process. Evolutionary responses of both natives and exotics occur not only in response to the direct influence of species on each other, but also to the changed conditions of the ecosystem induced by exotics. These responses mean that exotics, and especially those with strong ecosystem effects, have changed North American biotas in ways that can never be undone.

Chapter 19

Exotics and Evolution: Assimilation or Conquest?

> Once established, however, there is evolution of the invader over time, with genetic changes taking place both in the invader itself and in all likelihood also in the organisms on which it has an impact. This evolution takes place over longer time scales than the process of establishment itself, although it should be realized that the scales are relative to the generation times of the particular organisms.
>
> —Scott Carroll and Hugh Dingle, "The Biology of Post-Invasion Events"*

Ribwort plantain was one of the early weeds brought to North America by European settlers (Foy et al. 1983) (Table 19.1). This common perennial weed of lawns, fields, parks, and open woodlands is an obligately out-crossing plant; its individuals must be cross-pollinated. It only rarely reproduces by vegetative means. It is genetically variable in many characteristics, and populations in North America have been found to differ in many morphological and physiological characteristics (e.g., Tetramura and Strain 1979), indicating evolutionary adjustment to local physical and biotic conditions of habitat. This genetic adaptability has probably been a major factor in its spread throughout almost all of temperate North America. Ribwort plantain is an invader that is continually adapting to new conditions.

---

*Scott Carroll and Hugh Dingle (1996). "The Biology of Post-Invasion Events." *Biological Conservation* 78:207–214.

Table 19.1. Exotics involved in evolutionary change after introductions to North America.

PLANTS	
Asian buckthorn, *Rhamnus utilis*	Asia
black knapweed, *Centaurea nigra*	Eurasia
brown knapweed, *Centaurea jacea*	Eurasia
bulbous bluegrass, *Poa bulbosa*	Europe
California cordgrass, *Spartina foliosa*	native
cheatgrass, *Bromus tectorum*	Eurasia
common dandelion, *Taraxicum officinale*	Europe
Dalmatian toadflax, *Linaria dalmatica*	Eurasia
European buckthorn, *Rhamnus cathartica*	Europe
golden rain tree (small fruit), *Kolreuteria elegans*	Asia
golden rain tree (large fruit), *Kolreuteria paniculata*	Asia
heartseed vine, *Cardiospermum halicacabum*	South America
hybrid honeysuckle, *Lonicera x bella*	Eurasia*
Japanese chestnut, *Castanea crenata*	Asia
leafy spurge, *Euphorbia esula*	Eurasia
oriental white oak, *Quercus acutissima*	Asia
ribwort plantain, *Plantago lanceolata*	Europe
rose clover, *Trifolium hirtum*	Europe
slender wild oat, *Avena barbata*	Europe
smooth cordgrass, *Spartina alterniflora*	eastern North America
white poplar, *Populus alba*	Europe
witchweed, *Striga asiatica*	Asia
yellow star thistle, *Centaurea solsticialis*	Europe
AQUATIC INVERTEBRATES	
zebra mussel, *Dreissena polymorpha*	Europe
INSECTS	
soapberry bug, *Jadera haematoloma*	native
FISH	
American shad, *Alosa sapidissima*	eastern North America
BIRDS AND MAMMALS	
house sparrow, *Passer domesticus*	Europe
DISEASE ORGANISMS	
Dutch elm disease, *Ophiostoma ulmi*	Asia

*Both parental forms from Eurasia.

The ribwort plantain story has another important side. In parts of the high country of western Nevada, a native butterfly, Edith's checkerspot (*Euphydryas editha*), has evolved to prefer the exotic ribwort plantain over its native host, a herbaceous wildflower known commonly as Chinese houses (*Collinsia parviflora*). In 1982, in one location, only 5 percent of individual checkerspot butterflies preferred ribwort plantain, but by 1990, the preference had increased to nearly 65 percent (Singer et al. 1993). This change was not just due to increased plantain availability. The altered preference was shown even by butterflies reared in the laboratory from butterfly stock obtained from the field in 1983 and 1993. In nature, as well, the fraction of butterflies laying eggs on the two plants changed in the same fashion over the nine-year period. In 1990, several checkerspots were found that refused to oviposit on Chinese lanterns, even when confined with this original host plant for several days. In nearby locations, where ribwort plantain has never been present, however, checkerspots showed no preference for plantain over Chinese houses. Thus, some populations of this native butterfly have evolved preference for, and even dependence on, an exotic plant host. Exactly when ribwort plantain first reached the mountain meadows inhabited by Edith's checkerspot is uncertain, but this time period is probably on the order of only 100–150 years ago, or 100–150 butterfly generations.

Evolutionary Responses of Exotics and Natives

The plantain–checkerspot example shows, first of all, that in a vast new region an exotic species such as ribwort plantain will likely encounter habitats to which it is not already well adapted as a result of evolution in its native region. It may also fail to find habitats to which it is adapted. Thus, given appropriate genetic variation, selection will tend to adapt the exotic to the physical and biotic conditions of the new array of habitats. Physical conditions such as climate and soils will differ to some extent from the conditions to which the exotic was adapted in its native region. The exotic may be freed of many of its enemies, and thus selection may tend to eliminate specific defenses against them. On the other hand, new predators, parasites, and disease agents may attack it, and selection will favor adaptations to counter these enemies. The exotic may encounter new competitors, leading to evolutionary adjustment to them. It is likely that ribwort plantain has adjusted to many of these physical and biotic conditions of its new habitats.

Second, the example shows that an exotic species, as a result of the food web relationships it establishes, acts as a selective force on the other species with which it interacts. These species might be prey or predators, hosts or parasites, disease agents, or competitors. Consequently, these native species make evolutionary adjustments to an invading exotic, enabling them to utilize it to their benefit or to reduce its negative impacts on them. The sum total of such adjustments by members of the native community is termed counteradaptation. Ribwort plantain has experienced counteradaptation by Edith's checkerspot butterfly, which has evolved to feed on it.

Earlier chapters have presented many examples of exotics that flourish and spread rapidly in North America in part because they have been freed from diseases, parasites, predators, and competitors in their native region. These species have, in fact, entered North American communities whose members lack counteradaptation to them.

Thus, evolutionary processes, both on the part of an exotic and of the species with which it interacts, are set in motion when an exotic enters a new region (Carroll and Dingle 1996). Evolutionary theory thus suggests that, in time, exotics will be integrated into the community, and their most negative influences tempered. As an exotic species adapts to its new physical and biotic environment, members of the invaded community evolve counteradaptations to it.

Over long periods of evolutionary time—centuries to millennia—the processes of invasion, adaptation, and counteradaptation may create a pattern known as the taxon cycle (Wilson 1961). A taxon cycle begins when a species invades a new geographical region and spreads through it, aided by the lack of counteradaptation by native species. Local populations of the invader then gradually become differentiated, as selection adjusts them to local physical and biotic conditions. As time goes on, the invading species may become increasingly specialized, and some of its populations eliminated by counteradaptation and changing habitat conditions. Ultimately, the original invader may become a highly specialized form with a restricted distribution. Taxon cycles can be discerned in many kinds of organisms that have invaded new regions. They have most often been seen in taxa that have invaded island archipelagos (Ricklefs and Cox 1972), but they can also be seen in taxa expanding from tropical centers of origin into higher latitudes (Cox 1985).

Genetic Variability in Exotic Species

What is the evidence that processes of adaptation and counteradaptation are at work in the relatively short period of evolutionary time that humans have become active agents of dispersal of species to new areas? To consider this question, we shall first examine patterns of hereditary variability shown by exotic species. Then we shall look at patterns of evolutionary change shown by exotics and the species with which they interact.

Evolutionary adaptation depends on the availability of hereditary variation on which selection can act. Hereditary variation is created by a variety of processes that affect the genetic constitution of an organism. These include mutation, recombination, chromosomal changes, and hybridization. Protein electrophoresis and several techniques for detecting variability in mitochondrial, chloroplast, and nuclear DNA now make it possible to assess genetic variability within populations of exotics. These are now providing insight into the degree of genetic variability of invaders and how this relates to their success.

Analyses of the genetics of several exotic plants show diverse patterns of variation within and between populations. Some exotics are primarily self-pollinating, apomictic, or clonal and usually show little intra- or interpopulation variability. For such species, the level of genetic variability appears to depend largely on the number of introductions to North America and the interpopulation variability in their source areas. Others are largely outcrossing and show varying degrees of genetic variability. For such species, as well, genetic variability is often enhanced by multiple introductions from different source populations.

Several examples illustrate these patterns of genetic variability. Witchweed, for example, is a self-pollinating root parasite of several cultivated grain crops. It is widespread in Asia and Africa, where it shows several genetic races (Werth et al. 1984). In North America, where the plant has been known since 1956, individuals from two populations were found to be monomorphic for all thirty-two genetic loci examined. This suggests that witchweed was introduced from a single source area in the Old World and that genetic variability is likely to be slow to appear in North American populations.

Cheatgrass, another self-pollinating annual plant, also shows low genetic variability in North America. In the Old World, cheatgrass occupies a wide range from western Europe and North Africa east to Tibet and Pakistan. Over this enormous range, more genes are polymorphic and more alleles exist per gene than in North America. Within individual populations in the Old World, however, variability is low, so that most of this genetic variation reflects differences in the genetic structure of populations in different regions (Novak and Mack 1993). In North America, fewer genes show polymorphism, and the number of alleles per gene is less than in the Old World. Variability is low both within and among North American populations (Novak et al. 1991). Examination of where particular alleles appear in North America and the Old World suggests that North American cheatgrass was established from five or six independent introductions from locations ranging from Spain to Afghanistan (Novak et al. 1993). The success of cheatgrass in North America seems to reflect its broad adaptation to disturbance and phenotypic plasticity, rather than evolutionary adaptability.

Common dandelion, an apomict, shows a number of clonal forms, several of which may occur in a given population (Lyman and Ellstrand 1984). These forms are not variable and appear to represent independent introductions from the Old World. Their frequencies do not correlate with environmental differences throughout their enormous North American range. Again, phenotypic plasticity is probably responsible for the success of this species in a wide range of habitats.

A different pattern is shown by bulbous bluegrass, a species introduced into the United States in 1906, probably as a contaminant of alfalfa seed (Novak and Welfley 1997). It is now widespread in temperate North America. Although the North American populations are clonal, reproducing primarily by asexually produced bulblets, they show a high percentage of polymorphic loci, nearly 28 percent, and high heterozygosity, about 20 percent. This pattern is also consistent with multiple introductions.

For outcrossing species, the pattern is usually quite different; genetic variation within populations is usually high. Rowe et al. (1997), for example, examined genetic variation in five North American populations of leafy spurge, an outcrossing species that also reproduces vegetatively. High genetic variability characterizes this aggressive exotic. Analysis of chloroplast DNA, which is maternally inherited and does not experience recombination during sexual reproduction, revealed thirteen types, of which one to several occurred in individual populations, much like the case with the common dandelion. In contrast, nuclear DNA analyses revealed that each of the 123 individuals studied was unique. Whether this high genetic variability reflects a high variability in the source population in Russia or the Ukraine, or is the product of multiple introductions, is uncertain. High variability in source material might have introduced many chloroplast types to an original establishment area, with these having been sorted out by selection as the plant spread across North America. Spurge seed also might have been brought to different North American sites from different Old World sources. Of course, some combination of both patterns is also possible.

Yellow star thistle, a primarily outcrossing annual, also shows a high degree of genetic variability. Analysis of twenty-two populations from California, Washington, and Idaho showed that 56 percent of all loci were polymorphic, with polymorphism within local populations averaging 43 percent (Sun 1997). Heterozygosity within populations was also high, averaging 38 percent. Despite the wide geographical range of the populations sampled, little divergence was apparent in their genetic makeup. Yellow star thistle has been established in western North America for 100 years or more. Thus, it may possess a "general purpose" genotype that enables success through phenotypic plasticity rather than evolutionary adaptation.

Similarly, slender wild oat, a grass that is primarily outcrossing, shows a pattern of genetic variability similar to that seen in the Old World (Clegg and Allard 1972). All seven loci examined for individuals from sixteen populations in California were polymorphic, as were these loci in nine populations from the Mediterranean region of the Old World. Several specific alleles differed, however, between the California and the Old World populations.

Genetic Variability Related to Habitat Conditions

In the case of slender wild oat, however, genetic variability correlates with environmental conditions in North America. In California, populations of slender wild oat fall into two groups: one occupying the hot, dry Central Valley and southern coast, the other the milder Sierran foothills and coast ranges (Clegg and Allard 1972). The Central Valley populations show nearly complete fixation at all seven loci, whereas the foothill and coast range populations show greater variability and often a predominance or fixation of different alleles. These patterns can only reasonably be interpreted as resulting from selection for adaptation of the different climates of the Central Valley and surrounding foothills and coastal mountains. Although the species was first introduced to California very early, perhaps 400 years ago, it has certainly been reintroduced many times, and the average age of populations is probably of the order of 150 years.

Rose clover, an outcrossing legume, shows a closer relationship between genetic variability and habitat characteristics (Jain and Martins 1979). This clover was introduced to California as a range plant in 1946. More recently, rose clover has begun to invade roadsides, suggesting that adaptive change has occurred. Comparison of rangeland and roadside colonies of rose clover showed that the roadside populations tended to have high genetic variability and increased outcrossing rates. The roadside colonies also showed higher germination rates, earlier flowering, and a greater allocation of biomass to reproduction. The size and morphology of the calyx also tended to be different for roadside plants. Thus, subtle genetic changes seemed to have occurred in rose clover in only shortly over thirty years, increasing the invasiveness of some populations.

Hybridization and Variability of Exotics

Hybridization is a process that may restructure the genetics of invasive exotics. Introduced species frequently encounter close relatives, either natives or other exotics, with resultant hybridization. In several cases, invasive "species" may actually be forms derived from hybridization in nature. In other cases, hybrids may be created deliberately by plant and animal breeders, and the hybrid forms introduced into the wild. Thousands of plant hybrids are created annually by plant breeders, and many animal hybrids are also created. Mungall and Sheffield (1994), for example, list 143 ungulate hybrids that have produced living young. A number of these hybrids are commonly raised on Texas ranches, and some of these enter free-ranging populations.

Hybrids among exotics brought to North America from other continents are common. For example, many of the shrub honeysuckles invading northeastern forests are hybrids between two Eurasian species (*Lonicera tatarica* and *L. morrowi*) (Woods 1993). European and Asian buckthorns have hybridized near Ann Arbor, Michigan, to create a highly variable, fully fertile population of invasive shrubs (Gil-ad and Reznicek 1997). Dalmatian toadflax hybridizes with a number of other species in its genus (Vujnovik and Wein 1997). Black and brown knapweeds hybridize to produce a fertile form, known as meadow knapweed, that disperses more effectively and is weedier than either parent (Roché and Roché 1991). The extremely invasive leafy spurge in North America may, in fact, be a hybrid between true leafy spurge and another Old World spurge (Louda and Masters 1993).

Hybrids between exotics and North American natives are probably less frequent, but perhaps more serious in their impacts. In some cases, such hybrids may displace their North American parent. The exotic white poplar, for example, hybridizes with big-toothed and trembling aspens (*Populus grandidentata* and *P. tremuloides,* respectively), the progeny being clonal trees that spread aggressively (Spurr 1980).

Another hybridization between an exotic and a native has resulted from the introduction of smooth cordgrass to various Pacific coast estuaries, including San Francisco Bay (see Chapter 5). Recent studies (Daehler and Strong 1997b) show that it hybridizes with the native California cordgrass. Smooth cordgrass produces much more pollen than California cordgrass, and the seed set of California cordgrass plants pollinated by smooth cordgrass is about eightfold that of plants pollinated by their own species (Anttila et al. 1998). The hybrids appear to be intermediate in size between their parents, but they are not easily distinguished by morphology. They recruit abundantly from seed and show vigorous clonal spread. The hybrids also appear to grow lower in the intertidal zone than does California cordgrass, thus contributing to the loss of open mudflats that are important to many native organisms and migratory birds. Spread of smooth cordgrass to other locations along the Pacific coast is likely to expand the region of hybridization of these species.

Other cases of hybridization between exotics and natives involve aquatic animals. Electrophoretic study of enzyme variants in the water flea *Daphnia galeata* in the Great Lakes, for example, has revealed that European forms of this species had been introduced and that extensive hybridization has resulted (Taylor and Hebert 1993). As noted in Chapter 5, exotic and native forms of *Mytilus* species are now hybridizing in bays and estuaries on the Pacific coast.

Introductions from one part of North America to another may also lead to hybridization with local plants and animals. The danger here is not so much the creation of a new, invasive pest as the extinction of the local species by genetic swamping (Rhymer and Simberloff 1966). Hybridization of this sort was implicated as a factor in the extinction of 38 percent of the species or subspecies of fish that have disappeared in North America (Miller et al. 1989). As noted in Chapter 7, extensive hybridization is occurring between introduced and native pupfish in the Pecos River in Texas. Levin et al. (1996) also noted that hybridization with introduced plants can lead to the extinction of rare endemic plants.

In addition, hybrid fish have been created by captive breeding and introduced into new waters, where they may act as competitors for or predators on native faunas. At least nineteen such hybrid exotics are known (Boydstun et al. 1995). Chapter 7 noted the high incidence of hybrid fish that have been introduced into western streams and rivers.

These complex patterns of hybridization are increasing in frequency as exotics spread and increase in abundance. They insert a "wild card" variable in the evolutionary process. Hybridization provides a rich source of genetic variability for invasive exotics. In some cases, it may enable new invasive forms, such as meadow knapweed, to appear suddenly. In other cases, it can so alter the genetics of a native species, such as the Pecos River pupfish, that it effectively becomes extinct.

Hereditary Variability and Adaptive Change

Thus, hereditary variability of various sorts clearly exists in North American populations of many exotic species, because of both intrinsic genetic variability and hybridization. In several cases, this variability correlates with environmental variation. Furthermore, although evolutionary adaptation is often quite slow, several clear examples show that adaptive change has occurred since the introduction of species to North America.

In several instances, exotics have shown adaptive change to habitat conditions in their new North American ranges. The house sparrow, for example, has apparently shown an evolutionary response to selection over its extremely broad climatic range in North America (Johnson and Klitz 1977). Populations show a latitudinal gradient from large body size in northern localities to small size in southern areas. This pattern follows Bergman's Rule—increasing body size of warm-blooded animals in colder regions—and is usually interpreted as adaptive in aiding body temperature regulation. This gradient appears to have a strong genetic component. Among plants, the patterns of genetic adjustment by slender wild oat and rose clover, described earlier, are probably adaptive to the different climates and soil conditions of sites they occupy.

Similarly, the American shad, which is native to rivers in eastern North America, has been introduced to several river systems along the Pacific coast (see Chapter 7). Shad are anadromous fish that live in coastal waters and estuaries, entering the rivers, to spawn. Demographic differences, probably with a genetic basis, now exist between East Coast and West Coast populations. In their native eastern rivers, the fish are slightly longer lived, with only 20–40 percent of individuals spawning in more than one year. In Pacific rivers, the fish are shorter lived, but 32–77 percent spawned in more than one year, so that their lifetime fecundity was considerably higher than that of eastern fish (Dingle 1980).

Other patterns of evolutionary adaptation by exotics are related to their new biotic environment. Exotics may evolve in their ability to use new hosts. One well-documented example involves Dutch elm disease. The Dutch elm disease fungus has not only killed many native elms, it has also evolved more virulent strains, some of which have been reintroduced to Europe. These strains also pose a growing threat to other native North American elms. Other examples may involve phytophagous insects introduced to North America for biological control. About 16 percent of such species also attack native, nonpest species (Hawkins and Marino 1997), although whether or not genetic changes have aided such host shifts is uncertain.

Evolutionary adjustments may also occur when an exotic escapes many of its natural enemies. Selection may then favor the reduction of energy-expensive defenses and the allocation of energy to other activities, such as growth and reproduction. Populations of smooth cordgrass in Pacific coast estuaries, for example, now show differential evolutionary adjustments to release from herbivory (Daehler and Strong 1997b). In Willapa Bay, Washington, the smooth cordgrass population has been free of herbivory from the spartina leafhopper for at least a century. The San Francisco Bay population, introduced from Maryland about twenty years ago, comes from an area with spartina leafhoppers and still coexists with them. Greenhouse tests of the growth of smooth cordgrass plants from these three populations show that Maryland and San Francisco Bay plants are little affected by leafhopper herbivory. Plants from Willapa Bay, however, produced only 30 percent of normal biomass at the end of the first year of exposure to leafhoppers, and only 12 percent after two years. After two years, in fact, over a third of the Willapa Bay plants had died. Although the founder plants in the Willapa Bay population might have been more attractive to leafhoppers, experimental results indicated that they had lost a degree of herbivore tolerance, as well as gaining characteristics making them attractive to leafhoppers.

A similar pattern can be seen in purple loosestrife. Blossey and Notzold (1995) compared the growth of plants from seed obtained near Ithaca, New York, and a location in Switzerland. After one growing season, the Ithaca plants were four to five times greater in biomass than those from Switzerland. Furthermore, when root-feeding beetle larvae were grown on plants from the two locations, their survival was more than three times higher on the Ithaca plants, and their maximum body mass nearly 30 percent greater. Thus, it appears that the Ithaca plants had reduced their allocation to herbivore defense and had increased the allocation to growth.

Exotics and Evolutionary Change in Native Species

Exotics also influence the evolution of species with which they interact. Vermeij (1982), for example, found that invasion of the East Coast of North America by the predatory green crab led to an evolutionary change in the shell of the native dog whelk (*Nucella lapilla*). Dog whelks that co-occur with green crabs develop shells that are slenderer and heavier and that possess a smaller, thicker-lipped opening than do whelks in areas not invaded by green crabs. That these changes are adaptive is indicated by the fact that scarred whelk shells—evidence of attempted predation—have increased in frequency following appearance of the crab. The adaptive response by the dog whelk is believed to have been favored by the fact that this mollusk has larvae that are locally dispersed. Thus, local selection was able to influence the characteristics of the subsequent generation of settling whelks.

Counteradaptation by Natives

The long-term result of these evolutionary interactions, together with ecological adjustments, appears to be the establishment of counteradaptation by members of the native community. As shown in the introductory example, herbivorous native species such as Edith's checkerspot butterfly evolve to use exotic host plants. Native predators and parasites evolve and learn to utilize exotic herbivores. Native species acquire improved defenses and competitive abilities. This suggests that many exotic species will eventually be integrated into the community as "well-behaved" members.

One of the clearest examples of counteradaptive evolution by a native species to exotic invaders is provided by the soapberry bug, a true bug (Hemiptera) whose North American host is the soapberry tree (*Sapindus saponaria*) in the south-central United States and the balloon vine (*Cardiospermum corindum*) in Florida (Carroll and Dingle 1996). The bug uses its tubular beak to pierce the wall of the fruits of these species and feeds on the seeds, which it liquifies and sucks out. The length of the beak is thus related to the thickness of the fruit wall. Populations of the soapberry bug living on soapberry and balloon vine differ in average beak lengths (Table 19.2) in a fashion that corresponds closely to the difference in average radius of the fruits of these two plants.

Table 19.2. Radii of native and exotic sapindaceous fruits and beak length of the soapberry bugs feeding on them.

Fruit	*Location*	*Fruit Radius (mm)*	*Beak Length (mm)*
NATIVES			
soapberry	SC USA	6.05	6.68
balloon vine	FL	11.92	9.32
EXOTICS			
golden rain tree (small fruit)	FL	2.82	6.93
golden rain tree (large fruit)	SC USA + FL	7.09	7.23
heartseed vine	SC USA + FL	8.54	7.80

Three species of exotic plants of the family Sapindaceae have been introduced to the south-central United States and Florida, and soapberry bugs have found them suitable hosts. On the small-fruited golden rain tree, which occurs in Florida, soapberry bugs show beak lengths slightly over 25 percent shorter than those of bugs feeding on Florida balloon vines. On a second species of golden rain tree, which has larger fruits than those of the soapberry, the bugs show an average beak length about 8 percent greater than those on soapberry. On the heartseed vine, with fruits still larger, the bugs show average beak lengths almost 17 percent greater than those on soapberry. Laboratory breeding experiments show that beak length is inherited, so that rapid evolution of differences in beak lengths has evidently occurred. Populations of soapberry bugs living on the exotic small-fruited golden rain tree also produce smaller eggs and show greater fecundity on this host than on the native balloon vine, indicating that genetic differences in life-history traits have also evolved (Carroll et al. 1998).

Counteradaptation, both through evolution and the establishment of ecological interactions, is also shown by the numerous cases of rapidly increased use of exotic plants by native herbivorous insects and of exotic herbivorous insects by native parasitoids. Ecological interactions are often established quickly, perhaps in just a few generations, but these interactions are the basis for evolutionary changes over hundreds to thousands of generations.

The native insects that first attack exotic plants tend to be generalist plant feeders, and their accumulation is often rapid. Sugarcane (*Saccharum officinarum*), for example, has been introduced to many parts of the tropics and subtropics at various times. The number of herbivorous arthropods now associated with the plant bears no relation to the date of introduction but is strongly related to the area of sugarcane cultivation (Strong et al. 1977). This suggests that herbivores that are preadapted to feed on the plant colonize it very quickly. In another case, involving exotic trees and native herbivores, Auerbach and Simberloff (1988) examined the accumulation of leaf-mining insects on exotic trees in northern Florida. In 1967, two exotic trees, Japanese chestnut and oriental white oak, were introduced to a locality well isolated from other localities at which these species already occurred. Within less than twenty years, both exotics acquired a fauna of leaf-reproducing miners nearly as diverse as that on the native black oak (*Quercus nigra*).

The similarity of an exotic plant to species in its new environment is also an important consideration in its acceptance by native herbivores. Surveys of the accumulation of leaf miners on exotic trees indicate that most come from closely related native trees (Connor et al. 1980). The taxonomic distinctiveness of an exotic plant strongly influences how fast it accumulates herbivorous insects. Eucalyptus trees in California, for example, have apparently not been attacked by any native herbivores, although they have in Costa Rica (Strong 1979).

Exotic insect herbivores also accumulate native predators and parasites. Cornell and Hawkins (1993) analyzed the parasitoid complexes of eighty-seven species, both in their native regions and in regions where they were exotic. On average, exotics had accumulated about four species of parasitoids, whereas natives had nearly eight parasitoids. However, some exotics had accumulated more parasitoids than attacked them in their native region. Some insect herbivores were simply more vulnerable to parasitoids and were attacked by many species both in regions to which they were native and in those to which they were exotic. For exotic herbivores, the number of parasitoids tended to increase with time since colonization, which ranged from 1 to 150 years, but this effect was relatively weak.

Close evolutionary relationships between native species and exotics require longer periods, but they do appear. In Hawaii, for example, at least five species of moths have evolved specific feeding relationships with the banana plant (*Musa* sp.), introduced 1,000 or more years ago by Polynesian colonists (Zimmerman 1960). These moths are related to species that occur on native palms. Edith's checkerspot butterfly and the soapberry bug examples discussed earlier are incipient cases of such evolution.

Ecological assimilation is also suggested by the boom-and-bust pattern frequently shown by exotics. Exotics often show enormous outbreaks, sometimes immediately upon introduction and sometimes only after a lag period. The outbreak state is often not maintained, however, and the exotic may decline in abundance and impact. In some cases, as in the introduction of herbivorous mammals to predator-free islands, the boom-and-bust pattern results from population growth to the point that food resources are exhausted. In other cases, it may result from ecological and evolutionary adjustments of the native biota.

The spectacular invasion of North American freshwaters by zebra mussels (see Chapter 6), for example, has led to their utilization by a number of native fish, waterfowl, and even blue crabs. Other invertebrates, such as freshwater sponges, utilize the mussels as attachment substrates, sometimes overgrowing and killing them. Thus, the enormous early densities shown by zebra mussels are likely to decline as other members of the freshwater biota adapt to them.

Over the Short Term and the Long Term

Thus, members of the native biota make both ecological and evolutionary adjustments to exotics. Although ecological adjustments can be rapid, evolutionary adjustments are slow, requiring hundreds to thousands of generations. In time, the prognosis for assimilating exotics into the ecological community is good. Ultimately, the richness of species and diversity of evolutionary and ecological interactions may even be enhanced by the human-aided dispersal of organisms to new biogeographic regions. This long-term gain, however, is likely to come at high short-term costs: an accelerated extinction of native species and the disruption of ecosystem function over many future human generations.

Working as agents of dispersal, humans have clearly speeded up the pace of evolution. Introductions of exotics are contributing to a global mass extinction that will likely rival the mass extinctions of the geological past. Counteradaptation by native species is occurring and is aiding in the assimilation of exotics into North American ecosystems. To protect our distinctive biota and ecosystems, however, will require greatly improved scientific understanding of the invasion process, refined technological ability to control exotics, and an expanded public will to address the invasion of exotic species.

Part V

Policy Perspectives

Chapter 20

Living with Exotics: The Ecological Economics of Exotics

> I doubt that we will ever arrive at a concrete measure of the benefit of the biological control of any weed. However, we can say that the plant used to be a pest and that we are not bothered by it anymore. The plaque erected by ranchers in the west to commemorate the control of Klamath weed by imported leaf-feeding beetles is one of the best indications of a successful project.
>
> —John Drea, Jr., "The Philosophy, Procedures, and Cost of Developing a Classical Biological Control of Weeds Project"*

Klamath weed, a broad-leaved, perennial herb of the Saint John's wort genus, was introduced to North America from Europe in 1793, and it reached California and British Columbia in the late 1800s (Groves 1989) (Table 20.1). Later, it was introduced to Hawaii. It proved to be an extremely invasive range weed, and one toxic to livestock, as well. It reproduces sexually, each plant producing about 30,000 seeds, but also spreads vegetatively. By the 1940s, nearly 5 million acres of rangeland in the northwestern United States and western Canada were seriously infested with Klamath weed. In Hawaiian rangelands, it also became a serious

*John Drea, Jr. (1991). "The Philosophy, Procedures, and Cost of Developing a Classical Biological Control of Weeds Project." *Natural Areas Journal* 11:143–147.

Table 20.1. Scientific names and native regions of exotics with major economic and ecological impacts.

PLANTS	
Australian paperbark, *Melaleuca quinquenervia*	Australia
Brazilian pepper, *Schinus terebinthefolius*	South America
diffuse knapweed, *Centaurea diffusa*	Eurasia
Eurasian water milfoil, *Myriophyllum spicatum*	Eurasia
Klamath weed, *Hypericum perforatum*	Eurasia
lantana, *Lantana camara*	Central America
leafy spurge, *Euphorbia esula*	Eurasia
puncturevine, *Tribulus terrestris*	Eurasia
saltcedar, *Tamarix* spp.	Eurasia
spotted knapweed, *Centaurea maculosa*	Eurasia
water hyacinth, *Eichhornia crassipes*	South America
AQUATIC INVERTEBRATES	
zebra mussel, *Dreissena polymorpha*	Eurasia
TERRESTRIAL INVERTEBRATES	
giant African snail, *Achatina fulica*	Africa
rosy wolfsnail, *Euglandina rosea*	Florida, USA
INSECTS	
Argentine ant, *Iridomyrmex humilis*	Argentina
boll weevil, *Anthonomus grandis*	Mexico, Central America
cottony cushion scale, *Icerya purchasi*	Australia
European spruce sawfly, *Diprion hercyniae*	Europe
flowerhead weevil, *Rhinocyllus conicus*	Europe
gypsy moth, *Lymantria dispar*	Eurasia
Japanese beetle, *Popillia japonica*	Asia
Mediterranean fruit fly, *Ceratitus capitata*	Europe
winter moth, *Operophtera brumata*	Europe
FISH	
Chinese grass carp, *Ctenopharyngodon idella*	Asia
European ruffe, *Gymnocephalus cernuus*	Europe
BIRDS AND MAMMALS	
cattle egret, *Bubuculus ibis*	Africa
domestic cat, *Felis cattus*	Africa
domestic goat, *Capra hircus*	Eurasia
domestic hog, *Sus scrofa*	Eurasia
European rabbit, *Oryctolagus cuniculus*	Europe
DISEASE ORGANISMS	
citrus canker, *Xanthomonas campestris*	Asia
Dutch elm disease, *Ophiostoma ulmi*	Asia

weed. Cattle ranchers in these regions were seriously affected by the spread of this toxic weed.

Klamath weed had already become a serious problem in Australia, New Zealand, Chile, and South Africa. Efforts to find biological controls had been begun in 1919. Two species of leaf-feeding beetles that had proved effective in Australia were introduced into infested rangeland in California in 1945 and 1946. The more effective of these was *Chrysolina quadrigemina.* Adults of this beetle defoliate the plant in spring and early summer, and the larvae attack the buds and young leaves in fall and winter. A root-feeder and a leafbud gall-former were imported from France in 1950. Within ten years, these insects had reduced Klamath weed to less than 1 percent of its outbreak density in northern California and considerably reduced its importance elsewhere. In rangeland infested with Klamath weed, its biological control led to an increase of 35 percent in plant species diversity (Huffaker and Kennett 1959). In California, the annual benefit of this control was estimated to equal $3.5 million. This was the first successful biological control effort for a weed in North America. In Hawaii, a quite different insect, a gall midge (*Zeuxidiplosis giardi*), was found to be an equally effective biological control agent.

Economic Impacts of Exotic Species

The economic and ecological costs of exotics such as Klamath weed can clearly be enormous. Dollar estimates of the costs attributable to exotics, however, are largely limited to their impacts on crop and livestock agriculture, horticulture, commercial forestry, fisheries, and navigable waters. Recent rough estimates by Pimentel et al. (1999) suggest that exotic pests, not including microbial agents of human disease, create annual economic costs of over $115 billion in the United States. This includes over $4.3 billion in damage and control costs of forest insects and diseases. Almost $5.5 billion of damage and control costs are attributed to freshwater and marine plants, invertebrates, and fish. Exotic vertebrates are estimated to create economic costs of about $27.5 billion, largely to agricultural interests. The remaining $78 billion represents costs of weeds, insects, and diseases to agriculture and horticulture. Crop losses due to exotic weeds, together with weed control expenses, were estimated to equal $29 billion. Other estimates of the costs of exotics are somewhat lower. Peter Jenkins, of the University of New Mexico Center for Wildlife Law, has estimated that invasive exotics of all types produce damages equaling $5.5–7.5 billion annually in the United States (Baskin 1996). This is certainly a low estimate. In crop agriculture, the Office of Technology Assessment estimated that exotic weeds caused annual losses and incurred direct and indirect treatment costs equal to $4.6–6.4 billion in 1992 (OTA 1993). Control of aquatic weeds that clog navigable waterways in the United States costs about $100 million annually (OTA 1993).

Individual exotics can create enormous losses and treatment costs. One crop pest, the boll weevil, is estimated to cost farmers $250–400 million annually in crop losses and treatment costs (Smith 1998). Over the past century, these boll weevil costs to farmers exceed $15 billion. Damage to forests by the gypsy moth alone, in addition, amounts to $764 million annually (Stein and Flack 1996). Economic losses attributed to Dutch elm disease over the years total several billion dollars (Sinclair and Campana 1975). Individual exotics also cause major losses in aquatic ecosystems. The zebra mussel alone is projected to cause damage to lakes and streams totaling $5 billion by A.D. 2002 (Stein and Flack 1996).

Many other detrimental impacts of exotics, those that might be termed ecological costs, are real but are much harder to evaluate in dollar terms. Many of these constitute damage to what have been termed "nature's services," the role of ecological systems in protecting watersheds, regulating biological productivity, processing wastes of human activity, and giving stability and resilience to both human and natural ecosystems. Still other services relate to aesthetic properties of nature that many people value: beauty, wildness, uniqueness, diversity, and the like. For the biosphere as a whole, all of these services have been estimated to value about $33 trillion annually (Costanza et al. 1997). No estimates are available of the extent to which exotics have disrupted these services globally, or for North America. One suspects, however, that such damage ranges into the billions of dollars.

On the other side of the coin, the economic benefits of preventing the invasion of exotics, or of controlling those that have become established, are not only difficult to assess but are also frequently overlooked. Prevention of exotic invasions deals with a "great unknown," how the costs of prevention would stack up against the unknown damage from the invasions that were prevented. For control of known pests, the benefits are more apparent. Control of the Klamath weed, described earlier, cost about $750,000 but has yielded benefits of more than $100 million—benefits that are still accumulating (Batra 1982).

Pest Control Technologies

The economic and ecological damage caused by plant and animal pests and diseases has spawned a wide range of control technologies—mechanical, chemical, biological, cultural, and combinations of these approaches. The application of these technologies is still driven overwhelmingly by the economic impact of pests on agriculture, horticulture, forestry, and fisheries. More recently, a few exotics that primarily impact natural ecosystems have been subjected to control efforts.

The need to control pests and invasive exotics is clear. How to apply the available strategies in this effort remains controversial, however. At the core of this controversy is the degree of reliance placed on chemical pesticides versus that placed on biological control and other control strategies that minimize reliance on chemical pesticides. Despite the view among environmentalists that chemical pesticides are ecologically disruptive, they remain the most extensively applied control agents for pests that cause direct economic damage to agriculture and other productive enterprises. They are also being used judiciously in many programs aimed at controlling invasive exotics in natural ecosystems.

Chemical Pesticides

Sole reliance on chemical insecticides and herbicides for control of exotic pests in forests, grasslands, and other natural ecosystems, however, can lead to damage to native biodiversity. These chemicals, some of them applied to millions of acres of wildlands, often suppress nontarget, native species. In North America, for example, persistent broad-spectrum herbicides are being used against infestations of diffuse knapweed, spotted knapweed, and leafy spurge (Harris 1988) in situations where many native plants are likely to be affected. Similarly, between 1980 and 1995, in efforts to control the gypsy moth, 4.2 million acres of forest in eastern North America were sprayed with microbial pesticides or insecticides based on insect hormones. These pesticides are not specific for the gypsy moth, and their use has resulted in reductions of nontarget insects, spiders, and aquatic arthropods (Butler et al. 1997).

Chemical pesticides continue to show serious pollution side effects through food-chain transfers, as well. Even nonpersistent pesticides may result in mortality of animals that feed on target insects killed by their use. Various carbamate and organophosphate pesticides have caused mortality of birds feeding on plants or dead insects in treated areas, or even ingesting granular pesticides as grit (Cox 1997). A remarkable recent example of accidental vertebrate mortality from an insecticide involves the organophosphate monocrotophos. Although not used in either Canada or the United States, this insecticide was found to have killed an estimated 20,000 Swainson's hawks (*Buteo swainsoni*) in Argentina (Line 1996).

Biological Control

For many exotic pests, biological control offers an alternative to reliance on chemical pesticides. Biological control is the creation of a strong, negative biological interaction against an exotic species. Biological control is sometimes portrayed as "reestablishing the balance of nature," but its goal is really to create a strong "imbalance of nature" that reduces an undesirable species to such a low density that its impact is negligible. For example, to control a pest, one strives to encourage the enemies of the pest, but not the parasites and predators of these enemy species.

Biological control employs several different strategies. One strategy is simple protection, encouragement, or augmentation of native predators, parasites, and disease species. This can often be done by the selective use of pesticides in ways that do not kill natural enemies and by the improvement of microhabitats and resources on which existing natural enemies depend.

A second strategy is the release of native biological control agents at critical times, either to reestablish populations in habitats where they do not persist well or to augment natural populations in order to overwhelm the target pest at a critical stage of its life cycle. Augmentation is usually not economically practical for biological control agents in extensive forests, grasslands, and other natural areas, however.

The third strategy, known as classical biological control, is introduction of exotic biological control species, either from the native area of the pest species or from some other location. Classic biological control, the introduction of an exotic's enemies, is often more desirable than eradication, because as long as the exotic survives at low abundance, its control agent usually also survives. If a pest is eradicated locally, its enemies may also disappear, and reinvasion of the pest is likely to occur from a source area.

In the broadest sense, biological control also includes a variety of new strategies, such as the use of microbial agents that are distributed much as chemical pesticides are. Another new biological strategy is the release of artificially reared, sterilized individuals of a pest species to reduce the breeding success of pests in field populations.

Cultural and Integrated Control

A second alternative to chemical control is cultural control. In the agricultural context, cultural control applies to techniques of management of the phenology and environment of crop plants to minimize damage by pests. As an approach to control invasive exotics in natural ecosystems, cultural control refers to the use of fire, weeding, grazing, control of forest-stand structure, regulation of stream flow, and other general manipulations to disfavor exotics and encourage natives. For example, flood discharges from reservoirs on regulated rivers might be used to create open, wet, sand and mudflats suitable for germination of seeds of native cottonwoods and willows during the period when these trees shed their seeds. In forests, the maintenance of a mixed species-and-age structure and thinning to prevent overcrowding can reduce susceptibility of trees to exotic pests. Maintenance of snags can provide nest cavities for birds, and the encouragement of a herbaceous and shrubby ground layer can favor small mammals and understory birds that may reduce populations of exotics such as gypsy moths (Dahlsten 1986). Installing nest boxes for songbirds can serve a similar function.

All of these strategies, chemical and nonchemical, are increasingly being combined in a strategy known as integrated control or integrated pest management. This approach strives to combine chemical, biological, and cultural control in a fashion that employs chemical control only when needed to throw the burden of regulation of a pest species back to natural or nonchemical processes. Integrated pest management programs have been developed for many agricultural pests, but they have yet to be applied extensively to invasive exotics in natural ecosystems.

Overall Prevention and Control Strategies

Integrated pest management may be thought of as the capstone of a hierarchical prevention and control strategy. The first level of defense is prevention or delay of the entry of exotics into new geographical areas along routes created by humans. This strategy includes restrictions on importation, inspection and quarantine of biological materials at ports of entry to detect unwelcome associates, and treatment of cargos and carriers to kill hitchhiking invaders. These preventative aspects of regulation of exotics are examined in more detail in Chapter 21.

Secondary lines of defense must be instituted once a potentially invasive exotic becomes established within a region. If the population is very localized, eradication can sometimes be achieved. In a few cases, localized species that have threatened native biodiversity have been eradicated. The giant African snail was introduced to Miami, Florida, in 1966 by tourists who brought only three individuals from Hawaii and released them in their garden (Simberloff 1997). From this garden the population spread to several other locations in the greater Miami area. The infestation was recognized in 1969. Quarantine of affected properties, distribution of chemical baits, and human collection of snails accomplished eradication by 1973, although baiting and searching continued for several years. The cost of this effort was about $1 million. The removal of goats from San Clemente Island, off southern California (see Chapter 16), and of exotic fish populations and aquatic weeds from small lakes are other examples of eradication programs that have benefited natural ecosystems.

Efforts can often be made to control the spread of newly established invasive exotics. Quarantine and inspection programs can be instituted to prevent the movement of an exotic into new regions. Programs of this sort have been used, for example, to try to reduce the risk of gypsy moth invasion of the Pacific states. In California, inspections have discovered hundreds of gypsy moths and their eggs annually, most on cars, trucks, and other vehicles coming from the eastern states. In all probability, however, such programs intercept, at best, 10 percent of the invaders (Dahlsten 1986).

Direct control of the exotic or its biological vectors can often be used to slow the spread of an invasive species. In Hawaii, for example, this approach is being used against the Argentine ant. The Argentine ant has invaded Haleakala National Park on Maui and Hawaii Volcanos National Park on Hawaii, where it threatens many native arthropods (see Chapter 13). Efforts are being made to control the spread of this ant, which has flightless queens and spreads by the division of colonies, with the use of toxic baits (Stone and Loope 1987).

Another approach to dealing with newly established and spreading exotics is to locate and provide special protection to pristine areas. Small islands lying off the coastlines of continental areas or larger islands are often the last places to be invaded by exotic species. In regions with volcanic activity, such as Hawaii, islands of native vegetation, known as kipukas, often exist where lava flows have isolated them from the main landscape. Fencing, monitoring, and weeding such pristine sites may enable them to be maintained with relatively small effort.

Once a destructive exotic becomes widely established, still other strategies must be employed to reduce its impacts. Exotics can often be removed from local areas, and their reinvasion prevented. In Hawaii, exclosures for goats and hogs have been created in fifty-one sites to protect examples of native ecosystems or to promote recovery of native vegetation (Stone and Loope 1987). In some exclosures, plants thought to have gone extinct locally have reappeared, apparently from long-lived seeds in the soil. The extreme of this approach is the protection of individuals. Some exclosures in Hawaii protect the last individual plants from destruction by goats or hogs (Stone and Loope 1987). Chemical repellents have also been used in

Hawaii to deter rats from climbing into canopies of some native trees to feed on flowers and fruits, as a means of increasing the reproduction of endangered species.

For destructive exotics that threaten to spread over large areas, the first reaction to establishment is often to propose eradication. Eradication is usually very difficult, often impossible, however, once invaders have become widespread or have entered habitats that possess wilderness characteristics. Most eradication programs have been aimed at agricultural pests, and most have targeted outbreaks in urban areas, islands, or other localized situations. In California, for example, of thirty eradication efforts deemed successful, twenty-five were in urban areas (Dahlsten 1986).

Eradication has been successful for a few widely established exotics, however. A successful eradication effort was directed at citrus canker, a bacterial disease of citrus trees, in the southeastern United States. Citrus canker was introduced to the Gulf states about 1910, which then spread west to Texas and east to Florida within a few years. Quarantine of affected areas, coupled with chemical control, enabled its eradication by 1916 (Simberloff 1997). Inspection of imported citrus plants and fruit is still conducted to prevent its reintroduction. About $2.5 million was spent in this eradication program.

The most ambitious eradication program undertaken in North America is being directed at the boll weevil (Smith 1998). Native to southern Mexico and Central America, this pest has invaded cotton-growing areas throughout most of northern Mexico and the United States. After preliminary trials, an eradication program was begun in 1983, using a combination of intensive pesticide applications, cultural controls, and massive releases of sterilized weevils. This effort has been successful in eliminating the boll weevil from much of its range in the United States. The current cost of the eradication program itself is about $152 per acre, with a small ongoing cost for surveillance to detect reestablishment of the species. This eradication program has suffered a number of problems related to secondary outbreaks of other cotton pests due to pesticide treatments that released them from natural biological control. Because of this, some regions in Texas and Mississippi have refused to participate, creating difficulties in achieving total eradication of the boll weevil from the United States. In any case, this effort shows that eradication of a widespread exotic is possible, but that it is likely to be slow, very costly, logistically complex, and complicated by side effects.

In many cases, however, eradication programs, even for pests with enormous direct destructive potential, have been unsuccessful. Eradication efforts are often directed at isolated colonies of actively spreading species such as the gypsy moth and Japanese beetle. Even if successful, the likelihood of reinvasion is very great, and the money spent largely wasted (Dahlsten 1986). In southern California, for example, efforts have been made to eradicate the Mediterranean fruit fly when small outbreaks occurred between 1975 and 1994 (Carey 1996). These efforts, based on the trapping of 1 to 400 or so individuals, have used chemical insecticides and releases of sterile males at a cost of millions of dollars. On several occasions, successful eradication has been claimed. Examination of the locational history of outbreaks, however, indicates that the insect is probably established in several locations and that identified outbreaks probably spring from these populations.

If eradication or containment is impossible, general population reduction may be possible. Reduction of keystone exotics over large areas has been attempted for only a few species. Many techniques of control have been used, including the killing or capturing of animals, the weeding of undesirable plants, the use of chemical and biological pesticides, biological control, and cultural control.

Direct killing or removal of exotics is usually practical only on small areas. Cutting and weeding of invasive exotic plants have been carried out at several Nature Conservancy preserves in Florida, Arizona, and California, for example (Randall 1993). Efforts have also been made to reduce feral goat and hog populations in preserves in the Hawaiian Islands by hunting, both by citizen hunters and by hired hunters (Stone and Loope 1987).

Selective chemical control may play a useful role in restoration of native communities in at least some circumstances. For example, Getsinger et al. (1997) conducted an experiment on an area of the Pend Oreille River, Washington, that was infested by Eurasian water milfoil. They applied a short-life herbicide, triclopyr, to areas heavily infested by milfoil but supporting a remnant community of native aquatic plants. Although the biomass of all species was reduced immediately after the treatment, the diversity of native plants increased in the following year, and native plant biomass increased five- to tenfold. This strong recovery of native species inhibited the regrowth of milfoil for three years.

Classical biological control has had some spectacular successes, such as control of the Klamath weed, as described at the outset of this chapter. In California, the cottony cushion scale, introduced in 1860, threatened to destroy the emerging citrus industry of the state. A search for possible controls in Australia, the scale's home, was begun in 1888. This effort turned up several possible control species, including the vedalia beetle (*Rodalia cardinalis*) and a parasitic fly (*Cryptochaetum iceryae*). These species were released in California in 1888 and 1889 and have since given effective control of the scale throughout California (Cox and Atkins 1979).

Classical biological control has proven effective for terrestrial plants, invertebrates and vertebrates, and for freshwater plants and vertebrates. Among terrestrial plants, successful programs have been carried out for annual and perennial herbaceous plants, cacti, and woody shrubs. In the United States, as of 1987, releases of biological control insects have been made for control of twenty-eight species of plants (DeLoach 1991). In Hawaii alone, nine species of weedy plants have been completely or substantially controlled (DeLoach 1991). The exotic puncturevine, an annual herb, is among these Hawaiian successes. Also in Hawaii, substantial-to-complete control of lantana, a sprawling, weedy shrub, has been achieved, depending on location. Efforts to control this shrub began in 1902, and twenty-two species of potential biological control insects were introduced, several of these eventually contributing to successful control.

The potential exists for biological controls of invasive trees, as well (DeLoach 1991). In Florida, several Brazilian insects are being examined as possible controls for Brazilian peppertree, and several Australian insects for control of Australian paperbark trees. Several herbivorous insects from the Middle East and central Asia are also under study for control of saltcedars in the southwestern United States.

Many herbivorous insects and mites have been controlled effectively by imported biological agents. DeBach (1974) noted that, worldwide, 120 insect pests have been partially to completely controlled. Most of these species are agricultural pests, but a few forest insect pests have also been successfully controlled (Dahlsten 1986). An exotic forest insect, the European spruce sawfly, began to cause extensive defoliation of forests in eastern Quebec in the 1930s. Between 1933 and 1951, twenty-seven parasitic insects were imported and released. Three of these, and a virus possible released along with them, have given nearly complete control of this exotic. Another forest insect, the winter moth, appeared in Nova Scotia in 1949 and threatened to become a destructive defoliator of oak trees across much of North America. Between 1955 and 1960, six species of its parasites were introduced from Europe. Two became established, and these, together with a virus introduced in 1961, now give effective control.

Classical biological control can be applied to at least some terrestrial vertebrates (Usher 1989), although no such programs have been attempted in North America or Hawaii. Microbial agents have been used to control European rabbits in Australia and feral house cats on remote oceanic islands. Use of such agents in a continental situation is impractical because of the danger to both domestic animals and wildlife.

Classical biological control usually takes considerable research to identify, test, and introduce a safe and effective agent. Thus, for a long time the monetary investment may appear to be considerably in excess of the benefit. Once implemented biological control can be a highly cost-effective strategy. DeBach (1974) estimated that the chances of substantial-to-complete control of an established insect pest are about 4 in 10, and that the benefit-to-cost ratio for investment in biological control is about 30:1. Investments in chemical pest control are estimated to have a benefit-to-cost ratio of only 3:1 (Pimentel 1973). Still, the costs of developing a biological control program are uncertain in a given case and are immediate and out-of-pocket. Thus, they appear difficult to weigh against benefits that extend into the future and, in the eyes of most economists, must be discounted at some rate.

The uncertainty about the economic benefits of biological control tends to promote risky control strategies. Short-term economic considerations tend to favor the least expensive strategy that gives immediate relief. In many cases, this tends to be a chemical attack on an exotic, a tactic that may, in the long run, exacerbate the problem by damaging natural biotic controls and promoting the evolution of resistance. Worldwide, over 500 species of insects, mites, and other animal pests have evolved resistance to pesticides (McGaughey and Whalon 1992). Worldwide, 124 species of weeds have evolved herbicide resistance, in some cases to two or more classes of herbicides (Heap 1997). The increasing numbers of resistant species, and especially of species with multiple resistance, make future success in chemical control of arthropods and weeds very uncertain.

Biological Insecticides

A new branch of biological control involves the use of biological insecticides, or biopesticides, for control of agricultural insect pests and disease vectors. Biological insecticides include a variety of fungi, viruses, bacteria, and protozoa. The most effective of these are the insecticidal crystal proteins (ICPs) derived from the bacterium *Bacillus thuringiensis,* commonly known as Bt (Lambert and Peferoen 1992). Bt is found in most soils, and it probably parasitizes soil insects. It reproduces normally by simple cell division, but when environmental conditions become stressful, it forms spores. In spore formation, ICPs are also formed in considerable quantity. When ingested, these ICPs are toxic to many insects. Many genetic strains of Bt exist, perhaps tens of thousands. Some twenty-five or more ICPs, of related chemical structure but different toxic capabilities, are now recognized. A given Bt strain produces several ICPs.

Bt pesticides can be mass-produced and applied much as traditional pesticides. The genes for ICPs can also be introduced into other organisms, to facilitate production of the ICPs or possibly to give systemic protection to a species of value. Bt biopesticides are not highly specific and can sometimes impact nontarget species, especially moths and butterflies (e.g., Herms et al. 1997). Concern has arisen recently that wind-carried pollen from corn bearing Bt genes may be deposited on leaves of food plants of butterflies and moths, killing the larvae of these insects. Nevertheless, they may represent an improvement over synthetic chemical pesticides, many of which have serious side effects on nontarget animals, often including vertebrates. Bt pesticides are also vulnerable to evolution of resistance (McGaughey and Whalon 1992). So far, these biological pesticides have not been applied to control exotics in other agricultural arenas. But they might, for example, be employed in the control of invasive insects, such as mosquitoes that carry avian malaria in Hawaii.

Shortcomings of Classic Biological Control

Although biological control offers the greatest potential for reducing the magnitude of problems caused by invasive exotics, intercontinental introductions must be done with extreme care. Species with broad feeding habits can be destructive to native plants and animals. An exotic fish, the Chinese grass carp, has been introduced to freshwaters in Florida, Arkansas, and Texas for control of exotic aquatic weeds (see Chapter 9). In addition to consuming exotics, however, it also feeds very heavily on native aquatic plants (DeLoach 1991). The cattle egret, introduced to Hawaii in 1959 to control insects in cattle pastures, has proved to be a competitor with endangered Hawaiian waterbirds, as well as a predator on prawns in mariculture ponds. Even some of the most successful control agents for exotic plants such as Klamath weed and water hyacinth have proven to attack some ornamental plants, themselves exotics, and some native plants (Harris 1988).

Several cases of detrimental impacts of introduced biological control species have been documented. In Hawaii, native insects of several groups—pentatomid bugs, damselflies, and moths—are believed to have declined in abundance because of the impacts of introduced biological control agents (Howarth 1991) (see Chapter 13). Fish introduced to

Hawaii for control of mosquitoes and aquatic weeds have also eliminated native shrimp from some coastal ponds. Insects introduced to Hawaii to control the exotic puncturevine have also greatly reduced a native species of the same genus.

Two additional examples show the degree to which well-intentioned biological introductions can go astray. The exotic rosy wolfsnail, a predatory snail introduced to Hawaii for biological control of the giant African snail, has spread into mountain forests. There it has become a devastating predator on endemic snails of the genus *Achatinella,* all of which are now designated as federally endangered (see Chapter 13).

More recently, in the late 1960s and early 1970s, the flowerhead weevil was introduced to various states and Canadian provinces for control of exotic thistles of the genus *Carduus,* particularly the musk thistle. Its larvae feed in the flowerheads, destroying the seeds. This species, and a second weevil that feeds on the young rosettes, have produced dramatic declines of musk thistle in some places. Unfortunately, the flowerhead weevil also attacks some native thistles that are not problem weeds. From the initial introduction sites, the flowerhead weevil has been redistributed throughout most of the central and northern portion of the United States. These introductions and redistributions were done in spite of the fact that tests showed that potential hosts for this weevil included several genera of thistles native to North America. The decision to do this was evidently based on the observation that the weevils showed stronger preferences for *Carduus* than for the native genera.

Louda et al. (1997) have discovered that flowerhead weevils are now using several species of native thistles of the genus *Cirsium* as hosts. The effects of the weevils on seed production by these species were severe. One of the species, the Platte thistle (*Cirsium canescens*), is an endemic of the Nebraska sandhills, and is a form very closely related to Pitcher's thistle (*Cirsium pitcheri*), a federally threatened species that inhabits sand dunes around the Great Lakes.

These sorts of difficulties have led several ecologists to question the adequacy of current procedures for testing host or prey specificity of candidate biological control species (Simberloff and Stiling 1996; Cowie and Howarth 1998). Even when high specificity is demonstrated, the possibility that an introduced biological control species will adapt or evolve to utilize native species cannot be completely discounted. This issue is examined further in Chapter 21.

Unfortunately, some groups of organisms do not offer good opportunity for classic biological control. Among terrestrial plants, most grasses do not have specific herbivores, being defended by general features of structure and growth habit. The buds of perennial exotic grasses are usually located at or below the ground surface, where grazing animals usually do not destroy them. Their leaves often possess mechanical defenses, such as silica spicules, that deter grazers in general. Thus, biological control of invasive grasses in the western states and Florida, except by augmentation of native grass competitors, does not appear likely.

No examples of classic biological control exist for freshwater or marine invertebrates or fish. The difficulty of finding specific control agents in aquatic ecosystems appears to be greater than in terrestrial ecosystems. Only recently, however, have efforts been made to find such agents. Fish and aquatic invertebrates are hosts to many parasites and diseases, and some of these may have the specificity required for use as biological controls. Furthermore, many marine organisms have unique defensive chemicals and thus might have unique coevolved parasites and predators.

Classic biological control efforts in aquatic environments have involved the introduction of exotic fish that feed on mosquitoes or aquatic plants. As noted, these have tended to have undesirable side effects. Other biological control efforts involving the augmentation of native fish populations to control mollusks, such as zebra mussels, or small fish, such as the European ruffe, have been proposed. The possibility of introduction of exotic fish for these purposes has also been suggested, but it appears very risky.

Hierarchical Prevention and Control Strategy

To deal effectively with the challenges of the invasive exotics now flooding North America and Hawaii requires a much more integrated, responsive, and hierarchically structured system than now exists. A five-tiered system can be envisioned:

1. *Vulnerability analysis:* Conduct a comprehensive analysis of potential new exotic invaders and the routes by which they might enter North America. Identify the ecosystems, communities, and species likely to be detrimentally impacted by these exotics and evaluate the possible impact that new exotics might have. Assess possible ways to provide protection for critical systems or species.
2. *Exclusion and quarantine:* Develop more effective systems of reducing the volume of exotic propagules entering and moving from place to place in North American lands, freshwaters, and coastal marine waters. These must be based on a realistic assessment of the major routes and must apply measures that are effective in reducing exotic propagules.

3. *Discovery and eradication:* Establish a system of monitoring establishment zones, so that early detection of new exotics can be achieved. Develop a protocol to assess the potential for eradication of newly established exotics.
4. *Containment and research:* If eradication is not feasible, apply procedures to slow the spread of the exotic and institute research on the full array of controls that might be brought to bear to reduce the detrimental impacts of the exotic.
5. *Integrated exotic management:* Establish an integrated management program for exotics that have become widely established, the objective being to transform them into nondisruptive components of North American ecosystems.

An Emerging Field: Exotic Species Management

Invasive exotics are now causing serious economic and ecological damage to terrestrial, freshwater, and marine ecosystems throughout North America and Hawaii. Estimates of economic damage are still very rough and vary greatly according to their source. A thorough review of production losses and treatment costs in plant and animal agriculture, horticulture, aquaculture, and forestry, together with management expenses for control of exotics that impact industry, navigation, transportation, natural environments, and human health, is badly needed. Realistic estimates of the degradation of native ecosystems by exotic invaders, using techniques such as contingent valuation to capture costs that are not measured by marketplace transactions, are also needed (Hanemann 1994).

Stewards of our natural ecosystems, managers of productive wildlands, and guardians of local preserves throughout North America are facing increased expenses for control of exotics. They must also accept the fact that the threat of invasion by new exotics will always exist. The vulnerability of natural areas to exotic invasions is clearly a feature that must be taken into account in their designing and management. For example, connectivity of preserves, a feature accepted as basic by many conservation biologists, may promote the movement of exotics from one unit of a preserve system to another. Isolation of preserves may thus be desirable. In any case, ways must be found to reduce the rate of entry of exotics into North America and to reduce the vulnerability of regional and local ecosystems to invasion. Exotics that have already invaded, as well as future invaders, must be managed aggressively.

The challenge of exotics must be met by a new branch of applied science: exotic species management. Scientists in this field must be able to address all phases of the invasion process, evaluate the risks and costs of invasive exotics, and devise ecologically sound techniques of prevention and control of invasives.

Chapter 21

Exotics and Public Policy: Are All Exotics Undesirable?

> The total number of [nonindigenous species] and their cumulative impacts are creating a growing burden for the country. We cannot completely stop the tide of new harmful introductions. Perfect screening, detection, and control are technically impossible and will remain so for the foreseeable future. Nevertheless, the Federal and State policies designed to protect us from the worst species are not safeguarding our national interests in important areas.
>
> —Roger C. Herdman, in *Harmful Nonindigenous Species in the United States**

James T. Carlton (1996b) describes a close-to-life scenario, as follows. The month is May. In San Francisco Bay, the waters teem with the larvae of marine invertebrates, including dozens of exotics. Among them are larvae of Asian clams, Atlantic green crabs, and Chinese mitten crabs (see Table 21.1). A cargo ship, recently in from Japan, has unloaded its cargo and is readying for a run to Seattle to pick up cargo for the return trip. The ship is ballasting up, pumping bay water into its special ballast tanks and the main cargo hold, perhaps 25,000–30,000 thousand tons of water. In two days, this water will be discharged into Puget Sound, and the larvae of three exotic marine invertebrates will be introduced into new coastal marine waters. Is this ballasting procedure legal? Are modified procedures

*Office of Technology Assessment, U.S. Congress (1993). *Harmful Nonindigenous Species in the United States*. OTA-F-565. U.S. Government Printing Office, Washington, DC.

Table 21.1. Scientific names and native regions of exotics that pose major issues of public policy.

PLANTS	
African rue, *Peganum harmala*	North Africa
Russian olive, *Eleagnus angustifolia*	Asia
AQUATIC INVERTEBRATES	
Asian clam, *Potamocorbula amurensis*	Asia
Chinese mitten crab, *Eriocheir sinensis*	China
European green crab, *Carcinus maenas*	Europe
zebra mussel, *Dreissena polymorpha*	Eurasia
FISH	
northern pike, *Esox lucius*	eastern North America
REPTILES	
brown tree snake, *Boiga irregularis*	western Pacific Islands
MAMMALS	
feral horse, *Equus caballus*	Asia

available that would eliminate the transfer of estuarine larvae? Can they be implemented at reasonable cost? Can compliance with a requirement for modified ballasting procedures be monitored?

Another scenario illustrates how easily an invasive exotic plant can be introduced to new areas. African rue, a weedy shrub native to North Africa and southern Asia, has become established in desert areas of western Texas, southern New Mexico, and scattered across other locations in the Southwest. The invasiveness of this plant has led to its listing as a noxious weed in New Mexico, Arizona, Colorado, and other states. However, word has also spread through the recreational drug community that African rue contains psychoactive alkaloids. An Internet search quickly reveals that seeds of African rue are available from several sources, along with recipes for extracting and using the alkaloids. Seeds can be ordered, planted on vacant desert areas in the West and Southwest, and this invasive shrub introduced to new localities. Soon, no doubt, African rue will become established throughout its potential climatic range in North America. Is the sale of seeds of a species designated a noxious weed in many states legal? Is it practical to regulate the interstate shipment of such material? How can the spread of an invasive species such as this be prevented?

Complexity of the Invasion Challenge

Similar considerations exist for dozens of other mechanisms by which exotics reach new regions. Interregional trade in plant and animal food products, pets, horticultural plants, wood, and wood products is enormous. Exotics can be carried by interstate and international movements of trains, ships, airplanes, commercial truckers, military groups, and private citizens. Shipments of seeds and other propagules can be sent by public and private mail systems. Deliberate introductions of exotics are made by game and fish agencies, sportsmen's groups, agricultural organizations, and aquacultural enterprises. Unsuspected hitchhikers ride in legal shipments of plant and animal materials. Illegal releases occur by pet and aquarium dealers and by bored pet and aquarium owners. Escapes occur accidentally from zoos, game ranches, and aquacultural pens. Natural dispersal spreads exotics along routes of commerce. Do current laws apply to these actions? Are they adequate? Can they be improved? What will improved regulations cost? How will they be enforced? These questions, and more, must be answered as we attempt to stem the flood of exotics now invading North America.

A growing economic and ecological burden is clearly being placed on the North American public by invasive exotics. For exotics that directly impact sectors of the economy such as agriculture and forestry, in which costs are most easily measured, we know that the damage is enormous (see Chapter 20). And, although estimates are difficult to obtain, we know that North Americans attach high existence values to native species and natural ecosystems (Cox 1997). Collectively, the survival of pristine ecosystems and endangered species is valued in billions of dollars a year. Do we have the technological capability to deal with the threats posed by invasive exotics? Do we have the will to impose the restrictions and fund the programs necessary to slow the invasion of new exotics and reduce the impacts of those now established?

The Hierarchy of Public Concern

Exotic species pose extraordinarily difficult policy issues and regulatory difficulties. Exotics, first of all, range from highly beneficial to extremely detrimental. Exotics form the economic basis of plant and animal agriculture, aquaculture, and horticulture. Exotics are economically important in forestry and wildlife management. Exotics that serve as biological control species have contributed beneficially to lessening damage by other destructive exotics. Even some of the exotics that have "gone wild," such as the Russian olive in southwestern riparian ecosystems and feral horses in the Great Basin, are considered by many people to have great economic, aesthetic, and existence values. Against these species with valuable or redeeming qualities are arrayed a growing number of exotics with largely detrimental impacts.

To the public at large, however, even detrimental exotics range widely in perceived seriousness of their impacts. Although no surveys are available on ranking exotics, those considered most deleterious are certainly those with detrimental impacts on human health: disease agents and their vectors. Exotics that are destructive to consumable goods such as foods, fibers, and forest products, thereby increasing their cost, rank high in public concern. Exotics that cause visual impairment of valued landscapes in urban areas or parklands are perhaps next on the scale of public concern. Those that affect recreational resources, such as game and fish, would be lower in perceived seriousness. To most of the public, exotics that have negative impacts on biodiversity and ecosystem function probably rank lowest in concern.

This ranking of awareness and concern is evident in the public response to programs instituted to control exotics. In California, for example, the actions of the Department of Fish and Game in 1997 to eliminate northern pike from Lake Davis, a reservoir on the middle fork of the Feather River, drew heavy opposition from local residents. To wildlife ecologists and fisheries scientists, the pike was considered a serious threat to aquatic biodiversity in California freshwaters. Rotenone and other chemicals were therefore used to kill the pike, and also all other fish, in Lake Davis. This treatment, together with the restocking of trout, themselves exotics, cost about $2 million. Killing the fish temporarily eliminated trout fishing in Lake Davis, a major aspect of the economy of the town of Portola and its population of about 2,300 people. It also prevented the town from taking water for domestic use from the reservoir for several months. These impacts prompted Portola and many of its private citizens to file substantial claims against the state for economic losses. To the residents of Portola, the economic benefits from exotic fish clearly outweighed the danger of the pike to aquatic biodiversity in California freshwaters.

History of Regulation of Exotics in North America

This mix of benefits and detriments associated with exotics has contributed heavily to the fact that the history of federal law relating to exotics has been one of piecemeal reaction to immediate threats. The history of federal efforts in the United States and Canada is long, but most law has applied to, or been interpreted to apply to, introductions that threaten plant and animal agriculture, horticulture, and forestry. Concern in both countries about the regulation of species that threaten native biodiversity has been very recent.

The first federal effort to regulate exotics in the United States was the Lacey Act of 1900, which enabled the secretary of agriculture to ban the import of birds or mammals injurious to agriculture and horticulture. A related act in 1926 extended this regulation to fish. In 1981, the Lacey Act was amended and strengthened. The revised act covered foreign animals of all types, as well as some plants. It prohibited the import of a blacklist of species, and it established the requirement of a permit for the importation of any foreign species. The blacklist now bans the importation of species in twelve genera of mammals, four genera of birds, and two families of fish (Ruesink et al. 1995). The U.S. Fish and Wildlife Service was charged with determining if species proposed for importation are harmful. Although covering a very wide assortment of organisms, the Lacey Act suffers from a number of weaknesses. In practice, the greatest weakness is that most species cannot be recognized as harmful until they are well established. Thus, the result is that blacklisted species are mostly those that are already established and causing problems. A second serious shortcoming is that the act does nothing to regulate interstate transport of exotic species.

The Federal Plant Quarantine Act of 1912 gave authority to the Department of Agriculture to regulate the importation and interstate shipment of plant materials so as to exclude or prevent the spread of diseases and insects injurious to plants. Although this law has been interpreted to apply to plants of direct economic value, such as agricultural crops, the wording of the law does not make this restriction (Campbell 1993). This law provides the basis for inspections of personal baggage and shipments of commodities that enter the United States.

The Federal Seed Act of 1939 was passed to require that imported seed be free of noxious weeds. The Federal Plant Pest Act of 1957 was designed to prevent the import of agricultural pest species and to regulate their interstate transport. These acts, together with the Federal Noxious Weed Act of 1974, are the basis for federal regulation of entry and spread of noxious weeds, arthropods, and plant pathogens. The Federal Noxious Weed Act strengthened the ability of the secretary of agriculture to prohibit the importation of weeds that have detrimental effects on agriculture, navigation, public health, or fish and wildlife resources. Although broad in its objectives, the act has been applied largely to restrict importation of agricultural weeds. As a result of this act, a federal noxious weed list, including both aquatic and terrestrial plants, was developed. This list of noxious banned weeds currently includes ninety-four species. The Animal and Plant Health Inspection Service (APHIS) is responsible for port-of-entry inspections to prevent the entry of these species. However, although the act empowers the secretary of agriculture to prohibit interstate transport of designated noxious weeds, no mechanism to effect such action has been created.

In 1977, with concern beginning to focus on effects of exotic species on natural ecosystems, President Carter signed Executive Order 11987 requiring U.S. federal agencies to restrict the introduction of exotic species into natural ecosystems of federal lands and waters. Exotics determined not to have adverse impacts were exempted. Unfortunately, the various agencies have not developed guidelines to implement this order. This, together with the fact that it applies only to federal lands, has made it effectively powerless.

The U.S. Nonindigenous Aquatic Nuisance Prevention and Control Act of 1990 was a hurried response to invasion of the Great Lakes by zebra mussels and other exotic species. It charged the Fish and Wildlife Service and the National Marine Fisheries Service with controlling the introduction and spread of exotic aquatic pests. The act established an Aquatic Nuisance Species Task Force to develop a control program for such species. It also specifically created a National Ballast Water Control Program to prevent further introductions of exotics to the Great Lakes. This program, administered by the U.S. Coast Guard, applies to all vessels entering the Great Lakes, whether bound to Canadian or U.S. ports. These vessels must carry no ballast, or if they are ballasted, they either must have exchanged ballast water in the open ocean or must retain their ballast throughout their voyage in the Great Lakes (Reeves 1997a). These programs were budgeted at slightly over $30 million dollars per year.

In 1996, the U.S. National Invasive Species Act reauthorized the 1990 Nonindigenous Aquatic Nuisance Prevention and Control Act and amended it to make the ballast management procedures used in the Great Lakes applicable to all U.S. ports, beginning July 1, 1999. It also authorized funds for research of ways to reduce ballast water introductions of nuisance species and to control aquatic nuisance species that have become established.

Many U.S. states and Canadian provinces have created laws and agencies to deal with exotic pests. Again, most of these are directed at economic pests of agriculture, horticulture, and forestry. In California, however, the California Exotic Pest Plant Council (CalEPPC) focuses on invasive plants in the state's wildlands (Wade 1997). CalEPPC functions as a coordinating group that maintains a database of invasive exotic plants, develops protocols for invasive plant control, and sponsors conferences and workshops on invasive plants.

History of Global Efforts to Regulate Exotics

Ultimately, the problem of invasive exotics is a global problem, but international efforts to stem the flow of exotics are still rudimentary. As in North America, the greatest attention has been paid to agricultural pests, livestock diseases, and the parasites and pathogens of aquacultural fish and shellfish. The economic damage resulting from ballast water introductions to North American freshwaters of zebra and quagga mussels, of the ctenophore *Mnemiopsis leidyi,* from the western Atlantic to the Black Sea and of toxic marine algae to Australian waters has recently created international concern about other exotics. Canada and the United States were the first countries to apply regulations to water ballasting of ocean ships entering the Great Lakes. Regulation of ballast water discharge is being instituted in several European countries, Japan, Chile, Australia, and New Zealand (Reeves 1997a).

The first international effort to regulate exotic species was the International Plant Protection Convention, signed by the United States in 1972. Some ninety-four nations are party to this agreement. Administered by the Food and Agriculture Organization of the United Nations, this convention requires signatory nations to establish an agency to regulate agricultural plant pests, including certification that materials being exported are pest-free. In addition to being quite limited in scope, its implementation is very weak in many nations.

For the marine environment, the International Council for the Exploration of the Sea (ICES) has developed a "code of practice" relating to introductions of nonindigenous species (Sindermann 1986). This code, adopted in a revised form in 1979, applies primarily to species involved in aquaculture and is designed to prevent accidental introductions of parasites, pathogens, and other detrimental exotics with translocated fish, shellfish, or other aquacultural species. The code is strictly advisory and has served mainly to make member nations of ICES aware of the serious consequences that can result from the unintentional introduction of problem exotics.

The Convention on Biological Diversity, drafted at the United Nations Conference on Environment and Development in Brazil in 1992, commits its signatory parties to strive only to exclude, control, and eradicate exotic species that threaten ecosystems, habitats, or species. Although the United States participated in the Brazil conference and has continued to participate in activities of the organization, it has not formally ratified this convention. Canada has ratified the convention.

International scientific efforts are now being organized to address the problems of invasive exotics. In 1996, the United Nations sponsored a Conference on Alien Species in Trondheim, Norway. As noted in Chapter 1, a new SCOPE Project on Invasive Species is now under way. The Species Survival Commission of the World Conservation Union (IUCN) has established an Invasive Species Specialist Group. The mission of this group is to publicize the threats posed by exotics and to promote the spread of information on means of controlling or eradicating them. The emphasis of this group is on exotics that impact conservation values. The group publishes a newsletter, *Aliens,* that appears once or twice yearly.

These diverse international efforts have only begun to address the overall challenge of invasive exotics. They promise, however, to lead toward the development of a worldwide database on invasive species, their potential to invade various world regions, and the successes and failures that have occurred in combating them. Effective international controls on the movement of exotics, however, are probably far in the future.

Protecting North America Against Invasive Exotics

Even in North America, regulation of activities that allow the entry and spread of exotics is inadequate. In both the United States and Canada, policy relating to importation of exotics has been very permissive. Both countries, however, now face difficult policy issues in regard to strengthening the regulation and management of detrimental exotics. These relate, first of all, to preventing the invasion of exotics to North America, and second, to reducing the negative impacts of exotics that have become established.

Development of policy relating to introductions must recognize that exotics enter North America by deliberate legal importation, by illegal importation, and by accidental means (Ruesink et al. 1995). Introductions by all importation routes can be reduced by procedures used to screen for undesirable species or prevent their accidental movement into North America. Such procedures need to be adjusted closely to the risks involved. Explicit risk assessment procedures have largely been lacking in exotic management policy in the United States and Canada.

Unfortunately, even the most basic strategy for excluding exotics is a major issue. In the past, the policy of North American nations, states, and provinces has been to permit free movement of biotic materials unless they have been determined to be dangerous or deleterious. The burden of proof is thus on sectors of society that might be hurt by problem exotics rather than on the sectors of society whose activities introduce exotics. Often, those who are ultimately hurt are not aware of the possibility of introduction of detrimental species. These superficial importation controls and unregulated transport procedures guarantee that many new exotics will reach North America.

Nevertheless, for deliberate introductions, stricter requirements can clearly be established for the importation and release of exotics. The first approach to regulating exotics was the creation of a blacklist of species whose import or release is prohibited. Many scientists concerned about exotic invasions have recommended the alternative policy of establishing a "white" list of species proven safe. A white list might be accompanied by a "gray" list of species whose safety is under study. In its broadest sense, a white-list policy would require that agents of potential transport of exotics adopt measures to assure that their vehicles are free of exotics. "Exotic-free" ships, planes, and motor vehicles will necessitate major advances in technologies of treatment, inspection, and quarantine.

In the United States, introduction of these species could also be made subject to an environmental impact statement (EIS) under the National Environmental Policy Act. Many sets of criteria have already been proposed to evaluate the risks of deliberate introductions of exotic species (Ruesink et al. 1995). Many of these protocols, often aimed at specific groups of organisms, are inadequately conceived and ecologically naive. Many protocols do not consider characteristics that biologists have associated with high invasion potential. Most proposed protocols for assessing risk, for example, have not considered the species' intrinsic potential for population growth, the range of habitats they can occupy, their response to disturbance, and their invasiveness in other world regions. Protocols for assessing the risks of deliberate introduction have also been applied very unevenly (Ruesink et al. 1995). For example, potential biological control agents have been subject to very rigorous study before their release, but most horticultural plants are screened only for the possibility that they are carrying disease agents or hitchhiking arthropods.

A comprehensive set of criteria for consideration by an EIS for deliberate introductions needs to be developed. For organisms that are not intended for release into the environment, the EIS must evaluate the risk of escape and the consequences that might result if escape occurs. For organisms intended for release into the environment, an EIS must consider the geographical range that the species might achieve, the range of habitats that it might occupy, and the influence the species might have on native species and ecosystems. For deliberate introductions, the costs of evaluating a proposed importation should also be shifted to the group or segment of society that is likely to benefit from the new exotic.

For illegal and unintentional routes of entry of exotics, the problems of reducing the inflow of exotics are more difficult. The number of species entering illegally and unintentionally has increased as international travel and commerce have grown (Ruesink et al. 1995). Procedures to control such introductions tend to conflict with policies relating to international trade and travel. The most effective way to reduce the entry of exotics would be to reduce trade and travel, an approach that is politically unrealistic. If increased international trade and travel are encouraged, as is the worldwide trend, the vectors of movement of exotics will obviously increase in number and variety. Can such increases be managed to prevent further increase in or, better, to reduce the number of exotics reaching North America?

Sets of criteria have also been proposed to evaluate the risks of unplanned introductions of exotic species (Ruesink et al. 1995). These have tended to emphasize suitability of physical and biotic conditions for persistence of species that might reach North America. As for deliberate introductions, protocols for risk analysis of unplanned introductions have also tended to ignore basic demographic features of exotics and often their performance in other invaded world regions. Risk assessment of unplanned introductions must also consider the frequency of introductions, the numbers of individuals introduced, the difficulty of detecting early introductions, and the practicality of eradication of early invasions.

Technological improvements are needed both for reducing the numbers of exotics moving along various entry routes and for detecting those that reach North America. Improvements in technologies of international shipment of goods are needed to reduce the chance of accidental transport of exotics. For ocean and Great Lakes cargo vessels, current ballasting regulations are not fully effective. For example, ships entering the Great Lakes without ballast usually contain a substantial quantity of unpumpable water and sediment in their ballast tanks. If these ships take on and discharge ballast within the Great Lakes, material of foreign origin can still be discharged into the Great Lakes system. Problems such as this have led to a search for alternative means of ballast water. These include loading of sterilized ballast water, discharge of ballast water to shore facilities, filtration of ballast water, and sterilization of ballast water by various agents (Reeves 1997b). Redesign of ballast tanks to allow complete water exchange is also under consideration.

Detecting New Invaders

Regardless of improvement in risk assessment protocols and monitoring procedures for international travel and commerce, some exotics will continue to invade North America. Improved means of detecting entry and incipient establishments are needed. Dogs, or "canine sniffers," are used to detect species such as the brown tree snake at airports in Hawaii. Electronic "sniffers" that can detect a wide range of organisms and materials that may harbor organisms are under development.

Greater attention must also be given to detecting newly established invaders. Indeed, as Richard Mack (OTA 1993) has suggested, a better use of funds may be in early detection of newly established exotics rather than in greatly intensified inspections of passenger and commercial vehicles, travelers' baggage, and commercial cargo.

Early detection implies a program of rapid response to eradicate or contain new exotics. For localized or newly established populations of exotics, eradication appears to be most practical, especially for terrestrial organisms. Even in these cases, however, a realistic evaluation of the chances of killing all individuals of the exotic is required.

Managing Established Exotics

Major issues also exist about strategies for managing established exotics. These center on programs of eradication and on reliance on biological and chemical control. In the United States, so-called "eradication" programs have been undertaken for a number of agricultural and forest pests deemed highly destructive or potentially so. Although a few well-established serious pests of agriculture have been eradicated, most others have not (Myers et al. 1998). Eradication programs are politically popular, especially when conducted by governmental agencies, and are often justified by cost-benefit analyses that underestimate the costs of full eradication and overestimate the benefits that could result. Expensive eradication programs often lead to claims that eradication has been achieved, when subsequent resurgence of the pest shows that it was not. In some cases, eradication efforts also employ heavy pesticide applications that can lead to detrimental side effects. In many cases, therefore, integrated regional pest management programs are likely to be more effective than intensive eradication efforts (Myers et al. 1998).

A second issue in management of established exotics relates to the degree of dependence that can be placed on biological control as a technique in regional pest management programs. Classical biological control has been instituted against fifty-four species of weeds in the United States and eighteen in Canada (McFadyen 1998), and it is generally regarded as the safest and most cost-effective technique of controlling exotic weeds. In spite of the control effectiveness of biological agents, little evidence of the evolution of resistance to these agents by the target species exists (Holt and Hochberg 1997).

Nevertheless, enough cases of unanticipated impacts of biological control agents now exist (see Chapters 13 and 20) to demonstrate that classical biological control is not without risk. Many of these date from the early 1900s, when "shotgun" introductions of many possible control species were done with only limited screening (Samways 1997). Simberloff and Stiling (1996) argued that biological control is much riskier than generally believed, largely because little or no monitoring of nontarget species is carried out once biocontrol agents have been released. They believe that known and potential cases of nontarget effects are evidence that present procedures for evaluating biocontrol agents are inadequate. Frank (1998) and others contend that the risks of introducing biocontrol agents must be weighed against those of alternatives that are likely to be used in the absence of biological control. These include damage by the target species itself and by alternative control efforts, such as chemical pesticide applications, that are likely to be applied. Clearly, rigorous screening of proposed biological control species is certainly justified, and this screening must evaluate the risks to nontarget species as fully as practical.

Another issue relates to established exotics. Should all such exotics be discouraged in parks and nature reserves? In some situations, exotic plants have become important habitat elements for native species, and in some cases for threatened or endangered ones (Westman 1990). In California, for example, most wintering sites for monarch butterflies (*Danaus plexippus*) are in groves of eucalyptus or planted conifers. In other cases, the disturbance associated with removal of exotics, such as trees or large shrubs, might lead to invasion or spread of other exotics of even worse character or to destructive soil erosion. The costs of attempted eradication of established exotics may be very high, as well, and money spent in such efforts may be diverted from even more critical management activities. Clearly, control of exotics must be incorporated in an overall set of ecological management priorities.

Managing Exotics in the Era of Global Change

The policies that are formulated for exotics must also take into account global environmental change. Increasing atmospheric CO_2 concentration, climatic warming, and increased nitrogen deposition, together with altered disturbance regimes and habitat fragmentation, are likely to favor the spread and increase in abundance of many exotic species (Dukes and Mooney 1999). A warming climate, for example, will likely increase the suitability of higher latitudes for many tropical diseases and pests. Several serious human parasites and diseases, including malaria, dengue, viral encephalitis, and Chagas' disease, could spread into higher North

American latitudes. Many pests and pathogens of agriculture and commercial forests are also likely to spread to higher latitudes. Tropical and subtropical plants and animals capable of invading natural ecosystems, as they are doing on a large scale in Florida and Hawaii, are also likely to spread northward in the United States. Changing climate will also likely induce ecological imbalances in ecosystems at all latitudes—in effect, large-scale disturbance—that may favor weedy exotic species.

Global change also raises a second issue, that of change in the geographical ranges in which species can exist. For most species, climatic change is likely to shift the suitable geographical range faster than species can migrate. In North America, climatic zones for common tree species may shift northward several hundred miles over the coming century (OTA 1993). Thus, deliberate translocation of species to enable them to keep pace with their optimal climatic zones may be necessary. For example, climatic warming might make areas of southern Canada suitable for species of plants and animals of more southern distribution. If these species are slow to disperse into these regions, should humans assist in their translocation? Or should these species be treated as undesirable exotics and translocations opposed? Similarly, if species migrate on their own, should their spread be permitted, or should they be treated as exotics and their spread opposed? These questions will likely have to be addressed for individual species, because the consequences of changing range will vary. For example, natural or assisted spread of some species might lead to extinctions of other species that are pushed upward in elevation or northward in latitude until their habitat is extinguished.

Global change is also likely to increase the frequency of endangered species, as climatic limits force species upward in elevation, poleward in latitude, or toward barriers they cannot cross. Survival of such species may only be possible in botanical gardens, zoos, or aquaria or as exotics translocated to areas of suitable habitat in other world regions. Biodiversity preserves designed explicitly for exotics may be necessary to enable the survival of the full range of species threatened by shifting climatic zones.

Strengthening North American Institutions

Beyond the specific issues that we have identified are broader issues relating to governmental institutions concerned with managing exotic species. The United States and Canada need to strengthen existing institutions and improve their coordination in order to deal effectively with this challenge. New national, state, and provincial institutions may also be required. A strong international organization is also needed to address the problem of invasive exotics adequately.

Much can be done by strengthening laws and institutions already in place. For example, in the United States, the Lacey Act could be strengthened in a number of ways (OTA 1993). The list of injurious plant and animal species could be lengthened substantially, and the procedure for adding species to the list streamlined. An emergency listing capability could be created so that the import of species could be prohibited until evaluation of their danger is completed. The act could also be amended to allow the prohibition of interstate transport of injurious species. These functions could be accomplished by increasing the role of the U.S. Fish and Wildlife Service in exotics management.

Improved coordination of federal and state agencies that are involved with exotic species is badly needed. In the United States, twenty-one federal agencies are involved with regulation of exotic species or management of lands on which exotics exist. Some agencies, such as the Department of Energy, have no policies on management of exotics (OTA 1993). Regulation of exotics by states also varies greatly. A few states, including Hawaii, prohibit the import or release of all species except those on an approved ("white") list. Many others, such as Florida, have a blacklist of prohibited species, and many others, including California, have no prohibition against import of exotic species.

In the United States, the Federal Interagency Committee for Management of Noxious and Exotic Weeds (FICMNEW) has drafted a strategy for management of invasive plants (FICMNEW 1998). This strategy relies on voluntary cooperation of public and private groups to prevent invasions, control established exotics, and restore disturbed and invaded habitats. The Animal and Plant Health Inspection Service (APHIS) is also drafting a strategy for management of invasive exotics (APHIS 1997). This strategy, entitled "Campaign against Non-Native Invasive Species" sets goals for prevention and early detection of exotic invasions, identification of gaps in authority and organizational response to invasions, and improvements in policy and action. One specific goal is creation of a system for early detection of new invaders and for prompt action to prevent their spread and reduce their impacts. Identified gaps in authority include a lack of emergency response capability, inability to bill costs of emergency responses to those responsible for invasive introductions, inability to charge import fees to pay for risk assessments of exotics proposed for import, and weak coordination of exotics management programs among federal, state, and private sectors. Improvements in policy and action include increasing research, improving data availability and sharing, and increasing public awareness about exotics.

At the national level, informal proposals have also been made to create laws and institutions to address the problem of exotics in a comprehensive manner. Schmitz (1996), for example, has proposed a national center for biological pollution control and prevention, somewhat along the lines of the existing Center for Disease Control and Prevention. Such a center could act to coordinate efforts to prevent exotic invasions with international agencies, foreign governments, different national and state agencies, and private organizations.

Wade (1995) has proposed an exotic species act that establishes an exotic species service, akin to the Fish and Wildlife Service. The act would consolidate federal programs relating to exotic species in the exotic species service and give the service emergency powers to deal quickly with critical invasion problems. An environmental impact analysis would be required for importation of exotic organisms and a white list of species cleared established. The exotic species service would oversee and coordinate federal and state efforts at regulation of exotic species. It would also coordinate national research, eradication, and control efforts.

Strengthening Global Institutions

Exotics are indeed an international problem, and the need for international cooperation in their regulation is great. In North America, just as the need for continental management of migratory birds led to a migratory bird treaty involving Canada, the United States, and Mexico, the need exists for a continental agreement on exotic species. Many examples exist of deleterious exotics crossing international borders by human agency or natural dispersal.

The World Conservation Union, best known by the acronym IUCN, is also drafting a set of guidelines for the prevention of biodiversity loss due to biological invasion. These guidelines deal with reducing the risks of both intentional and unintentional introductions. For intentional introductions, risk analysis and environmental impact assessment (EIA) are recommended. A detailed list of items that should be addressed in an EIA is provided. For unintentional introductions, attention is given to reducing risks both in the source areas of exotics and in recipient areas. Specific recommendations are made for reducing introductions in baggage, cargos, and ballast water. Specific recommendations are also made for management of invasive species that do become established. Early detection and eradication are recommended for incipient invasions, and general strategies for reducing the impacts of widespread exotics are suggested.

Challenge to the Next Century

Exotics will always be with us. Thousands of species have already invaded North America and Hawaii, and their impacts are growing exponentially. More are sure to come. Like other environmental challenges in the twentieth century, to ignore the issue is to accept massive economic, ecological, and aesthetic damage to the North American environment. Just as strong national, state, and provincial programs have been established to protect soils, waters, forests and grasslands, endangered species, and the chemical health of air, water, and land, we must address this challenge with a strong exotic species management program.

Literature Cited

Abbott, I. 1980. Aboriginal man as an exterminator of wallaby and kangaroo populations on islands round Australia. *Oecologia* 44: 347–354.

Abernethy, K. 1994. The establishment of a hybrid zone between red and sika deer (genus *Cervus*). *Molecular Ecology* 3:551–562.

Alberts, A. C., A. D. Richman, D. Tran, R. Sauvajot, C. McCalvin, and D. T. Bolger. 1993. Effects of habitat fragmentation on native and exotic plants in southern California coastal scrub. Pp. 103–110 in J. E. Keeley (Ed.), *Interface between ecology and land development in California.* California Academy of Sciences, Los Angeles, CA.

Allen, C. R., R. S. Lutz, and S. Demerais. 1998. Ecological effects of the invasive nonindigenous ant, *Solenopsis invicta,* on native vertebrates: The wheels of the bus. Transactions of the North American Wildlife and Natural Resource Conference 63:56–65.

Allen, D. W. 1996. Habitat relationships of vertebrates on Naval Air Station Miramar, San Diego, California. Master's Thesis, San Diego State University, San Diego, CA.

Allen, J. A. 1998. Mangroves as alien species: The case of Hawaii. *Global Ecology and Biogeography Letters* 7:61–71.

Allan, J. D., and A. S. Flecker. 1993. Biodiversity conservation in running waters. *BioScience* 43:32–43.

Allendorf, F. W., and R. F. Leary. 1988. Conservation and distribution of genetic variation in a polytypic species, the cutthroat trout. *Conservation Biology* 2:170–184.

Anable, M. E., M. P. McClaran, and G. B. Ruyle. 1992. Spread of introduced Lehmann lovegrass *Eragrostis lehmanniana* Nees. in southern Arizona, USA. *Biological Conservation* 61:181–188.

Anderson, B. W., A. Higgins, and R. D. Ohmart. 1977. Avian use of saltcedar communities in the lower Colorado River. Pp. 128–136 in R. R. Johnson and D. A. Jones (Eds.), *Importance, preservation and management of riparian habitat: A symposium.* USDA Forest Service General Technical Report RM-43.

Anderson, B. W., and R. D. Ohmart. 1985. Riparian revegetation as a mitigating process in stream and river restoration. Pp. 41–80 in J. A. Gore (Ed.), *The restoration of rivers and streams: Theories and experience.* Butterworth Publishers, Boston, MA.

Anderson, M. G. 1995. Interactions between *Lythrum salicaria* and native organisms: A critical review. *Environmental Management* 19:225–231.

Anonymous. 1998. Eagle count declining as grasses disappear. *The New Mexican* (Santa Fe). January 3, page B4.

Anttila, C. K., C. C. Daehler, N. E. Rank, and D. R. Strong. 1998. Greater male fitness of a rare invader (*Spartina alterniflora,* Poaceae) threatens a common native *(Spartina foliosa)* with hybridization. *American Journal of Botany* 85:1597–1601.

APHIS. 1997. Campaign against non-native invasive species (draft). Internet: http://www.aphis.usda.gov/HyperNews/get/nis.html.

Athens, J. S. 1997. Hawaiian native lowland vegetation in prehistory. Pp. 248–270 in P. V. Kirch and T. L. Hunt (Eds.), *Historical ecology in the Pacific islands.* Yale University Press, New Haven, CT.

Auerbach, M., and D. Simberloff. 1988. Rapid leaf-miner colonization of introduced trees and shifts in sources of herbivore mortality. *Oikos* 52:41–50.

Baker, H. G. 1965. Characteristics and modes of origin of weeds. Pp. 147–169 in H. G. Baker and C. L. Stebbins (Eds.), *The genetics of colonizing species.* Academic Press, New York, NY.

Baltz, D. M. 1991. Introduced fishes in marine systems and inland seas. *Biological Conservation* 56:151–177.

Bartolome, J. W., and B. Gemmill. 1981. The ecological status of *Stipa pulchra* (Poaceae) in California. *Madroño* 28:172–184.

Bartolome, J. W., S. E. Klukkert, and W. J. Barry. 1986. Opal phytoliths as evidence for displacements in California annual grassland. *Madroño* 33:217–222.

Baskin, Y. 1996. Curbing undesirable invaders. *BioScience* 46:732–736.

Batra, S. W. T. 1982. Biological control in agroecosystems. *Science* 215:134–139.

Beck, K. G. 1993. How do weeds affect us all? Eighth Grazing Lands Forum, Washington, DC.

Beerling, D. J. 1995. General aspects of plant invasions: An overview. Pp. 237–247 in P. Pysek, K. Prach, M. Rejmánek, and W. Wade (Eds.), *Plant invasions: General aspects and special problems.* SPB Academic Publishing, Amsterdam, The Netherlands.

Belcher, J. W., and S. D. Wilson. 1989. Leafy spurge and the species composition of a mixed-grass prairie. *Journal of Range Management* 42:172–175.

Belsky, A. J., and D. M. Blumenthal. 1996. Effects of livestock grazing on stand dynamics and soils in upland forests of the interior West. *Conservation Biology* 10:1–14.

Bergersen, E. P., and D. E. Anderson. 1997. The distribution and spread of *Myxobolus cerebralis* in the United States. *Fisheries* 22(8):6–7.

Bertness, M. D. 1984. Habitat and community modification by an introduced herbivorous snail. *Ecology* 65:370–381.

Billings, W. D. 1990. *Bromus tectorum,* a biotic cause of ecosystem impoverishment in the Great Basin. Pp. 301–322 in G. M. Woodwell (Ed.), *The earth in transition: Patterns and processes of biotic impoverishment.* Cambridge University Press, Cambridge, England.

Biswell, H. H. 1956. Ecology of California grasslands. *Journal of Range Management* 9:19–24.

Bjergo, C., C. Boydstun, M. Crosby, S. Kokkanakis, and R. Sayers, Jr. 1995. Nonnative aquatic species in the United States and coastal waters. Pp. 428–431 in E. T. LaRoe (Ed.), *Our living resources.* USDI National Biological Service, Washington, DC.

Blackburn, W. H., R. W. Knight, and J. L. Schuster. 1982. Saltcedar influence on sedimentation in the Brazos River. *Journal of Soil and Water Conservation* 37:298–301.

Blair, R. M., and M. J. Langlinais. 1960. Nutria and swamp rabbits damage baldcypress seedlings. *Journal of Forestry* 58:388–389.

Bland, J. D., and S. A. Temple. 1993. The Himalayan snowcock: North America's newest exotic bird. Pp. 149–155 in B. N. McKnight (Ed.), *Biological pollution: The control and impact of invasive exotic species.* Indiana Academy of Science, Indianapolis, IN.

Blossey, B., and R. Notzold. 1995. Evolution of increased competitive ability in invasive nonindigenous plants: A hypothesis. *Journal of Ecology* 83:887–889.

Bock, C. E., J. H. Bock, K. L. Jepson, and J. C. Ortega. 1986. Ecological effects of planting African lovegrasses in Arizona. *National Geographic Research* 2:456–463.

Bock, C. E., and L. W. Lepthien. 1976. Population growth in the cattle egret. *Auk* 93:164–166.

Bock, J. H., and C. E. Bock. 1995. The challenges of grassland conservation. Pp. 199–222 in A. Joern and K. H. Keeler (Eds.), *The changing prairie: North American grasslands.* Oxford University Press, New York, NY.

Bodle, M. J., A. P. Ferriter, and D. D. Thayer. 1994. The biology, distribution, and ecological consequences of *Melaleuca quinquenervia* in the Everglades. Pp. 341–355 in S. M. Davis and J. C. Ogden (Eds.), *Everglades: The system and its restoration.* St. Lucie Press, Delray Beach, FL.

Boileau, M. G. 1985. The expansion of the white perch, *Morone americana,* in the lower Great Lakes. *Fisheries* 10(1):6–10.

Bolze, D. 1992. *The wild bird trade.* Wildlife Conservation International Policy Report No. 2. New York Zoological Society, Bronx, NY.

Bossard, C. C. 1991. The role of habitat disturbance, seed predation and ant dispersal on establishment of the exotic shrub *Cytisus scoparius* in California. *American Midland Naturalist* 126:1–13.

Bottoms, R. M., and T. D. Whitson. 1998. A systems approach for the management of Russian knapweed *(Centaurea repens). Weed Technology* 12:363–366.

Boyd, D. 1985. Gorse. *Fremontia* 12(4):16–17.

Boydstun, C., P. Fuller, and J. D. Williams. 1995. Nonindigenous fish. Pp. 431–433 in E. T. LaRoe (Ed.), *Our living resources.* USDI National Biological Service, Washington, DC.

Bradford, D. F. 1989. Allotopic distribution of native frogs and introduced fishes in high Sierra Nevada lakes of California: Implications of the negative effect of fish introductions. *Copeia* 1989:775–778.

Bradford, D. F., S. E. Franson, A. C. Neale, D. T. Heggem, G. R. Miller, and G. E. Canterbury. 1998. Bird species assemblages as indicators of biological integrity in Great Basin rangeland. *Environmental Monitoring and Assessment* 49:1–22.

Bradford, D. F., F. Tabatabai, and D. M. Graber. 1993. Isolation of remaining populations of the native frog, *Rana mucosa,* by introduced fishes in Sequoia and Kings Canyon National Parks, California. *Conservation Biology* 7:882–888.

Bratton, S. P. 1975. The effect of the European wild boar, *Sus scrofa,* on gray beech forest in the Great Smoky Mountains. *Ecology* 56:1356–1366.

Brenton, B., and R. Klinger. 1994. Modeling the expansion and control of fennel *(Foeniculum vulgare)* on the Channel Islands. Pp. 497–504 in W. L. Halvorson and G. J. Maender (Eds.), *The fourth California islands symposium: Update on the status of resources.* Santa Barbara Museum of Natural History, Santa Barbara, CA.

Brock, J. H., and M. C. Farkas. 1997. Alien woody plants in a Sonoran Desert urban riparian corridor: An early warning system about invasiveness? Pp. 19–35 in J. H. Brock, M. Wade, P. Pysek, and D. Green (Eds.), *Plant invasions: Studies from North America and Europe.* Backhuys Publishers, Leiden, The Netherlands.

Brock, J. H., M. Wade, P. Pysek, and D. Green (Eds.). 1997. *Plant invasions: Studies from North America and Europe.* Backhuys Publishers, Leiden, The Netherlands.

Brothers, T. S., and A. Spingarn. 1992. Forest fragmentation and alien plant invasion of central Indiana old-growth forests. *Conservation Biology* 6:91–100.

Brown, B. J., and C. E. Wickstrom. 1997. Adventitious root production and survival of purple loosestrife *(Lythrum salicaria)* shoot sections. *Ohio Journal of Science* 97:2–4.

Brown, B. T. 1990. Ecology and management of riparian breeding birds in tamarisk habitats along the Colorado River in Grand Canyon National Park, Arizona. Pp. 68–73 in M. R. Kunzmann, R. R. Johnson, and P. S. Bennett (Eds.), *Tamarisk control in the southwestern United States.* USDI, National Park Service, Special Rep. No. 9.

Brown, B. T., and M. W. Trosset. 1989. Nesting-habitat relationships of riparian birds along the Colorado River in Grand Canyon, Arizona. *Southwestern Naturalist* 34:260–270.

Brown, J. H. 1989. Patterns, modes and extents of invasions by vertebrates. Pp. 85–109 in J. A. Drake, H. A. Mooney, F. di Castri, R. H. Groves, F. J. Kruger, M. Rejmánek, and M. Williamson (Eds.), *Biological invasions: A global perspective.* John Wiley and Sons, New York, NY.

Brown, L. R., and P. B. Moyle. 1997. Invading species in the Eel River, California: Successes, failures, and relationships with resident species. *Environmental Biology of Fishes* 49:271–291.

Brown, N. S., and G. I. Wilson. 1975. A comparison of the ectoparasites of the house sparrow *(Passer domesticus)* from North America and Europe. *American Midland Naturalist* 94:154–165.

Bruce, K. A., G. N. Cameron, and P. A. Harcombe. 1995. Initiation of a new woodland type on the Texas coastal prairie by the Chinese tallow tree *(Sapium sebiferum* (L.) Roxb.). *Bulletin of the Torrey Botanical Club* 122:215–225.

Bruce, K. A., G. N. Cameron, P. A. Harcombe, and G. Jubinsky. 1997. Introduction, impact on native habitats, and management of a woody invader, the Chinese tallow tree, *Sapium sebiferum* (L.) Roxb. *Natural Areas Journal* 17: 255–260.

Bucher, E. H. 1992. The causes of extinction of the passenger pigeon. *Current Ornithology* 9:1–36.

Buchmann, S. L., and G. P. Nabhan. 1996. *The forgotten pollinators.* Island Press, Washington, DC.

Bump, G. 1968. Exotics and the role of the State-Federal Foreign Game Introduction Program. Pp. 5–8 in *Symposium on introduction of exotic animals: Ecological and socioeconomic considerations.* Cesar Kleberg Research Program in Wildlife Ecology. Texas A&M University, College Station.

Bury, R. B., and R. A. Luckenbach. 1976. Introduced amphibians and reptiles in California. *Biological Conservation* 10:1–14.

Busch, D. E., N. L. Ingraham, and S. D. Smith. 1992. Water uptake in woody riparian phreatophytes of the southwestern United States: A stable isotope study. *Ecological Applications* 2:450–459.

Busch, D. E., and M. L. Scott. 1995. Western riparian ecosystems. Pp. 286–290 in E. T. LaRoe (Ed.), *Our living resources.* USDI National Biological Service, Washington, DC.

Busch, D. E., and S. D. Smith. 1993. Effects of fire on water and salinity relations of riparian woody taxa. *Oecologia* 94:186–194.

———. 1995. Mechanisms associated with decline of woody species in riparian ecosystems of the southwestern U.S. *Ecological Monographs* 65:347–370.

Busiahn, T. R. 1997. Ruffe control: A case study of an aquatic nuisance species control program. Pp. 69–86 in F. M. D'Itri (Ed.), *Zebra mussels and aquatic nuisance species.* Ann Arbor Press, Chelsea, MI.

Butler, L., V. Kondo, and D. Blue. 1997. Effects of tebufenozide (RH-5992) for gypsy moth (Lepidoptera: Lymantriidae) suppression on nontarget canopy arthropods. *Environmental Entomology* 26:1009–1015.

Byrd, G. V., E. P. Bailey, and W. Stahl. 1997. Restoration of island populations of black oystercatchers and pigeon guillemots by removing introduced foxes. *Colonial Waterbirds* 20:253–260.

Cameron, G. N., and S. R. Spencer. 1989. Rapid leaf decay and nutrient release in a Chinese tallow forest. *Oecologia* 80:222–228.

Campbell, F. T. 1993. Legal avenues for controlling exotics. Pp. 243–250 in B. N. McKnight (Ed.), *Biological pollution: The control and impact of invasive exotic species.* Indiana Academy of Science, Indianapolis, IN.

Campbell, F. T., and S. E. Schlarbaum. 1994. *Fading forests: North American trees and the threat of exotic pests.* Natural Resources Defense Council, New York, NY.

Carey, J. R. 1996. The incipient Mediterranean fruit fly population in California: Implications for invasion biology. *Ecology* 77:1690–1697.

Carlquist, S. 1980. *Hawaii: A natural history.* Pacific Tropical Botanic Garden, Honolulu, HI.

Carlton, J. T. 1989. Man's role in changing the face of the ocean: Biological invasions and implications for conservation of near-shore environments. *Conservation Biology* 3:265–273.

———. 1992a. Dispersal of living organisms into aquatic ecosystems as mediated by aquaculture and fisheries activities. Pp. 13–45 in A. Rosenfeld and R. Mann (Eds.), *Dispersal of living organisms into aquatic ecosystems.* Maryland Sea Grant Publication, College Park, MD.

———. 1992b. Introduced marine and estuarine mollusks of North America: An end-of-the-20th-century perspective. *Journal of Shellfish Research* 11:489–505.

———. 1996a. Biological invasions and cryptogenic species. *Ecology* 77:1653–1655.

———. 1996b. Marine bioinvasions: The alteration of marine ecosystems by nonindigenous species. *Oceanography* 9:36–43.

Carlton, J. T., and J. B. Geller. 1993. Ecological roulette: The global transport of nonindigenous marine organisms. *Science* 261:78–82.

Carlton, J. T., J. K. Thompson, L. E. Schemel, and F. H. Nichols. 1990. Remarkable invasion of San Francisco Bay (California, USA) by the Asian clam *Potamocorbula amurensis.* I. Introduction and dispersal. *Marine Ecology Progress Series* 66:81–94.

Carmichael, G. J., J. N. Hanson, M. E. Schmidt, and D. C. Morizot. 1993. Introgression among Apache, cutthroat, and rainbow trout in Arizona. *Transactions of the American Fisheries Society* 122:121–130.

Carothers, S. W., and R. R. Johnson. 1983. Status of the Colorado River Ecosystem in Grand Canyon National Park and Glen Canyon National Recreation Area. Pp. 139–160 in V. D. Adams and V. A. Lamarra (Eds.), *Aquatic resources management of the Colorado River Ecosystem.* Ann Arbor Science Publishers, Ann Arbor, MI.

Carothers, S. W., M. E. Stitt, and R. R. Johnson. 1976. Feral asses on public lands: An analysis of biotic impact, legal considerations and management alternatives. *North American Wildlife Conference* 41:396–405.

Carroll, S. P., and H. Dingle. 1996. The biology of post-invasion events. *Biological Conservation* 78:207–214.

Carroll, S. P., S. P. Klassen, and H. Dingle. 1998. Rapidly evolving adaptations to host ecology and nutrition in the soapberry bug. *Evolutionary Ecology* 12:955–968.

Carson, H. L., and D. A. Clague. 1995. Geology and biogeography of the Hawaiian Islands. Pp. 14–29 in W. L. Wagner and V. A. Funk (Eds.), *Hawaiian biogeography.* Smithsonian Institution Press, Washington, DC.

Carson, R. 1962. *Silent spring.* Houghton-Mifflin, New York, NY.

Case, T. J., and D. T. Bolger. 1991. The role of introduced species in shaping the distribution and abundance of island reptiles. *Evolutionary Ecology* 5:272–290.

Case, T. J., A. Suarez, and D. T. Bolger. 1997. The interaction of habitat fragmentation and invasion in southern California ant communities. Society for Conservation Biology Annual Meeting, Victoria, British Columbia, Canada.

Causey, M. K., and C. A. Cude. 1978. Feral dog predation of gopher tortoise, *Gopherus polyphemus* (Reptilia, Testudines, Testudinidae), in southeast Alabama. *Herpetological Review* 9:94–95.

Cavers, P. B., M. I. Heagy, and R. F. Kokron. 1979. The biology of Canadian weeds. 35. *Alliaria petiolata* (M. Bieb.) Cavara and Grande. *Canadian Journal of Plant Science* 59:217–229.

Center, T. D., J. H. Frank, and F. A. Dray, Jr. 1997. Biological control. Pp. 245–263 in D. Simberloff, D. C. Schmitz, and T. C. Brown (Eds.), *Strangers in paradise.* Island Press, Washington, DC.

Chace, J. F., and A. Cruz. 1998. Range of the brown-headed cowbird in Colorado: Past and present. *Great Basin Naturalist* 58:245–249.

Chandler, M. 1997. The species turnover in the fish fauna over a forty-year period in Massachusetts lakes. Society for Conservation Biology Annual Meeting, Victoria, British Columbia, Canada.

Chapuis, J. L. 1994. Alien animals, impact and management in the French subantarctic islands. *Biological Conservation* 67:97–104.

Charlebois, P. M., and G. A. Lamberti. 1996. Invading crayfish in a Michigan stream: Direct and indirect effects on periphyton and macroinvertebrates. *Journal of the North American Benthological Society* 15:551–563.

Charlebois, P. M., J. E. Marsden, R. G. Goettel, R. K. Wolfe, D. J. Jude, and S. Rudnika. 1997. *The round goby.* Illinois-Indiana Sea Grant Program and Illinois Natural History Survey. INHS Special Publication No. 20.

Chicoine, T. K., P. K. Fay, and G. A. Nielsen. 1985. Predicting weed migration from soil and climate maps. *Weed Science* 34:57–61.

Christie, W. J. 1974. Changes in the fish species composition of the Great Lakes. *Journal of the Fisheries Research Board of Canada* 31:827–854.

Clarkson, R. W., and J. C. Rorabaugh. 1989. Status of leopard frogs *(Rana pipiens* complex: Ranidae) in Arizona and southeastern California. *Southwestern Naturalist* 34:531–538.

Clegg, M. T., and R. W. Allard. 1972. Patterns of genetic differentiation in the slender wild oat species *Avena barbata. Proceedings of the National Academy of Science USA* 69:1820–1824.

Cleverly, J. R., S. D. Smith, A. Sala, and D. A. Devitt. 1997. Invasive capacity of *Tamarix ramosissima* in a Mojave Desert floodplain: The role of drought. *Oecologia* 111:12–18.

Cloern, J. E., and A. E. Alpine. 1991. *Potamocorbula amurensis,* a recently introduced Asian clam, has had dramatic effects on the phytoplankton biomass and production in northern San Francisco Bay. *Journal of Shellfisheries Research* 10:258–259.

Coblenz, B. E. 1978. The effects of feral goats *(Capra hircus)* on island ecosystems. *Biological Conservation* 13:279–286.

Cohen, A. N., and J. T. Carlton. 1995. *Nonindigenous aquatic species in a United*

States estuary: A case study of the biological invasions of the San Francisco Bay and Delta. U.S. Fish and Wildlife Service, Washington, DC.

———. 1998. Accelerating invasion rate in a highly invaded estuary. *Science* 279:555–558.

Cole, F. R., L. L. Loope, A. C. Medeiros, J. A. Raikes, and C. S. Wood. 1995. Conservation implications of introduced game birds in high-elevation Hawaiian shrubland. *Conservation Biology* 9:306–313.

Cole, F. R., A. C. Medeiros, L. L. Loope, and W. W. Zuehlke. 1992. Effects of the Argentine ant on arthropod fauna of Hawaiian high-elevation shrubland. *Ecology* 73:1313–1322.

Coleman, J. S., S. A. Temple, and S. R. Craven. 1996. Cats and wildlife: A conservation dilemma. Internet: http://www.wisc.edu/wildlife/extension/catfly3.htm.

Conner, W. H., and J. R. Toliver. 1990. Long-term trends in the bald-cypress *(Taxodium distichum)* resource in Louisiana (U.S.A.). *Forest Ecology and Management* 33/34:543–557.

Connor, E. F., S. H. Faeth, D. Simberloff, and P. A. Opler. 1980. Taxonomic isolation and the accumulation of herbivorous insects: A comparison of introduced and native trees. *Ecological Entomology* 5:205–211.

Cope, W. G., M. R. Bartsch, and R. R. Hayden. 1997. Longitudinal patterns in abundance of the zebra mussel *(Dreissena polymorpha)* in the upper Mississippi River. *Journal of Freshwater Biology* 12:235–238.

Cordo, H. A., C. J. DeLoach, and R. Ferrer. 1981. Biological studies on two weevils, *Oechetina bruchi* and *Onychylis cretanus,* collected from *Pistia* and other aquatic plants in Argentina. *Annals of the Entomological Society of America* 74:363–369.

Cornell, H. V., and B. A. Hawkins. 1993. Accumulation of native parasitoid species on introduced herbivores: A comparison of hosts as natives and hosts as invaders. *American Naturalist* 141: 847–865.

Costanza, R., R. D'Arge, R. de Groot, S. Farber, M. Grasso, B. Hannon, K. Limburg, S. Naeem, R. V. O'Neill, J. Paruelo, R. G. Raskin, P. Sutton, and M. van den Belt. 1997. The value of the world's ecosystem services and natural capital. *Nature* 387:253–260.

Counts, C. L. 1986. The zoogeography and history of the invasion of the United States by *Corbicula fluminea* (Bivalvia: Corbiculidae). *American Malacological Bulletin,* Special Edition #2:7–39.

Courtenay, W. R., Jr. 1997. Nonindigenous fishes. Pp. 109–122 in D. Simberloff, D. C. Schmitz, and T. C. Brown (Eds.), *Strangers in paradise.* Island Press, Washington, DC.

Courtenay, W. R. Jr., D. A. Hensley, J. N. Taylor, and J. A. McCann. 1984. Distribution of exotic fishes in the continental United States. Pp. 41–77 in W. R. Courtenay, Jr. and J. R. Stauffer, Jr. (Eds.). 1984. *Distribution, biology, and management of exotic fishes.* Johns Hopkins University Press, Baltimore, MD.

Courtenay, W. R. Jr., and C. C. Kohler. 1986. Exotic fishes in North American fishery management. Pp 401–413 in R. H. Stroud (Ed.), *Fish culture in fisheries management.* American Fisheries Society, Bethesda, MD.

Courtenay, W. R., Jr., and J. R. Stauffer, Jr. (Eds.). 1984. *Distribution, biology, and management of exotic fishes.* Johns Hopkins University Press, Baltimore, MD.

Cowie, R. H. 1998. Patterns of introduction of non-indigenous non-marine snails and slugs in the Hawaiian Islands. *Biodiversity and Conservation* 7:349–368.

Cowie, R. H., N. L. Evenhuis, and C. C. Christensen. 1995. *Catalog of the native land and freshwater molluscs of the Hawaiian Islands.* Backhuys Publishers, Leiden, The Netherlands.

Cowie, R. H., and F. G. Howarth. 1998. Biological control: Disputing the indisputable. *Trends in Ecology and Evolution* 13:110.

Cox, G. W. 1985. The evolution of avian migration systems between temperate and tropical regions of the New World. *American Naturalist* 126:451–474.

———. 1997. *Conservation biology: Concepts and applications.* Second edition. Wm. C. Brown Company, Dubuque, IA.

Cox, G. W., and M. D. Atkins. 1979. *Agricultural ecology.* W. H. Freeman and Company, San Francisco, CA.

Craig, G. B., Jr. 1993. The diaspora of the Asian tiger mosquito. Pp. 101–120 in B. N. McKnight (Ed.), *Biological pollution: The control and impact of invasive exotic species.* Indiana Academy of Science, Indianapolis, IN.

Craven, R. B., D. A. Eliason, D. B. Francy, P. Reiter, and E. G. Campos. 1988. Importation of *Aedes albopictus* and other mosquito species into the United States in used tires from Asia. *Journal of the Mosquito Control Association* 4:138–142.

Crawford, C. S., L. M. Ellis, and M. C. Molles, Jr. 1996. The middle Rio Grande bosque: An endangered ecosystem. *New Mexico Journal of Science* 36:276–299.

Crawley, M. J. 1989. Chance and timing in biological invasions. Pp. 407–423 in J. A. Drake, H. A. Mooney, F. di Castri, R. H. Groves, F. J. Kruger, M. Rejmánek, and M. Williamson (Eds.), *Biological invasions: A global perspective.* John Wiley and Sons, New York, NY.

Cronk, Q. C. B., and J. L. Fuller. 1995. *Plant invaders: The threat to natural systems.* Chapman and Hall, London, England.

Crooks, J. A. 1996. The population ecology of an exotic mussel, *Musculista senhousia,* in a southern California bay. *Estuaries* 19:42–50.

———. 1997. Invasions and effects of exotic marine species: A perspective from southern California. American Fisheries Society Annual Meeting, Monterey, CA.

———. 1998. Habitat alteration and community-level effects of an exotic mussel, *Musculista senhousia. Marine Ecology Progress Series* 162:137–152.

Crooks, J. A., and M. E. Soulé. 1999. Lag times in population explosions of invasive species: Causes and implications. In press in D. T. Sandland (Ed.), *Invasive species and biodiversity management.* Chapman and Hall, New York, NY.

Crosby, A. W. 1986. *Ecological imperialism: The biological expansion of Europe 900–1900.* Cambridge University Press, Cambridge, England.

Crossman, E. J. 1984. Introduction of exotic fishes into Canada. Pp. 78–101 in W. R. Courtney, Jr. and J. R. Stauffer, Jr. (Eds), *Distribution, biology, and management of exotic fishes.* Johns Hopkins University Press, Baltimore, MD.

Daehler, C. C. 1998. The taxonomic distribution of invasive angiosperm plants: Ecological insights and comparison to agricultural weeds. *Biological Conservation* 84:167–180.

Daehler, C. C., and C. Anttila. 1997. Impacts of introduced smooth cordgrass *(Spartina alterniflora)* on Pacific estuaries. Ecological Society of America Annual Meeting, Albuquerque, NM.

Daehler, C. C., C. K. Anttila, D. R. Ayres, D. R. Strong, and J. P. Bailey. 1999. Evolution of a new ecotype of *Spartina alterniflora* (Poaceae) in San Francisco Bay, California, USA. *American Journal of Botany* 86:543–546.

Daehler, C. C., and D. R. Strong. 1994. Variable reproductive output among clones of *Spartina alterniflora* (Poaceae) invading San Francisco Bay, California: The influence of herbivory, pollination, and establishment site. *American Journal of Botany* 81:307–313.

———. 1996. Status, prediction and prevention of introduced cordgrass *Spartina* spp. invasions in Pacific estuaries, USA. *Biological Conservation* 78:51–58.

———. 1997a. Reduced herbivore resistance in introduced smooth cordgrass *(Spartina alterniflora)* after a century of herbivore-free growth. *Oecologia* 110:99–108.

———. 1997b. Hybridization between introduced smooth cordgrass *(Spartina alterniflora;* Poaceae) and native California cordgrass *(S. foliosa)* in San Francisco Bay, California, USA. *American Journal of Botany* 84:607–611.

Dahl, F. H., and R. B. McDonald. 1980. Effects of control of the sea lamprey *(Petromyzon marinus)* on migratory and resident fish populations. *Canadian Journal of Fisheries and Aquatic Sciences* 37:1886–1894.

Dahlsten, D. L. 1986. Control of invaders. Pp. 275–302 in H. A. Mooney and J. A. Drake (Eds.), *Ecology of biological invasions of North America and Hawaii.* Springer-Verlag, New York, NY.

Dalrymple, G. H. 1994. Nonindigenous amphibians and reptiles in Florida. Pp. 67–78 in D. C. Schmitz and T. C. Brown (Eds.), *An assessment of invasive non-indigenous species in Florida's public lands.* Florida Department of Environmental Protection, Tallahassee, FL.

D'Antonio, C. M., and P. M. Vitousek. 1992. Biological invasions by exotic grasses, the grass/fire cycle, and global change. *Annual Review of Ecology and Systematics* 23:63–87.

Dark, S. J., R. J. Gutierrez, and G. I. Gould, Jr. 1998. The barred owl *(Strix varia)* invasion in California. *Auk* 115:50–56.

Daughtrey, M. L., C. R. Hibben, K. O. Britton, M. T. Windham, and S. C. Redlin. 1996. Dogwood anthracnose: Understanding a disease new to North America. *Plant Disease* 80:349–358.

Davis, D. E. 1950. The growth of starling, *Sturnus vulgaris,* populations. *Auk* 67:460–465.

DeBach, P. 1974. *Biological control by natural enemies.* Cambridge University Press, London, England.

DeLoach, C. J. 1991. Past successes and current prospects in biological control of weeds in the United States and Canada. *Natural Areas Journal* 11:129–142.

Demarais, S., J. T. Baccus, and M. S. Traweek, Jr. 1998. Nonindigenous ungulates in Texas: Long-term population trends and possible competitive mechanisms. *Transactions of the North American Wildlife and Natural Resource Conference* 63:49–55.

Denney, R. N. 1974. The impact of uncontrolled dogs on wildlife and livestock. *Transactions of the North American Wildlife and Natural Resource Conference* 39:257–291.

Dermott, R., and M. Munawar. 1993. Invasion of Lake Erie offshore sediments by *Dreissena,* and its ecological implications. *Canadian Journal of Fisheries and Aquatic Sciences* 50:2298–2304.

Dermott, R., J. Witt, Y. M. Um, and M. González. 1998. Distribution of the Ponto-Caspian *Echinogammarus ischnius* in the Great Lakes and replacement of native *Gammarus fasciatus. Journal of Great Lakes Research* 24:442–452.

Devitt, D. A., J. M. Piorkowski, S. D. Smith, J. R. Cleverly, and A. Sala. 1997. Plant water relations of *Tamarix ramosissima* in response to the imposition and alleviation of soil moisture stress. *Journal of Arid Environments* 36:527–540.

Dexter, D. M., and J. A. Crooks. Submitted. The soft bottom benthos of an urbanized bay and its invasion by an exotic mussel: A 27-year history. Submitted to the *Bulletin of the Southern California Academy of Sciences.*

Dill, W. A., and A. J. Cordone. 1997. History and status of introduced fishes in California, 1871–1996: Conclusions. *Fisheries* 22(10):15–18, 35.

Dingle, H. 1980. The ecology and evolution of migration. Pp. 1–101 in S. A. Gauthreaux, Jr. (Ed.), *Animal migration, orientation, and navigation.* Academic Press, New York, NY.

DiTomaso, J. M. 1998. Impact, biology, and ecology of saltcedar (*Tamarix* spp.) in the southwestern United States. *Weed Technology* 12:326–336.

Dobkin, D. S., A. C. Rich, and W. H. Pyle. 1998. Habitat and avifaunal recovery from livestock grazing in a riparian meadow system of the northwestern Great Basin. *Conservation Biology* 12:209–221.

Donald, W. W. 1994. The biology of Canada thistle *(Cirsium arvense). Reviews of Weed Science* 6:77–101.

Dowell, R. V., and R. Gill. 1989. Exotic invertebrates and their effects on California. *Pan-Pacific Entomologist* 65:132–145.

Dowling, T. E., and M. R. Childs. 1992. Impact of hybridization on a threatened trout of the southwestern United States. *Conservation Biology* 6:355–364.

Drake, D. R. 1998. Relationships among the seed rain, seed bank, and vegetation of a Hawaiian forest. *Journal of Vegetation Science* 9:103–112.

Drake, J. A., H. A. Mooney, F. di Castri, R. H. Groves, F. J. Kruger, M. Rejmánek, and M. Williamson (Eds.). 1989. *Biological invasions: A global perspective.* John Wiley and Sons, New York, NY.

Drea, J. J., Jr. 1991. The philosophy, procedures, and cost of developing a classical biological control of weeds project. *Natural Areas Journal* 11:143–147.

Drees, B. M. 1994. Red imported fire ant predation on nestlings of colonial water birds. *Southwestern Entomologist* 19:355–359.

Drost, C. A., and G. M. Fellers. 1995. Non-native animals on public lands. Pp.

440–442 in E. T. LaRoe (Ed.), *Our living resources.* USDI National Biological Service, Washington, DC.

Dukes, J. S., and H. A. Mooney. 1999. Does global change increase the success of biological invaders? *Trends in Ecology and Evolution* 14:135–139.

Duncan, K. W. 1997. A case study in *Tamarix ramosissima* control: Spring Lake, New Mexico. Pp. 115–121 in J. H. Brock, M. Wade, P. Pysek, and D. Green (Eds.), *Plant invasions: Studies from North America and Europe.* Backhuys Publishers, Leiden, The Netherlands.

Dyer, A. R., and K. J. Rice. 1997. Intraspecific and diffuse competition: The response of *Nasella pulchra* in a California grassland. *Ecological Applications* 7:484–492.

Ebinger, J. E. 1996. *Acer ginnala,* Amur maple. P. 25 in J. M. Randall and J. Marinelli (Eds.), *Invasive plants: Weeds of the global garden.* Brooklyn Botanic Garden, Brooklyn, NY.

Edwards, R. J. 1979. A report of Guadaloupe bass *Micropterus treculi* x smallmouth bass *Micropterus dolomieui* hybrids from 2 localities in the Guadaloupe River, Texas USA. *Texas Journal of Science* 31:231–238.

Edwards, S. W. 1992. Observations on the prehistory and ecology of grazing in California. *Fremontia* 20:3–11.

Egerton, F. N. 1987. Pollution and aquatic life in Lake Erie: Early scientific studies. *Environmental Review* 11:189–205.

Ehrenfeld, J. G. 1997. Invasion of deciduous forest preserves in the New York metropolitan region by Japanese barberry (*Berberis thunbergii* DC.). *Journal of the Torrey Botanical Society* 124:210–215.

Ehrlich, P. R. 1989. Attributes of invaders and the invading process: Vertebrates. Pp. 315–328 in J. A. Drake, H. A. Mooney, F. di Castri, R. H. Groves, F. J. Kruger, M. Rejmánek, and M. Williamson (Eds.), *Biological invasions: A global perspective.* John Wiley and Sons, New York, NY.

Eldredge, L. G., and S. E. Miller. 1998. Numbers of Hawaiian species: Supplement 3, with notes on fossil species. *Bishop Museum Occasional Papers* 55:3–15.

Elliott, J. J., and R. S. Arbib, Jr. 1953. Origin and status of the house finch in the eastern United States. *Auk* 70:31–37.

Ellis, L. M. 1995. Bird use of saltcedar and cottonwood vegetation in the Middle Rio Grande Valley of New Mexico, U.S.A. *Journal of Arid Environments* 30:339–349.

Ellis, L. M., C. S. Crawford, and M. C. Molles, Jr. 1997. Rodent communities in native and exotic riparian vegetation in the Middle Rio Grande Valley of central New Mexico. *The Southwestern Naturalist* 42:13–19.

———. 1998. Comparison of litter dynamics in native and exotic riparian vegetation along the Middle Rio Grande Valley of central New Mexico, U.S.A. *Journal of Arid Environments* 38:283–296.

Elton, C. S. 1958. *The ecology of invasions by animals and plants.* Methuen & Co., London, England.

Engel S. 1995. Eurasian watermilfoil as a fishery management tool. *Fisheries* 20(3):20–27.

Evers, D. E., C. E. Sasser, J. G. Gosselink, D. A. Fuller, and J. M. Visser. 1998. The impact of vertebrate herbivores on wetland vegetation in Atchafalaya Bay, Louisiana. *Estuaries* 21:1–13.

Ewel, J. J. 1986. Invasibility: Lessons from South Florida. Pp. 215–230 in H. A. Mooney and J. A. Drake (Eds.), *Ecology of biological invasions of North America and Hawaii.* Springer-Verlag, New York, NY.

Fausch, K. D., and K. R. Bestgen. 1997. Ecology of fishes indigenous to the central and southwestern Great Plains. Pp. 131–166 in F. L. Knopf and F. B. Samson (Eds.), *Ecology and conservation of Great Plains vertebrates.* Springer-Verlag, New York, NY.

Feldhamer, G. A., J. A. Chapman, and R. L. Miller. 1978. Sika deer and white–tailed deer on Maryland's eastern shore. *Wildlife Society Bulletin* 6:155–157.

Fernald, M. L. 1950. *Gray's manual of botany.* Eighth edition. American Book Company, New York, NY.

Ferrell, M. A., T. D. Whitson, D. W. Koch, and A. E. Gade. 1998. Leafy spurge *(Euphorbia esula)* control with several grass species. *Weed Technology* 12: 374–380.

FICMNEW (Federal Interagency Committee for Management of Noxious and Exotic Weeds). 1998. *Pulling together: National strategy for invasive plant management.* 2nd edition. U.S. Government Printing Office, Washington, DC. (Internet: http://bluegoose.arw.r9.fws.gov/ficmnewfiles/NatlweedStrategytoc.html)

Fitch, L., and B. W. Adams. 1998. Can cows and fish co-exist? *Canadian Journal of Plant Science* 78:191–198.

Fitzgerald, B. M. 1988. Diet of domestic cats and their impact on prey populations. Pp. 123–147 in D. C. Turner and P. Bateson (Eds.), *The domestic cat: The biology of its behaviour.* Cambridge University Press, Cambridge, England.

Fleischner, T. L. 1994. Ecological costs of livestock grazing in western North America. *Conservation Biology* 8:629–644.

Flyger, V. 1960. Sika deer on islands in Maryland and Virginia. *Journal of Mammalogy* 41:140.

Fosberg, F. R. 1948. Derivation of the flora of the Hawaiian Islands. Pp. 107–119 in E. C. Zimmerman (Ed.), *Insects of Hawaii.* Univ. of Hawaii Press, Honolulu.

———. 1972. Man's effects on island ecosystems. Pp. 869–880 in M. T. Farvar and J. P. Milton (Eds.), *The careless technology: Ecology and international development.* The Natural History Press, Garden City, NY.

Fox, A. M., and C. T. Bryson. 1998. Wetland nightshade *(Solanum tampicense):* A threat to wetlands in the United States. *Weed Technology* 12:410–413.

Foy, C. L., D. R. Forney, and W. E. Cooley. 1983. History of weed introductions. Pp. 65–92 in C. L. Wilson and C. L. Graham (Ed.), *Exotic plant pests and North American agriculture.* Academic Press, San Diego, CA.

Frank, J. H. 1998. How risky is biological control? Comment. *Ecology* 79:1829–1834.

Frank, J. H., and E. D. McCoy. 1995a. Invasive adventive insects and other organisms in Florida. *Florida Entomologist* 78:1–15.

———. 1995b. Precinctive insect species in Florida. *Florida Entomologist* 78: 21–35.

Frank, J. H., and M. C. Thomas. 1994. *Metamasius callizona* (Chevrolat) (Coleoptera: Curculionidae), an immigrant pest, destroys bromeliads in Florida. *Canadian Entomologist* 128:673–682.

French, J. R. P. III. 1993. How well can fishes prey on zebra mussels in eastern North America? *Fisheries* 18(6):13–19.

Friedman, J. M., M. L. Scott, and G. T. Auble. 1997. Water management and cottonwood forest dynamics along prairie streams. Pp. 49–71 in F. L. Knopf and F. B. Samson (Eds.), *Ecology and conservation of Great Plains vertebrates.* Springer-Verlag, New York, NY.

Gagné, W. C. 1988. Conservation priorities in Hawaiian natural systems. *BioScience* 38:264–271.

Gamradt, S. C., and L. B. Kats. 1996. Effect of introduced crayfish and mosquitofish on California newts. *Conservation Biology* 10:1155–1162.

Gamradt, S. C., L. B. Kats, and C. Anzalone. 1997. Aggression by non-native crayfish deters breeding in California newts. *Conservation Biology* 11:793–796.

Garcia-Monteil, D. C., and D. Binkley. 1998. Effect of *Eucalyptus saligna* and *Albizia falcataria* on soil processes and nitrogen supply in Hawaii. *Oecologia* 113:547–556.

Getsinger, K. D., E. G. Turner, J. D. Madsen, and M. D. Netherland. 1997. Restoring native vegetation in a Eurasian water milfoil-dominated plant community using the herbicide Triclopyr. *Regulated Rivers: Research & Management* 13:357–375.

Gil-ad, N. L., and A. A. Reznicek. 1997. Evidence for hybridization of two Old World *Rhamnus* species—*R. cathartica* and *R. utilis* (Rhamnaceae)—in the New World. *Rhodora* 99:1–22.

Gipson, P. S., B. Hlavachick, and T. Berger. 1998. Range expansion by wild hogs across the central United States. *Wildlife Society Bulletin* 26:279–286.

Goodwin, B. J., A. J. McAllister, and L. Fahrig. 1999. Predicting invasiveness of plant species based on biological information. *Conservation Biology* 13: 422–426.

Goodwin, C. N., C. P. Hawkins, and J. L. Kershner. 1997. Riparian restoration in the western United States: Overview and perspective. *Restoration Ecology* 5: 4–14.

Gordon, D. R. 1998. Effects of invasive, non-indigenous plant species on ecosystem processes: Lessons from Florida. *Ecological Applications* 8:975–989.

Gordon, D. R., and K. P. Thomas. 1997. Florida's invasion by nonindigenous plants: History, screening and regulation. Pp. 21–37 in D. Simberloff, D. C. Schmitz, and T. C. Brown (Eds.), *Strangers in paradise.* Island Press, Washington, DC.

Gosliner, T. M. 1995. Introduction and spread of *Philine auriformis* (Gastropoda: Opisthobranchia) from New Zealand to San Francisco Bay and Bodega Harbor. *Marine Biology* 122:249–255.

Gottschalk, K. W. 1989. Gypsy moth effects on mast production. Pp. 42–50 in *Southern Appalachian mast management: Workshop proceedings.* USDA Forest Service and University of Tennessee, Knoxville, TN.

Graf, W. L. 1978. Fluvial adjustments to the spread of tamarisk in the Colorado Plateau region. *Geological Society of America Bulletin* 89:1491–1501.

Green, D. M. 1978. Northern leopard frogs and bullfrogs on Vancouver Island. *Canadian Field-Naturalist* 92:78–79.

Greenberg, N., R. L. Garthwaite, and D. C. Potts. 1996. Allozyme and morphological evidence for a newly introduced species of *Aurelia* in San Francisco Bay, California. *Marine Biology* 125:401–410.

Gregg, M. E. 1994. Invasive non-indigenous plant species impacting Florida's local-level governments and selected county-owned lands. Pp. 281–298 in D. C. Schmitz, and T. C. Brown (Eds.), *An assessment of invasive non-indigenous species in Florida's public lands.* Florida Department of Environmental Protection, Tallahassee, FL.

Griffith, J. T., and J. C. Griffith. 1999. Cowbird control and the endangered least Bell's vireo: A management success story. In press in J. N. M. Smith, T. L. Cook, S. I. Rothstein, S. G. Sealey, and S. R. Robinson (Eds.), *Ecology and management of cowbirds.* University of Texas Press, Austin, TX.

Grosholz, E. D. 1996. Contrasting rates of spread for introduced species in terrestrial and marine systems. *Ecology* 77:1680–1686.

———. 1997. Asymmetry, diffusion, and chance in the spread of introduced marine organisms. American Society of Limnology and Oceanography Annual Meeting, Santa Fe, NM.

Grosholz, E. D., and G. M. Ruiz. 1995. Spread and potential impact of the recently introduced European green crab, *Carcinus maenas,* in central California. *Marine Biology* 122:239–247.

———. 1996. Predicting the impact of introduced marine species: Lessons from the multiple invasions of the European green crab, *Carcinus maenas. Biological Conservation* 78:59–66.

———. 1997. The multitrophic level impacts of introduced and native predators on a marine food web. Society for Conservation Biology Annual Meeting, Victoria, British Columbia, Canada.

Groves, R. H. 1989. Ecological control of invasive terrestrial plants. Pp. 437–461 in J. A. Drake, H. A. Mooney, F. di Castri, R. H. Groves, F. J. Kruger, M. Rejmánek, and M. Williamson (Eds.), *Biological invasions: A global perspective.* John Wiley and Sons, New York, NY.

Groves, R. H., and J. J. Burdon (Eds.). 1986. *Ecology of biological invasions.* Cambridge University Press, New York, NY.

Gullion, G. W. 1965. A critique concerning foreign game bird introductions. *Wilson Bulletin* 77:409–414.

Gunderson, J. L., M. L. Klepinger, C. R. Bronte, and J. E. Marsden. 1998. Overview of the International Symposium on Eurasian Ruffe *(Gymnocephalus cernuus)* biology, impacts, and control. *Journal of Great Lakes Research* 24:165–169.

Haack, R. A., K. R. Law, V. C. Mastro, H. S. Ossenbruggen, and B. J. Raimo. 1997.

New York's battle with the Asian long-horned beetle. *Journal of Forestry* 95(12):11–15.

Hadfield, M. G. 1986. Extinction in Hawaiian Achatinelline snails. *Malacologia* 27:67–81.

Hadfield, M. G., S. E. Miller, and A. H. Carwile. 1993. The decimination of endemic Hawaiian tree snails by alien predators. *American Zoologist* 33:610–622.

Hager, H. A., and K. D. McCoy. 1998. Implications of accepting untested hypotheses: A review of the effects of purple loosestrife *(Lythrum salicaria)* in North America. *Biodiversity and Conservation* 7:1069–1079.

Hair, J. D., and D. J. Forrester. 1970. The helminth parasites of the starling *(Sturnus vulgaris)*. *American Midland Naturalist* 83:555–564.

Hajek, A. E. 1997. Fungal and viral epizootics in gypsy moth (Lepidoptera: Lymantriidae) populations in central New York. *Biological Control* 10:58–68.

Hallegraeff, G. M., and C. J. Bolch. 1991. Transport of toxic dinoflagellate cysts via ships' ballast water. *Marine Pollution Bulletin* 22:27–30.

Hammerson, G. A. 1982. Bullfrog eliminating leopard frogs in Colorado? *Herpetological Review* 13:115–116.

Hanemann, W. M. 1994. Valuing the environment through contingent valuation. *Journal of Economic Perspectives* 8(4):19–43.

Hanna, G. D. 1966. Introduced mollusks of western North America. *Occasional Papers of the California Academy of Sciences,* No. 48.

Hansen, R. W., R. D. Richard, P. E. Parker, and L. E. Wendel. 1997. Distribution of biological control agents of leafy spurge *(Euphorbia esula* L.) in the United States: 1988–1996. *Biological Control* 10:129–142.

Harcombe, P. A., G. N. Cameron, and E. G. Glumac. 1993. Above-ground net primary productivity in adjacent grassland and woodland on the coastal prairie of Texas, USA. *Journal of Vegetation Science* 4:521–530.

Harper, J. L. 1977. *The population biology of plants.* Academic Press, New York, NY.

Harrington, R. A., B. J. Brown, and P. B. Reich. 1989. Ecophysiology of exotic and native shrubs in southern Wisconsin. I. Relationship of leaf characteristics, resource availability, and phenology to seasonal patterns of carbon gain. *Oecologia* 80:356–367.

Harris, P. 1988. Environmental impact of weed-control insects. *BioScience* 38: 542–548.

Hart, A. 1978. The onslaught against Hawaii's tree snails. *Natural History* 87(10): 46–57.

Hartman, L. H., and D. S. Eastman. 1999. Distribution of introduced raccoons *Procyon lotor* on the Queen Charlotte Islands: Implications for burrow-nesting birds. *Biological Conservation* 88:1–13.

Hartman, L. H., A. G. Gaston, and D. S. Eastman. 1997. Raccoon predation on ancient murrelets on East Limestone Island, British Columbia. *Journal of Wildlife Management* 61:377–388.

Hastings, A. 1996. Models of spatial spread: Is the theory complete? *Ecology* 77: 1675–1679.

Havens, K. J., W. I. Priest III, and H. Berquist. 1997. Investigation and long-term monitoring of *Phragmites australis* within Virginia's constructed wetland sites. *Environmental Management* 21:599–605.

Hawkins, B. A., and P. C. Marino. 1997. The colonization of native phytophagous insects in North America by exotic parasitoids. *Oecologia* 112:566–571.

Hayes, M. P., and M. R. Jennings. 1986. Decline of Ranid frog species in western North America: Are bullfrogs *(Rana catesbiana)* responsible? *Journal of Herpetology* 20:490–509.

Heap, I. M. 1997. The occurrence of herbicide-resistant weeds worldwide. *Pesticide Science* 51:235–243.

Heath, D. D., S. Springer, and J. Mindel. 1997. Invasion dynamics of marine invertebrates: Marine mussels in British Columbia's waters. Society for Conservation Biology Annual Meeting, Victoria, British Columbia, Canada.

Hecnar, S. J., and R. T. M'Closkey. 1997. The effects of predatory fish on amphibian species richness and distribution. *Biological Conservation* 79:123–131.

Hendry, G. W. 1931. The adobe brick as a historical source. *Agricultural History* 5:125.

Herbold, B., and P. B. Moyle. 1989. *The ecology of the Sacramento–San Joaquin Delta: A community profile.* USDI Fish and Wildlife Service, Biological Report 85(7.22).

Herdman, R. C. 1993. *Harmful nonindigenous species in the United States.* Office of Technology Assessment, U.S. Congress. OTA-F-565. U.S. Government Printing Office, Washington, DC.

Herms, C. P., D. G. McCullough, L. S. Baue, R. A. Haack, D. L. Miller, and N. R. Dubois. 1997. Susceptibility of the endangered Karner blue butterfly (Lepidoptera: Lycaenidae) to *Bacillus thuringiensis* var. *kurstaki* used for gypsy moth suppression in Michigan. *Great Lakes Entomologist* 30:125–141.

Hesse, I. D., W. H. Conner, and J. M. Visser. 1997. Nutria: Another threat to Louisiana's vanishing coastal wetlands. *Aquatic Nuisance Species Digest* 2(1):4–5.

Heywood, V. H. 1989. Patterns, extents and modes of invasions by terrestrial plants. Pp. 31–60 in J. A. Drake, H. A. Mooney, F. di Castri, R. H. Groves, F. J. Kruger, M. Rejmánek, and M. Williamson (Eds.), *Biological invasions: A global perspective.* John Wiley and Sons, New York, NY.

Hickman, T. J. 1983. Effects of habitat alteration by energy resource developments in the upper Colorado River basin on endangered species. Pp. 537–550 in V. D. Adams and V. A. Lamarra (Eds.), *Aquatic resources management of the Colorado River ecosystem.* Ann Arbor Science Publishers, Ann Arbor, MI.

Hiers, J. K., and J. P. Evans. 1997. Effects of anthracnose on dogwood mortality and forest composition of the Cumberland Plateau (U.S.A.). *Conservation Biology* 11:1430–1435.

Hight, S. D., and J. J. Drea, Jr. 1991. Prospects for a classical biological control project against purple loosestrife *(Lythrum salicaria L.). Natural Areas Journal* 11:151–157.

Hilborn, R. 1992. Hatcheries and the future of salmon in the Northwest. *Fisheries* 17:5–8.

Hindar, K., N. Ryman, and F. Utter. 1991. Genetic effects of cultured fish on natural fish populations. *Canadian Journal of Fisheries and Aquatic Science* 48:945–957.

Hitchcock, A. S. 1950. *Manual of the grasses of the United States.* Second edition. U.S. Government Printing Office, Washington, DC.

Hobbs, R. J. 1989. The nature and effects of disturbance relative to invasion. Pp. 389–405 in J. A. Drake, H. A. Mooney, F. di Castri, R. H. Groves, F. J. Kruger, M. Rejmánek, and M. Williamson (Eds.), *Biological invasions: A global perspective.* John Wiley and Sons, New York, NY.

Hobbs, R. J., and L. F. Huenneke. 1992. Disturbance, diversity, and invasion: Implications for conservation. *Conservation Biology* 6:324–337.

Hobbs, R. J., and H. A. Mooney. 1998. Broadening the extinction debate: Population deletions and additions in California and Western Australia. *Conservation Biology* 12:271–283.

Hobdy, R. 1993. Lana'i—A case study: The loss of biodiversity on a small Hawaiian island. *Pacific Science* 47:201–210.

Holden, C. 1985. Hawaiian rain forest being felled. *Science* 228:1073–1074.

Holland, R. E, T. H. Johengen, and A. M. Beeton. 1995. Trends in nutrient concentrations in Hatchery Bay, western Lake Erie, before and after *Dreissena polymorpha. Canadian Journal of Fisheries and Aquatic Science* 52:1202–1209.

Holt, R. D., and M. E. Hockberg. 1997. When is biological control evolutionarily stable (or is it)? *Ecology* 78:1673–1683.

Hone, J., and C. P. Stone. 1989. A comparison and evaluation of feral pig management in two national parks. *Wildlife Society Bulletin* 17:419–425.

Horvitz, C. C., S. McMann, and A. Freedman. 1995. Exotics and hurricane damage in three hardwood hammocks in Dade County Parks, Florida. *Journal of Coastal Research* (Special Hurricane Andrew Issue) 18:145–158.

Houghton, D. C., J. J. Dimick, and R. V. Frie. 1998. Probable displacement of riffle-dwelling invertebrates by the introduced rusty crayfish, *Orconectes rusticus* (Decapoda: Cambaridae), in a north-central Wisconsin stream. *The Great Lakes Entomologist* 31:13–24.

Houston, D. N., V. Stevens, and B. B. Moorhead. 1994. Mountain goat population. Pp. 47–70 in D. B. Houston, E. G. Schreiner, and B. B. Moorhead (Eds.), *Mountain goats in Olympic National Park: Biology and management of an introduced species.* USDI National Park Service Scientific Monograph NPS/NROLYM/NRSM-94/25.

Howarth, F. G. 1991. Environmental impacts of classical biological control. *Annual Review of Entomology* 36:485–509.

Howarth, F. G., and A. C. Medeiros. 1989. Non-native invertebrates. Pp. 82–87 in C. P. Stone and D. B. Stone (Eds.), *Conservation biology in Hawaii.* University of Hawaii Cooperative National Park Resources Study Unit, Honolulu, HI.

Howarth, F. G., and G. W. Ramsey. 1991. The conservation of island insects and their habitats. Pp. 71–107 in N. M. Collins and J. A. Thomas (Eds.), *The conservation of insects and their habitats.* Academic Press, San Diego, CA.

Howe, W. H., and F. L. Knopf. 1991. On the imminent decline of Rio Grande cottonwoods in central New Mexico. *Southwestern Naturalist* 36:218–224.

Huenneke, L. F. 1997. No species is an island: Interactions between plant invasions and other aspects of global change. Ecological Society of America Annual Meeting, Albuquerque, NM.

Huffaker, C. B., and C. E. Kennett. 1959. A ten-year study of vegetational changes associated with the biological control of Klamath weed. *Journal of Range Management* 12:69–82.

Hughes, F., P. M. Vitousek, and T. Tunison. 1991. Alien grass invasion and fire in the seasonal submontane zone of Hawaii. *Ecology* 72:743–746.

Human, K. G., and D. M. Gordon. 1997. Effects of Argentine ants on invertebrate biodiversity in northern California. *Conservation Biology* 11:1242–1248.

Humphrey, N. 1994. History, status and management of Lehmann lovegrass. *Rangelands* 16:205–206.

Humphrey, S. R., and D. B. Barbour. 1981. Status and habitat of three subspecies of *Peromyscus polionotus* in Florida. *Journal of Mammalogy* 62:840–844.

Hunter, R. D., S. A. Toczylowski, and M. G. Janech. 1997. Zebra mussels in a small river: Impact on unionids. Pp. 161–186 in F. M. D'Itri (Ed.), *Zebra mussels and aquatic nuisance species.* Ann Arbor Press, Chelsea, MI.

Hunter, W. C., R. D. Ohmart, and B. W. Anderson. 1988. Use of exotic saltcedar *(Tamarix chinensis)* by birds in arid riparian systems. *Condor* 90: 113–123.

Hurlbert, S. H. 1997. Functional importance *vs.* keystoneness: Reformulating some questions in theoretical biocenology. *Australian Journal of Ecology* 22: 369–382.

Huryn, V. M. B. 1997. Ecological impacts of introduced honey bees. *Quarterly Review of Biology* 72:275–297.

Hutchinson, T. F., and J. L. Vankat. 1997. Invasibility and effects of Amur honeysuckle in southwestern Ohio forests. *Conservation Biology* 11:1117–1124.

Hyman, J., and S. Pruett-Jones 1995. Natural history of the monk parakeet in Hyde Park, Chicago. *Wilson Bulletin* 107:510–517.

Ingold, D. J. 1994. Influence of nest-site competition between European starlings and woodpeckers. *Wilson Bulletin* 106:227–241.

Jackson, L. E. 1985. Ecological origins of California's mediterranean grasses. *Journal of Biogeography* 12:349–361.

Jain, S. K., and P. S. Martins. 1979. Ecological genetics of the colonizing ability of rose clover (*Trifolium hirtum* All.). *American Journal of Botany* 66:361–366.

James, W. H., J. W. Balko, and H. L. Eakin. 1997. Nutrient regeneration by the zebra mussel *(Dreissena polymorpha). Journal of Freshwater Ecology* 12:209–216.

Jenkins P. T. 1996. Free trade and exotic species introductions. *Conservation Biology* 10:300–302.

Jennings, M. R., and M. P. Hayes. 1985. Pre-1900 overharvest of California red-legged frogs *(Rana aurora draytonii):* The inducement for bullfrog *(Rana catesbiana)* introduction. *Herpetologica* 41:94–103.

Johnsgard, P. A. 1967. Sympatry changes and hybridization incidence in mallards and black ducks. *American Midland Naturalist* 77:51–63.

Johnson, D. L. 1975. New evidence on the origin of the fox, *Urocyon littoralis clementae*, and feral goats on San Clemente Island, California. *Journal of Mammalogy* 56:925–927.

Johnson, L. E., and J. T. Carlton. 1996. Post-establishment spread in large-scale invasions: Dispersal mechanisms of the zebra mussel *Dreissena polymorpha. Ecology* 77:1686–1690.

Johnson, L. E., and D. K. Padilla. 1996. Geographic spread of exotic species: Ecological lessons and opportunities from the invasion of the zebra mussel *Dreissena polymorpha. Biological Conservation* 78:23–33.

Johnson, R. F., and W. J. Klitz. 1977. Variation and evolution in a granivorous bird: The house sparrow. Pp. 15–51 in J. Pinowski and S. C. Kendeigh (Eds.), *Granivorous birds in ecosystems.* Cambridge University Press, London, England.

Johnson, W. C. 1994. Woodland expansion in the Platte River, Nebraska: Patterns and causes. *Ecological Monographs* 64:45–84.

Jones, C. G., R. S. Ostfeld, M. P. Richard, E. M. Schauber, and J. O. Wolff. 1998. Chain reactions linking acorns to gypsy moth outbreaks and Lyme disease risk. *Science* 279:1023–1026.

Jones, D. T., and R. F. Doren. 1997. The distribution, biology and control of *Schinus terebinthefolius* in southern Florida, with special reference to Everglades National Park. Pp. 81–93 in J. H. Brock, M. Wade, P. Pysek, and D. Green (Eds.), *Plant invasions: Studies from North America and Europe.* Backhuys Publishers, Leiden, The Netherlands.

Joyce, J. C. 1992. Impact of *Eichornia* and *Hydrilla* in the United States. *ICES Marine Science Symposium* 194:106–109.

Jude, D. J., and F. J. Tesar. 1985. Recent changes in the inshore forage fish of Lake Michigan. *Canadian Journal of Fisheries and Aquatic Sciences* 42:1154–1157.

Jude, D. J., R. H. Reider, and G. R. Smith. 1992. Establishment of Gobiidae in the Great Lakes basin. *Canadian Journal of Fisheries and Aquatic Sciences* 49:416–421.

Kaufman, J. B., R. L. Beschta, N. Otting, and D. Lytjen. 1997. An ecological perspective of riparian and stream restoration in the western United States. *Fisheries* 22(5):12–24.

Kearns, C. A., and D. W. Inouye. 1997. Pollinators, flowering plants, and conservation biology. *BioScience* 47:297–307.

Keegan, D. R., B. E. Coblentz, and C. S. Winchell. 1994. Feral goat eradication on San Clemente Island, California. *Wildlife Society Bulletin* 22:56–61.

Keiper, R. R. 1985. Are sika deer responsible for the decline of white-tailed deer on Assateague Island, Maryland? *Wildlife Society Bulletin* 13:144–146.

Keiper, R. R., J. Stephens, and D. Baldwin. 1984. Sex, age, and dressed weights of sika and white-tailed deer from Assateague Island, Maryland. *Proceedings of the Pennsylvania Academy of Sciences* 58:101–102.

Kendeigh, S. C. 1982. Bird populations in east-central Illinois: Fluctuations, variations, and development over a half-century. *Illinois Biological Monographs* 52:1–136.

Kerpez, T. A., and N. S. Smith. 1987. *Saltcedar control for wildlife habitat improvement in the southwestern United States.* USDI Fish and Wildlife Service, Publication #169.

Kitayama, K., and D. Mueller-Dombois. 1995. Biological invasion on an oceanic island mountain: Do alien plant species have wider ecological ranges than native species? *Journal of Vegetation Science* 6:667–674.

Kittleson, P. M., and M. J. Boyd. 1997. Mechanisms of expansion of an introduced species of cordgrass, *Spartina densiflora,* in Humboldt Bay, California. *Estuaries* 20:770–778.

Klein, D. R. 1968. The introduction, increase, and crash of reindeer on St. Matthew Island. *Journal of Wildlife Management* 32:350–367.

———. 1980. Conflicts between domestic reindeer and their wild counterparts: A review of Eurasian and North American experience. *Arctic* 33:739–756.

Knopf, F. L., and T. E. Olson. 1984. Naturalization of Russian olive: Implications to Rocky Mountain wildlife. *Wildlife Society Bulletin* 12:289–298.

Knopf, F. L., and M. L. Scott. 1990. Altered flows and created landscapes in the Platte River headwaters, 1840–1990. Pp. 47–70 in J. M. Sweeney (Ed.), *Management of dynamic ecosystems.* North Central Section, The Wildlife Society, West Lafayette, IN.

Kondratieff, B. C., R. J. Bishop, and A. M. Brasher. 1997. The life cycle of an introduced caddisfly (*Cheumatopsyche pettiti* [Banks]; Trichoptera: Hydropsychidae) in Waikolu Stream, Molokai, Hawaii. *Hydrobiologia* 350:81–85.

Kornberg, Sir H., and M. H. Williamson (Eds.). 1987. Quantitative aspects of the ecology of biological invasions. *Philosophical Transactions of the Royal Society of London,* B 314:501–742.

Kotanen, P. M. 1995. Responses of vegetation to a changing regime of disturbance: Effects of feral pigs in a Californian coastal prairie. *Ecography* 18:190–199.

———. 1997. Effects of experimental soil disturbance on revegetation by natives and exotics in coastal California meadows. *Journal of Applied Ecology* 34: 631–644.

Kruger, F. J., G. J. Breytenbach, I. A. W. Macdonald, and D. M. Richardson. 1989. The characteristics of invaded mediterranean-climate regions. Pp. 181–213 in J. A. Drake, H. A. Mooney, F. di Castri, R. H. Groves, F. J. Kruger, M. Rejmánek, and M. Williamson (Eds.), *Biological invasions: A global perspective.* John Wiley and Sons, New York, NY.

Krysl, L. J., C. D. Simpson, and G. G. Gray. 1980. Dietary overlap of sympatric Barbary sheep and mule deer in Palo Duro Canyon, Texas. Pp. 97–103 in C. D. Simpson (Ed.), *Proceedings of the symposium on ecology and management of Barbary sheep.* Department of Range and Wildlife Management, Texas Tech University, Lubbock, TX.

Kupferberg, S. J. 1997. Bullfrog *(Rana catesbiana)* invasion of a California river: The role of larval competition. *Ecology* 78:1736–1751.

Kus, B. E. 1999. Impacts of brown-headed cowbird parasitism on productivity of the endangered least Bell's vireo. *Studies in Avian Biology* 18:160–166.

Lacey, J., P. Husby, and G. Handl. 1990. Observations on spotted and diffuse knapweed invasion into ungrazed bunchgrass communities in western Montana. *Rangelands* 12:30–32.

Lafferty, K. D., and A. M. Kuris. 1996. Biological control of marine pests. *Ecology* 77:1989–2000.

Lafferty, K. D., and C. J. Page. 1997. Predation on the endangered tidewater goby, *Eucyclogobius newberryi,* by the introduced African clawed frog, *Xenopus laevis,* with notes on the frog's parasites. *Copeia* 1997:589–592.

Lambert, B., and M. Peferoen. 1992. Insecticidal promise of *Bacillus thuringiensis. BioScience* 42:112–122.

Lane, J. N. 1997. Nonindigenous mammals. Pp. 157–186 in D. L. Simberloff, D. C. Schmitz, and T. C. Brown (Eds.), *Strangers in paradise.* Island Press, Washington, DC.

LaRosa, A. M., C. W. Smith, and D. E. Gardner. 1985. Role of alien and native birds in the dissemination of firetree (*Myrica faya* Ait.—Myricaceae) and associated plants in Hawaii. *Pacific Science* 39:372–378.

Larson, D. L., P. J. Anderson, and W. E. Newton. 1997. Exotic plant infestation in native mixed-grass prairie in North Dakota. Ecological Society of America Annual Meeting, Albuquerque, NM.

Laycock, G. 1966. *The alien animals.* Natural History Press, Garden City, NY.

Laycock, W. A. 1991. Stable states and thresholds of range condition on North American rangelands: A viewpoint. *Journal of Range Management* 44:427–433.

Layne, J. N. 1997. Nonindigenous mammals. Pp. 157–186 in D. Simberloff, D. C. Schmitz, and T. C. Brown (Eds.), *Strangers in paradise.* Island Press, Washington, DC.

Leary, R. F., F. W. Allendorf, and S. H. Forbes. 1993. Conservation genetics of bull trout in the Columbia and Klamath River drainages. *Conservation Biology* 7:856–865.

Lehman, J. T., and C. E. Cáceres. 1993. Food-web responses to species invasion by a predatory invertebrate: *Bythotrephes* in Lake Michigan. *Limnology and Oceanography* 38:879–891.

Leigh, P. 1998. Benefits and costs of the ruffe control program for the Great Lakes fishery. *Journal of Great Lakes Research* 24:351–360.

Leonard, Z. 1904. *Adventures of Zenas Leonard.* W. F. Wagner (Ed.). Burrows, Cleveland, OH.

Leopold, A. 1938. Chukaremia. *Outdoor America* 3:3.

Leopold, A. 1941. Cheat takes over. *The Land* 1:310–313.

Levin, D. A., J. Francisco-Ortega, and R. K. Jansen. 1996. Hybridization and the extinction of rare plant species. *Conservation Biology* 10:10–16.

Lewin, V., and G. Lewin. 1984. The Kalij pheasant, a newly established game bird on the island of Hawaii. *Wilson Bulletin* 96:634–646.

Li, H. W., and P. B. Moyle. 1981. Ecological analysis of species introductions into aquatic ecosystems. *Transactions of the American Fisheries Society* 110:772–782.

Lidicker, W. Z., Jr. 1991. Introduced mammals in California. Pp. 263–271 in R. H. Groves and F. DiCastri (Eds.), *Biogeography of mediterranean invasions.* Cambridge University Press, Cambridge, England.

Liebhold, A. M., K. W. Gottschalk, A. Mason, and R. R. Bush. 1997. Forest susceptibility to the gypsy moth. *Journal of Forestry* 95(5):20–24.

Liebhold, A. M., J. A. Halverson, and G. A. Elmes. 1992. Gypsy moth invasion in North America: A quantitative analysis. *Journal of Biogeography* 19:513–520.

Liebhold, A. M., W. L. MacDonald, D. Bergdahl, and V. C. Mastro. 1996. Invasion by exotic forest pests: A threat to forest ecosystems. *Forest Science Monographs* 30:1–49.

Line, L. 1996. Accord is reached to recall pesticide devastating hawk. *New York Times,* 15 October, p. B7.

———. 1997. The return of an American classic. *Audubon* 99(5):70–75.

Lippencott, C. L. 1997. Ecological consequences of invasion by cogongrass *(Imperata cylindrica),* a non-indigenous grass, in Florida sandhill ecosystems. Ecological Society of America Annual Meeting, Albuquerque, NM.

Locke, A., D. M. Reid, H. C. van Leeuwen, W. G. Sprules, and J. T. Carlton. 1993. Ballast water exchange as a means of controlling dispersal of freshwater organisms by ships. *Canadian Journal of Fisheries and Aquatic Sciences* 50:2086–2093.

Lodge, D. M. 1993. Species invasions and deletions: Community effects and responses to climate and habitat change. Pp. 367–387 in P. M. Kareiva, J. G. Kingsolver, and R. B. Huey (Eds.), *Biotic interactions and global change.* Sinauer Associates, Sunderland, MA.

Logan, D. J., E. L. Bibles, and D. F. Markle. 1996. Recent collections of exotic aquarium fishes in the freshwaters of Oregon and thermal tolerance of oriental weatherfish and pirapatinga. *California Fish and Game* 82:66–80.

Long, J. L. 1981. *Introduced birds of the world.* Universe Books, New York, NY.

Loope, L. L. 1994. Untitled. Aspen Global Change Institute, Aspen, CO.

Loope, L. L., and D. Mueller-Dombois. 1989. Characteristics of invaded islands, with special reference to Hawaii. Pp. 257–279 in J. A. Drake, H. A. Mooney, F. di Castri, R. H. Groves, F. J. Kruger, M. Rejmánek, and M. Williamson (Eds.), *Biological invasions: A global perspective.* John Wiley and Sons, New York, NY.

Loope, L. L., and P. G. Scowcroft. 1985. Vegetation response within exclosures in Hawaii: A review. Pp. 377–402 in C. P. Stone and J. M. Scott (Eds.), *Hawaii's terrestrial ecosystems: Preservation and management.* Univ. of Hawaii, Honolulu.

Louda, S. M., D. Kendall, J. Connor, and D. Simberloff. 1997. Ecological effects of an insect introduced for the biological control of weeds. *Science* 277:1088–1090.

Louda, S. M., and R. A. Masters. 1993. Biological control of weeds in Great Plains rangelands. *Great Plains Research* 3:215–247.

Ludwig, D., B. Walker, and C. S. Holling. 1997. Sustainability, stability, and resilience. *Conservation Ecology* [online] 1(1):8 (http://www.consecol.org).

Ludyanskiy, M. L., D. McDonald, and D. MacNeill. 1993. Impact of the zebra mussel, a bivalve invader. *BioScience* 43:533–544.

Luken, J. O., and D. T. Mattimiro. 1991. Habitat-specific resilience of the invasive shrub Amur honeysuckle *(Lonicera maackii)* during repeated clipping. *Ecological Applications* 1:104–109.

Lym, R. G. 1998. The biology and integrated management of leafy spurge *(Euphorbia esula)* on North Dakota rangeland. *Weed Technology* 12:367–373.

Lyman, J. C., and N. C. Ellstrand. 1984. Clonal diversity in *Taraxacum officinale* (Compositae), an apomict. *Heredity* 53:1–10.

MacIsaac, H. J., I. A. Grigorovich, J. A. Hoyle, N. D. Yan, and V. E. Panov. 1999.

Invasion of Lake Ontario by the Ponto-Caspian predatory cladoceran *Cercopagis pengoi. Canadian Journal of Fisheries and Aquatic Science* 56:1–5.

Mack, M. C., and C. M. D'Antonio. 1998. Impacts of biological invasions on disturbance regimes. *Trends in Ecology and Evolution* 13:195–198.

Mack, R. N. 1981. Invasion of *Bromus tectorum* L. into western North America: An ecological chronicle. *Agro-Ecosystems* 7:145–165.

———. 1986. Alien plant invasion into the intermountain West: A case history. Pp. 191–213 in H. A. Mooney and J. A. Drake (Eds.), *Ecology of biological invasions of North America and Hawaii.* Springer-Verlag, New York, NY.

———. 1989. Temperate grasslands vulnerable to plant invasions: Characteristics and consequences. Pp. 155–179 in J. A. Drake, H. A. Mooney, F. di Castri, R. H. Groves, F. J. Kruger, M. Rejmánek, and M. Williamson (Eds.), *Biological invasions: A global perspective.* John Wiley and Sons, New York, NY.

———. 1991. The commercial seed trade: An early dispenser of weeds in the United States. *Economic Botany* 45:257–273.

Mack, R. N., and J. N. Thompson. 1982. Evolution in steppe with few large, hooved mammals. *American Naturalist* 119:757–773.

Madenjian, C. P. 1995. Removal of algae by the zebra mussel *(Dreissena polymorpha)* population in western Lake Erie: A bioenergetics approach. *Canadian Journal of Fisheries and Aquatic Sciences* 52:381–390.

Magnuson, J. J. 1976. Managing with exotics—A game of chance. *Transactions of the American Fisheries Society* 105:1–9.

Makarewicz, J. C., and P. Bertram. 1991. Evidence for the restoration of the Lake Erie ecosystem. *BioScience* 41:216–223.

Makarewicz, J. C., T. W. Lewis, and P. Bertram. 1999. Phytoplankton composition and biomass in the offshore waters of Lake Erie: Pre- and post-*Dreissena* introduction (1983–1993). *Journal of Great Lakes Research* 25:135–148.

Mal, T. K., J. Lovett-Doust, and L. Lovett-Doust. 1997. Time-dependent competitive displacement of *Typha angustifolia* by *Lythrum salicaria. Oikos* 79:26–33.

Malecki, R. A., B. Blossey, S. D. Height, D. Schroeder, L. T. Kok, and J. R. Coulson. 1993. Biological control of purple loosestrife. *BioScience* 43:680–686.

Marlette, G. M., and J. E. Anderson. 1986. Seed banks and propagule dispersal in crested wheatgrass stands. *Journal of Applied Ecology* 23:161–175.

Marsh, P. C., and M. E. Douglas. 1997. Predation by introduced fishes on endangered humpback chub and other native fishes in the Little Colorado River, Arizona. *Transactions of the American Fisheries Society* 126:343–346.

Martin, P. S. 1970. Pleistocene niches for alien animals. *BioScience* 20:218–221.

———. 1973. The discovery of America. *Science* 179:969–974.

Mattson, W. J. 1996. Escalating anthropogenic stresses on forest ecosystems: Forcing benign plant–insect interactions into new interaction trajectories. Pp. 338–342 in E. Korpilahti, H. Mikkela, and T. Salonen (Eds.), *Caring for the forest: Research in a changing world.* Congress Rept. Vol. 2. Finn. IUFRO World Congress Organizing Committee.

Mayer, J. J., and I. L. Brisbin, Jr. 1991. *Wild pigs in the United States: Their history, comparative morphology, and current status.* Savannah River Ecology Laboratory, Aiken, SC.

Mayfield, H. 1965. The brown-headed cowbird, with old and new hosts. *Living Bird* 4:13–28.

Mcauley, D. G., D. A. Clugston, and J. R. Longcore. 1998. Outcome of aggressive interactions between American black ducks and mallards during the breeding season. *Journal of Wildlife Management* 62:134–141.

McCarthy, B. C. 1997. Response of a forest understory community to experimental removal of an invasive nonindigenous plant (*Alliaria petiolata*, Brassicaceae). Pp. 117–130 in J. O. Luken and J. W. Thieret (Eds.), *Assessment and management of plant invasions.* Springer-Verlag, New York, NY.

McClure, M. S. 1990. Role of wind, birds, deer, and humans in the dispersal of hemlock wooly adelgid (Homoptera: Adelgidae). *Environmental Entomology* 19:36–43.

McDonald, J. H., and R. K. Koehn. 1988. The mussels *Mytilus galloprovincialis* and *M. trossulus* on the Pacific coast of North America. *Marine Biology* 99:111–118.

McFadyen, R. E. C. 1998. Biological control of weeds. *Annual Review of Entomology* 43:369–393.

McGaughey, W. H., and M. E. Whalon. 1992. Managing insect resistance to *Bacillus thuringiensis* toxins. *Science* 258:1451–1455.

McKeown, S. 1996. *A field guide to the reptiles and amphibians of the Hawaiian Islands.* Diamond Head Publications, Los Osos, CA.

McKnight, B. (Ed.). 1993. *Biological pollution: The control and impact of invasive exotic species.* Indiana Academy of Science, Indianapolis, IN.

McLean, M., D. Ogle, and J. Gunderson. 1992. *Ruffe: A new threat to our fisheries.* Minnesota Sea Grant Program, University of Minnesota, St. Paul, MN.

McMahon, R. F. 1983. Ecology of the invasive pest bivalve *Corbicula.* Pp. 505–561 in *The Mollusca.* Vol. 6, *Ecology.* Academic Press, New York, NY.

Meffe, G. K., and R. C. Vrijenhoek. 1988. Conservation genetics in the management of desert fishes. *Conservation Biology* 2:157–169.

Mensing, S., and R. Byrne. 1998. Pre-mission invasion of *Erodium cicutarium* in California. *Journal of Biogeography* 25:757–762.

Miller, R. R., J. D. Williams, and J. E. Williams. 1989. Extinctions of North American fishes during the past century. *Fisheries* 14(6):22–38.

Mills, E. L., R. M. Dermott, E. F. Roseman, D. Distin, E. Mellina, D. B. Conn, and A. P. Spidle. 1993a. Colonization, ecology, and population structure of the "quagga" mussel (Bivalvia: Dreissenidae) in the lower Great Lakes. *Canadian Journal of Fisheries and Aquatic Sciences* 50:2305–2314.

Mills, E. L., J. H. Leach, J. T. Carlton, and C. L. Secor. 1993b. Exotic species in the Great Lakes: A history of biotic crises and anthropogenic introductions. *Journal of Great Lakes Research* 19:1–54.

Mills, C. E., and F. Sommer. 1995. Invertebrate introductions in marine habitats: Two species of hydromedusae (Cnidaria) native to the Black Sea, *Maeotias inexspectata* and *Blackfordia virginica,* invade San Francisco Bay. *Marine Biology* 122:279–288.

Minckley, W. L., and J. E. Deacon. 1968. Southwestern fishes and the enigma of "endangered species." *Science* 159:1424–1432.

Minckley, W. L., P. C. Marsh, J. E. Brooks, J. E. Johnson, and B. L. Jensen. 1991. Management toward recovery of the razorback sucker. Pp. 303–357 in W. L. Minckley and J. E. Deacon (Eds.), *Battle against extinction: Native fish management in the American West.* University of Arizona Press, Tucson, AZ.

Mitchell, D. S., and P. J. Ashton. 1989. Aquatic plants: Patterns and modes of invasion, attributes of invading species and assessment of control programmes. Pp. 111–154 in J. A. Drake, H. A. Mooney, F. di Castri, R. H. Groves, F. J. Kruger, M. Rejmánek, and M. Williamson (Eds.), *Biological invasions: A global perspective.* John Wiley and Sons, New York, NY.

Modin, J. 1998. Whirling disease in California: A review of its history, distribution, and impacts, 1965–1997. *Journal of Aquatic Animal Health* 10:132–142.

Modlin, R. F., and J. J. Orsi. 1997. *Acanthomysis bowmani,* a new species, and *A. aspera* Ii, Mysidae newly reported from the Sacramento–San Joaquin estuary, California. *Proceedings of the Biological Society of Washington* 110:439–446.

Molles, M. C., Jr., C. S. Crawford, L. M. Ellis, H. M. Valett, and C. N. Dahm. 1998. Managed flooding for riparian restoration. *BioScience* 48:749–756.

Molloy, D. P. 1998. The potential for using biological control technologies in the management of *Dreissena* spp. *Journal of Shellfish Research* 17:177–183.

Mooney, H. A., and J. A. Drake (Eds.). 1986. *Ecology of biological invasions of North America and Hawaii.* Springer-Verlag, New York, NY.

Mooney, H. A., S. P. Hamberg, and J. A. Drake. 1986. The invasions of plants and animals into California. Pp. 250–272 in H. A. Mooney and J. A. Drake (Eds.), *Ecology of biological invasions of North America and Hawaii.* Springer-Verlag, New York, NY.

Moore, R. J. 1975. The biology of Canadian weeds. 13. *Cirsium arvense* (L.) Scop. *Canadian Journal of Plant Science* 55:1033–1048.

Morgan, M. D., S. T. Threlkeld, and C. R. Goldman. 1978. Impact of the introduction of kokanee *(Oncorhynchus nerka)* and opossum shrimp *(Mysis relicta)* on a subalpine lake. *Journal of the Fisheries Research Board of Canada* 35:1572–1579.

Morrison, T. W., W. E. Lynch, Jr., and K. Dabrowski. 1997. Predation on zebra mussels by freshwater drum and yellow perch in western Lake Erie. *Journal of Great Lakes Research* 23:177–189.

Morton, B. 1997. The aquatic nuisance species problem: A global perspective and review. Pp. 1–54 in F. M. D'Itri (Ed.), *Zebra mussels and aquatic nuisance species.* Ann Arbor Press, Chelsea, MI.

Moulton, M. P., and S. L. Pimm. 1983. The introduced Hawaiian avifauna: Biogeographic evidence for competition. *American Naturalist* 121:669–690.

Moulton, M. P., and S. L. Pimm. 1986. Species introductions to Hawaii. Pp. 231–249 in H. A. Mooney and J. A. Drake (Eds.), *Ecology of biological invasions of North America and Hawaii.* Springer-Verlag, New York, NY.

Mountainspring, S., and J. M. Scott. 1985. Interspecific competition among Hawaiian forest birds. *Ecological Monographs* 55:219–239.

Moyle, P. B. 1973. Effects of introduced bullfrogs, *Rana catesbiana,* on the native frogs of the San Joaquin Valley, California. *Copeia* 1973:18–22.

———. 1976. Fish introductions in California: History and impact on native fishes. *Biological Conservation* 9:101–118.

———. 1986. Fish introductions into North America: Patterns and ecological impact. Pp. 27–43 in H. A. Mooney and J. A. Drake (Eds.), *Ecology of biological invasions of North America and Hawaii.* Springer-Verlag, New York, NY.

Moyle, P. B., H. W. Li, and B. A. Barton. 1986. The Frankenstein effect: Impact of introduced fishes on native fishes in North America. Pp. 415–426 in R. H. Stroud (Ed.), *Fish culture in fisheries management.* American Fisheries Society, Bethesda, MD.

Moyle, P. B., and T. Light. 1996a. Fish invasions in California: Do abiotic factors determine success? *Ecology* 77:1666–1670.

———. 1996b. Biological invasions of fresh water: Empirical rules and assembly theory. *Biological Conservation* 78:149–161.

Mueller-Dombois, D. 1973. A non-adapted vegetation interferes with water removal in a tropical rain forest area in Hawaii. *Tropical Ecology* 14:1–18.

Mueller-Dombois, D., and L. L. Loope. 1990. Some unique ecological aspects of oceanic island ecosystems. *Monographs in Systemic Botany of the Missouri Botanical Garden* 32:21–27.

Mullin, B. H. 1998. The biology and management of purple loosestrife *(Lythrum salicaria). Weed Technology* 12:397–401.

Mungall, E. C., and W. J. Sheffield. 1994. *Exotics on the range: The Texas example.* Texas A&M Press, College Station, TX.

Myers, J. H., A. Savoie, and E. van Randen. 1998. Eradication and pest management. *Annual Review of Entomology* 43:471–491.

Myers, R. L. 1983. Site susceptibility to invasion by the exotic tree *Melaleuca quinquenervia* in southern Florida. *Journal of Applied Ecology* 20:645–658.

———. 1984. Ecological compression of *Taxodium distichum* var. *nutans* by *Melaleuca quinquenervia* in southern Florida. Pp. 358–364 in K. C. Ewel and H. T. Odum (Eds.), *Cypress swamps.* University of Florida Press, Gainesville, FL.

Nalepa, T. F. 1994. Decline of native unionid bivalves in Lake St. Clair after infestation by the zebra mussel, *Dreissena polymorpha. Canadian Journal of Fisheries and Aquatic Sciences* 51:2227–2233.

National Research Council. 1986. *Pesticide resistance: Strategies and tactics for management.* National Academy Press, Washington, DC.

Nehlsen, W., J. E. Williams, and J. A. Lichatowich. 1991. Pacific salmon at the crossroads: Stocks at risk from California, Oregon, Idaho, and Washington. *Fisheries* 16(2):4–21.

Neill, W. M. 1983. *The tamarisk invasion of desert riparian areas.* Educational Bulletin 83–4, Desert Protective Council, Spring Valley, CA.

Nelson, J. C. 1917. The introduction of foreign weeds in ballast, as illustrated by the ballast plants at Linnton, Oregon. *Torreya* 17:151–160.

Newman, R. M., M. E. Borman, and S. W. Castro. 1997. Developmental performance of the weevil *Euhrychiopsis lecontei* on native and exotic watermilfoil host plants. *Journal of the North American Benthological Society* 16:627–634.

Nichols, F. H., J. E. Cloern, S. H. Luoma, and D. H. Peterson. 1986. The modification of an estuary. *Science* 231:567–573.

Nichols, F. H., and J. K. Thompson. 1985. Persistence of an introduced mudflat community in South San Francisco Bay, California. *Marine Ecology Progress Series* 24:83–97.

Nickens, E. 1996. Beyond the birds and the bees. *Audubon* 98(5):23–24.

Nico, L. G., and P. L. Fuller. 1999. Spatial and temporal patterns of nonindigenous fish introductions in the United States. *Fisheries* 24:16–27.

Niemela, P., and W. J. Mattson. 1996. Invasion of North American forests by European phytophagous insects. *BioScience* 46:741–753.

Noble, I. R. 1989. Attributes of invaders and the invading process: Terrestrial and vascular plants. Pp. 301–313 in J. A. Drake, H. A. Mooney, F. di Castri, R. H. Groves, F. J. Kruger, M. Rejmánek, and M. Williamson (Eds.), *Biological invasions: A global perspective.* John Wiley and Sons, New York, NY.

Norment, C., and C. L. Douglas. 1977. *Ecological studies of feral burros in Death Valley.* Cooperative National Park Resource Studies Unit, University of Nevada, Las Vegas, NV.

Novak, S. J., and R. N. Mack. 1993. Genetic variation in *Bromus tectorum* (Poaceae): Comparison between native and introduced populations. *Heredity* 71:167–176.

Novak, S. J., R. N. Mack, and D. E. Soltis. 1991. Genetic variation in *Bromus tectorum* (Poaceae): Population differentiation in its North American range. *American Journal of Botany* 78:1150–1161.

Novak, S. J., R. N. Mack, and P. S. Soltis. 1993. Genetic variation in *Bromus tectorum* (Poaceae): Introduction dynamics in North America. *Canadian Journal of Botany* 71:1441–1448.

Novak, S. J., and A. Y. Welfley. 1997. Genetic diversity in the introduced clonal grass *Poa bulbosa* (bulbous bluegrass). *Northwest Science* 71:271–280.

Nuzzo, V. A. 1991. Experimental control of garlic mustard [*Alliaria petiolata* (Bieb.) Cavara & Grande] in northern Illinois using fire, herbicide and cutting. *Natural Areas Journal* 11:158–167.

Oberbauer, T. A. 1989. Exploring San Clemente Island: A small world with a character all its own. *Environment Southwest* Winter–Spring:4–7.

Odum, W. E., C. C. McIvor, and T. J. Smith, Jr. 1982. *The ecology of the mangroves of South Florida: A community profile.* U.S. Fish and Wildlife Service, Office of Biological Services, Washington, DC FWS/OBS-81/24.

Ogle, D. H. 1998. A synopsis of the biology and life history of ruffe. *Journal of Great Lakes Research* 24:170–185.

Ohmart, R. D. 1996. Historical and present impacts of livestock grazing on fish and wildlife resources in western riparian habitats. Pp. 245–279 in P. R. Krausman (Ed.), *Rangeland wildlife.* The Society for Range Management, Denver, CO.

Oldemeyer, J. L. 1994. Livestock grazing and the desert tortoise in the Mojave

Desert. *USDI National Biological Service, Fish and Wildlife Research Report* 13:95–103.

Oliver, J. D. 1992. Carrotwood: An invasive plant new to Florida. *Aquatics* 14:4–9.

Olson, B. E., R. T. Wallander, and R. W. Scott. 1997. Recovery of leafy spurge seed from sheep. *Journal of Range Management* 50:10–15.

Olson, S. L., and H. F. James. 1984. The role of Polynesians in the extinction of the avifauna of the Hawaiian Islands. Pp. 768–780 in P. S. Martin and R. G. Klein (Eds.), *Quaternary extinctions.* University of Arizona Press, Tucson, AZ.

Olson, T. E., and F. L. Knopf. 1986a. Naturalization of Russian-olive in the western United States. *Western Journal of Applied Forestry* 1:65–69.

Olson, T. E., and F. L. Knopf. 1986b. Agency subsidization of a rapidly spreading exotic. *Wildlife Society Bulletin* 14:492–493.

Olson, T. M., D. M. Lodge, G. M. Capelli, and R. J. Houlihan. 1991. Mechanisms of impact of an introduced crayfish *(Orconectes rusticus)* on littoral congeners, snails, and macrophytes. *Canadian Journal of Fisheries and Aquatic Sciences* 48:1853–1861.

Orwig, D. A., and D. R. Foster. 1998. Forest response to the introduced hemlock wooly adelgid in southern New England, USA. *Journal of the Torrey Botanical Society* 125:60–73.

Osmund, D. 1997. Common buckthorn control using Roundup and girdling. *Natural Areas Journal* 17:203.

Osmundson, D. B., R. J. Ryel, and T. E. Mourning. 1997. Growth and survival of Colorado squawfish in the upper Colorado River. *Transactions of the American Fisheries Society* 126:687–698.

OTA (Office of Technology Assessment, U.S. Congress). 1993. *Harmful non-indigenous species in the United States.* OTA-F-565. U.S. Government Printing Office, Washington, DC.

Painter, E. L. 1995. Threats to the California flora: Ungulate grazers and browsers. *Madroño* 42:180–188.

Parker, I. M. 1997a. Pollinator limitation of *Cytisus scoparius* (Scotch broom), an invasive exotic shrub. *Ecology* 78:1457–1470.

———. 1997b. Using models to gain insight into the relative importance of different plant characteristics to invasive spread. Ecological Society of America Annual Meeting, Albuquerque, NM.

Parsons, D. J., and T. J. Stohlgren. 1989. Effects of varying fire regimes on annual grasslands in the southern Sierra Nevada of California. *Madroño* 36:154–168.

Patterson, D. T., and D. A. Mortenson. 1985. Effects of temperature and photoperiod on common crupina *(Crupina vulgaris). Weed Science* 33:333–338.

Peine, J. D., and J. A. Farmer. 1990. Wild hog management program at Great Smoky Mountains National Park. Pp. 221–227 in L. R. Davis and R. E. Marsh (Eds.), *Proceedings of the 14th Vertebrate Pest Conference.* University of California, Davis, CA.

Perrins, J., M. Williamson, and A. Fitter. 1992. Do annual weeds have predictable characters? *Acta Oecologica* 13:517–533.

Peterson, M. R. 1982. Predation on seabirds by red foxes at Shaiak Island, Alaska. *Canadian Field Naturalist* 96:41–45.

Philbrick, C. T., R. A. Aakjar, Jr., and R. L. Stuckey. 1998. Invasion and spread of *Callitriche stagnalis* (Callitrichidae) in North America. *Rhodora* 100:25–38.

Pierce, R. W., J. T. Carlton, D. A. Carlton, and J. B. Geller. 1997. Ballast water as a vector for tintinnid transport. *Marine Ecology Progress Series* 149:295–297.

Pimentel, D. 1973. Extent of pesticide use, food supply, and pollution. *Journal of the New York Entomological Society* 81:13–33.

Pimentel, D., L. Lach, R. Zuniga, and D. Morrison. 1999. Environmental and economic costs associated with non-indigenous species in the United States. *BioScience* 49: In press.

Pimm, S. L. 1989. Theories of predicting success and impact of introduced species. Pp. 351–367 in J. A. Drake, H. A. Mooney, F. di Castri, R. H. Groves, F. J. Kruger, M. Rejmánek, and M. Williamson (Eds.), *Biological invasions: A global perspective.* John Wiley and Sons, New York, NY.

Pogacnik, T. 1995. Wild horses and burros on public lands. Pp. 456–458 in E. T. LaRoe (Ed.), *Our living resources.* USDI National Biological Service, Washington, DC.

Porter, S. D., D. F. Williams, R. S. Patterson, and H. G. Fowler. 1997. Intercontinental differences in the abundance of *Solenopsis* fire ants (Hymenoptera: Formicidae): Escape from natural enemies? *Environmental Entomology* 26: 373–384.

Posey, M. H. 1988. Community changes associated with the spread of an introduced seagrass, *Zostera japonica. Ecology* 69:974–983.

Power, M. E. 1990. Effects of fish in river food webs. *Science* 250:811–814.

Pysek, P. 1998. Is there a taxonomic pattern to plant invasions? *Oikos* 82:282–294.

Pysek, P., K. Prach, M. Rejmánek, and W. Wade (Eds.). 1995. *Plant invasions: General aspects and special problems.* SPB Academic Publishing, Amsterdam, The Netherlands.

Rabenold, K. N., P. T. Fauth, B. W. Goodner, J. A. Sadowski, and P. G. Barker. 1998. Response of avian communities to disturbance by an exotic insect in spruce-fir forests of the southern Appalachians. *Conservation Biology* 12:177–189.

Race, M. S. 1982. Competitive displacement and predation between introduced and native mud snails. *Oecologia* 54:337–347.

Radonski, G. C., N. S. Prosser, R. G. Martin, and R. H. Stroud. 1984. Exotic fishes and sport fishing. Pp. 313–321 in W. R. Courtney, Jr. and J. R. Stauffer, Jr. (Eds.), *Distribution, biology, and management of exotic fishes.* Johns Hopkins University Press, Baltimore, MD.

Rai, K. S. 1991. *Aedes albopictus* in the Americans. *Annual Review of Entomology* 36:459–484.

Ramakrishnan, P. S., and P. M. Vitousek. 1989. Ecosystem-level processes and the consequences of biological invasions. Pp. 281–300 in J. A. Drake, H. A. Mooney, F. di Castri, R. H. Groves, F. J. Kruger, M. Rejmánek, and M. Williamson (Eds.), *Biological invasions: A global perspective.* John Wiley and Sons, New York, NY.

Randall, J. M. 1993. Exotic weeds in North American and Hawaiian natural areas: The Nature Conservancy's plan of attack. Pp. 159–172 in B. N. McKnight (Ed.), *Biological pollution: The control and impact of invasive exotic species.* Indiana Academy of Science, Indianapolis, IN.

Randall, J. M., M. Rejmánek, and J. C. Hunter. 1998. Characteristics of the exotic flora of California. *Fremontia* 26(4):2–12.

Ray, W. J., and L. D. Corkum. 1997. Predation of zebra mussels by round gobies, *Neogobius melanostomus. Environmental Biology of Fishes* 50:267–273.

Rayachhetry, M. B., and M. L. Elliott. 1997. Evaluation of fungus-chemical compatibility for melaleuca *(Melaleuca quinquenervia)* control. *Weed Technology* 11:64–69.

Reed, C. F. 1977. History and distribution of Eurasian watermilfoil in the United States and Canada. *Phytologia* 36:417–436.

Reese, J. G. 1975. Productivity and management of feral mute swans in Chesapeake Bay. *Journal of Wildlife Management* 39:280–286.

———. 1980. Demography of European mute swans in Chesapeake Bay. *Auk* 97:449–464.

Reeves, M. E. 1997a. Techniques for the protection of the Great Lakes from infection by exotic organisms in ballast water. Pp. 283–299 in F. M. D'Itri (Ed.), *Zebra mussels and aquatic nuisance species.* Ann Arbor Press, Chelsea, MI.

———. 1997b. *A general survey of proposals, studies, and research on means to prevent infection by exotic organisms in ballast water.* United States Coast Guard Ninth District Marine Safety and Analysis Branch, 1240 E. Ninth Street, Cleveland, OH.

Reichard, S. H., and C. W. Hamilton. 1997. Predicting invasions of woody plants introduced into North America. *Conservation Biology* 11:193–203.

Reid, W. V., and K. R. Miller. 1989. *Keeping options alive: The scientific basis for conserving biodiversity.* World Resources Institute, Washington, DC.

Reid, W. H., and G. R. Patrick. 1982. Gemsbok *(Oryx gazella)* in White Sands National Monument. *Southwestern Naturalist* 28:97–99.

Rejmánek, M. 1989. Invasibility of plant communities. Pp. 369–388 in J. A. Drake, H. A. Mooney, F. di Castri, R. H. Groves, F. J. Kruger, M. Rejmánek, and M. Williamson (Eds.), *Biological invasions: A global perspective.* John Wiley and Sons, New York, NY.

———. 1996. A theory of seed plant invasiveness: The first sketch. *Biological Conservation* 78:171–181.

———. 1998. Invasive plant species and invasible ecosystems. Pp. 79–102 in O. T. Sandlund, P. J. Schei, and A. Vilken (Eds.), *Invasive species and biodiversity management.* Kluwer Academic Publishers, Dordrecht, The Netherlands.

Rejmánek, M., and D. M. Richardson. 1996. What attributes make some plant species more invasive? *Ecology* 77:1655–1661.

Rejmánek, M., C. D. Thomsen, and I. D. Peters. 1991. Invasive vascular plants of California. Pp. 81–101 in R. H. Groves and F. Di Castri (Eds.), *Biogeography of Mediterranean invasions.* Cambridge University Press, Cambridge, England.

Reusch, T. B. H., and S. L. Williams. 1998. Variable responses of native eelgrass

Zostera marina to a non-indigenous bivalve *Musculista senhousia. Oecologia* 113:428–441.

Rhymer, J. M., and D. Simberloff. 1966. Extinction by hybridization and introgression. *Annual Review of Ecology and Systematics* 27:83–109.

Ricciardi, A., R. J. Neves, and J. B. Rasmussen. 1998. Impending extinctions of North American freshwater mussels (Unionidae) following the zebra mussel *(Dreissena polymorpha)* invasion. *Journal of Animal Ecology* 67:613–619.

Ricciardi, A., F. G. Whoriskey, and J. B. Rasmussen. 1996. Impact of the *Dreissena* invasion on native bivalves in the upper St. Lawrence River. *Canadian Journal of Fisheries and Aquatic Sciences* 53:1434–1444.

———. 1997. The role of the zebra mussel *(Dreissena polymorpha)* in structuring macroinvertebrate communities on hard substrata. *Canadian Journal of Fisheries and Aquatic Sciences* 54:2596–2608.

Rice, K. J., J. D. Gerlach, and A. R. Dyer. 1997. Closed canopies and open niches: Patterns and mechanisms of sequential invasions in California grasslands. Ecological Society of America Annual Meeting, Albuquerque, NM.

Rice, P. M., J. C. Toney, D. J. Bedunah, and C. E. Carlson. 1997. Elk winter forage enhancement by herbicide control of spotted knapweed. *Wildlife Society Bulletin* 25:627–633.

Rich, T. D. 1995. Cheatgrass invasion of shrubsteppe vegetation: Effect on breeding birds. Abstract. International Conference and Training Workshop on Conservation and Ecology of Grassland Birds. Tulsa, OK.

———. In preparation. Wildlife responses to changes from native sagebrush-steppe to exotic annual-dominated rangelands—A review.

Richter, B. D., D. P. Braun, M. A. Mendelson, and L. L. Master. 1997. Threats to imperiled freshwater fauna. *Conservation Biology* 11:1081–1093.

Ricklefs, R. E., and G. W. Cox. 1972. The taxon cycle in the West Indian avifauna. *American Naturalist* 106:195–219.

Robinson, S. K. 1997. The case of the missing songbirds. *Consequences* 3(1):3–15.

Robinson, W. L., and E. G. Bolen. 1989. *Wildlife ecology and management.* Second Edition. Macmillan Publishing Company, New York, NY.

Robles, M., and F. S. Chapin III. 1995. Comparison of the influence of two exotic communities on ecosystem processes in the Berkeley hills. *Madroño* 42: 349–357.

Roché, B. F., Jr., and C. T. Roché. 1991. Identification, introduction, distribution, and economics of *Centaurea* species. Pp. 274–291 in L. F. James, J. O. Evans, M. H. Ralphs, and R. D. Child (Eds.), *Noxious range weeds.* Westview Press, Boulder, CO.

Roché, B. F., Jr., C. T. Roché, and R. C. Chapman. 1994. Impacts of grassland habitat on yellow starthistle (*Centaurea solsticialis* L.) invasion. *Northwest Science* 68:86–96.

Roché, C. T., and B. F., Roché, Jr. 1988. Distribution and amount of four knapweed (*Centaurea* L.) species in eastern Washington. *Northwest Science* 62:242–253.

Roché, C. T., D. C. Thill, and B. Shafi. 1997. Prediction of flowering in common crupina *(Crupina vulgaris). Weed Science* 45:519–528.

Roe, S. L., and H. J. Macissac. 1997. Deepwater population structure and reproductive state of quagga mussels *(Dreissena bugensis)* in Lake Erie. *Canadian Journal of Fisheries and Aquatic Sciences* 54:2428–2433.

Rood, S. B., and J. M. Mahoney. 1990. Collapse of riparian poplar forests downstream from dams in western prairies: Probable causes and prospects for mitigation. *Environmental Management* 14:451–464.

Rosen, P. C., and C. R. Schwalbe. 1995. Bullfrogs *(Rana catesbiana):* Introduced predators in southwestern wetlands. Pp. 452–454 in E. T. LaRoe, G. S. Farris, C. E. Puckett, P. D. Doran, and M. J. Mac (Eds.), *Our living resources: A report to the nation on the distribution, abundance, and health of U.S. plants, animals, and ecosystems.* U.S. Department of Interior National Biological Service, Washington, DC.

Rowe, M. L., D. J. Lee, S. J. Nissen, and R. A. Masters. 1997. Genetic variation in North American leafy spurge *(Euphorbia esula)* determined by DNA markers. *Weed Science* 45:446–454.

Roy, J., M. L. Navas, and L. Sonié. 1991. Invasion by annual brome grasses: A case study challenging the homocline approach to invasions. Pp. 207–224 in R. H. Groves and F. Di Castri (Eds.), *Biogeography of Mediterranean invasions.* Cambridge University Press, Cambridge, England.

Ruesink, J. L., I. M. Parker, M. J. Groom, and P. M. Kareiva. 1995. Reducing the risks of nonindigenous species introductions. *BioScience* 45:465–477.

Ruiz, G. M., J. T. Carlton, E. D. Grosholz, and A. H. Hines. 1997. Global invasions of marine and estuarine habitats by non-indigenous species: Mechanisms, extent, and consequences. *American Zoologist* 37:621–632.

Russell, D. J. 1992. The ecological invasion of Hawaiian reefs by two marine red algae, *Acanthophora spicifera* (Vahl) Boerg. and *Hypnea musciformis* (Wulfen) J. Ag., and their association with two native species, *Laurencia ninifica* J. Ag. and *Hypnea cervicornis* J. Ag. *ICES Marine Science Symposium* 194:110–125.

Sailer, R. I. 1983. History of insect introductions. Pp. 15–38 in C. L. Wilson and C. L. Graham (Eds.), *Exotic plant pests and North American agriculture.* Academic Press, San Diego, CA.

Sala, A., S. D. Smith, and D. A. Devitt. 1996. Water use by *Tamarix ramosissima* and associated phreatophytes in a Mojave Desert floodplain. *Ecological Applications* 6:888–898.

Samways, M. J. 1997. Classical biological control and biodiversity conservation: What risks are we prepared to accept? *Biodiversity and Conservation* 6:1309–1316.

Sather, N. 1995a. Element stewardship abstract. *Poa pratensis* (Kentucky bluegrass). The Nature Conservancy, Arlington, VA. Online. (http://www.catalinas.net/seer/er/plants/poa_prat.htm)

———. 1995b. Element stewardship abstract. *Cirsium arvense* (Canada thistle). The Nature Conservancy, Arlington, VA. Online. (http://www.catalinas.net/seer/er/plants/poa_prat.htm)

Savidge, J. A. 1987. Extinction of an island forest avifauna by an introduced snake. *Ecology* 68:660–668.

Scavia, D., and G. L. Fahnenstiel. 1988. From picoplankton to fish. Pp. 85–97 in S. R. Carpenter (Ed.), *Complex interactions in lake communities.* Springer-Verlag, New York, NY.

Schaffer, W. M., D. B. Jensen, D. E. Hobbs, J. Gurevitch, J. R. Todd, and M. Valentine Schaffer. 1979. Competition, foraging energetics, and the cost of sociality in three species of bees. *Ecology* 60:976–987.

Schaffer, W. M., D. W. Zeh, S. L. Buchmann, S. Kleinhans, M. Valentine Schaffer, and J. Antrim. 1983. Competition for nectar between introduced honey bees and native North American bees and ants. *Ecology* 64:564–577.

Scheffer, V. B. 1993. The Olympic goat controversy: A perspective. *Conservation Biology* 7:916–919.

Schierenbeck, K. A., R. N. Mack, and R. R. Scharitz. 1994. Effects of herbivory on growth and biomass allocation of native and introduced species of *Lonicera. Ecology* 75:1661–1672.

Schiffman, P. M. 1994. Promotion of exotic weed establishment by endangered giant kangaroo rats *(Dipodomys ingens)* in a California grassland. *Biodiversity and Conservation* 3:524–537.

Schlarbaum, S. 1994. *Tree improvement as a strategy for coping with invasions.* Aspen Global Change Institute, Aspen, CO.

Schloesser, D. W., and T. F. Nalepa. 1994. Dramatic decline of unionid bivalves in offshore waters of western Lake Erie after infestation by the zebra mussel, *Dreissena polymorpha. Canadian Journal of Fisheries and Aquatic Sciences* 51:2234–2242.

———. 1996. Zebra mussel infestation of unionid bivalves (Unionidae) in North America. *American Zoologist* 36:300–310.

Schloesser, D. W., T. F. Nalepa, and G. L. Mackie. 1996. Zebra mussel infestation of unionid bivalves (Unionidae) in North America. *American Zoologist* 36:300–310.

Schmitz, D. C. 1996. Weed scientists need to start lobbying for a centre to combat biological pollution in the United States. *Aliens* No. 3:8–9.

Schmitz, D. C., D. Simberloff, R. H. Hoffstetter, W. Haller, and D. Sutton. 1997. The ecological impact of nonindigenous plants. Pp. 39–61 in D. Simberloff, D. C. Schmitz, and T. C. Brown (Eds.), *Strangers in paradise.* Island Press, Washington, DC.

Schoenherr, A. A. 1992. *A natural history of California.* University of California Press, Berkeley, CA.

Schultz, S. 1994. Review of the historical evidence relating to mountain goats in the Olympic Mountains before 1925. Pp. 256–293 in D. B. Houston, E. G. Schreiner, and B. B. Moorhead (Eds.), *Mountain goats in Olympic National Park: Biology and management of an introduced species.* USDI National Park Service Scientific Monograph NPS/NROLYM/NRSM-94/25.

Schwalbe, C. R., and P. C. Rosen. 1988. Preliminary report on effect of bullfrogs on wetland herpetofaunas in southeastern Arizona. Pp. 166–173 in *Management of amphibians, reptiles, and small mammals in North America. Proceedings of the Symposium*. General Technical Report RM-166, USDA Forest Service, Fort Collins, CO.

Scott, J. M., C. B. Kepler, C. van Riper III, and S. I. Fefer. 1988. Conservation of Hawaii's vanishing avifauna. *BioScience* 38:238–253.

Scott, J. M., S. Mountainspring, F. L. Ramsey, and C. B. Kepler. 1986. *Forest bird communities of the Hawaiian Islands: Their dynamics, ecology, and conservation.* Cooper Ornithological Society, Studies in Avian Biology, No. 9.

Scott, J. M., and J. L. Sincock. 1985. Hawaiian birds. Pp. 549–562 in R. L. Di Silvestro (Ed.), *Audubon wildlife report 1985.* National Audubon Society, New York.

Scott, M. L., G. T. Auble, and J. M. Friedman. 1997. Flood dependency of cottonwood establishment along the Missouri River, Montana, USA. *Ecological Applications* 7:677–690.

Scott, T. A., and M. L. Morrison. 1990. Natural history and management of the San Clemente loggerhead shrike. *Proceedings of the Western Foundation of Vertebrate Zoology* 4(2):23–57.

Scott, W. B., and W. J. Christie. 1963. The invasion of the lower Great Lakes by the white perch, *Roccus americanus* (Gmelin). *Journal of the Fisheries Research Board of Canada* 20:1189–1195.

Seegmiller, R. F., and R. D. Ohmart. 1981. *Ecological relationships of feral burros and desert bighorn sheep.* The Wildlife Society, Wildlife Monograph, No. 78.

Seiger, L. A. 1997. The status of *Fallopia japonica (Reynoutria japonica; Polygonum cuspidatum)* in North America. Pp. 95–102 in J. H. Brock, M. Wade. P. Pysek, and D. Green (Eds.), *Plant invasions: Studies from North America and Europe.* Backhuys Publishers, Leiden, The Netherlands.

Shafroth, P. B., G. T. Auble, and M. L. Scott. 1995. Germination and establishment of the native plains cottonwood (*Populus deltoides* Marshall subsp. *monilifera*) and the exotic Russian-olive (*Eleagnus angustifolia* L.). *Conservation Biology* 9:1169–1175.

Sharp, W. M. 1957. Social and range dominance in gallinaceous birds—pheasants and prairie grouse. *Journal of Wildlife Management* 21:242–244.

Sheley, R. L., B. E. Olson, and L. L. Larson. 1997. Effect of weed seed rate and grass defoliation level on diffuse knapweed. *Journal of Range Management* 50:39–43.

Simberloff, D. 1989. Which insect introductions succeed and which fail? Pp. 61–75 in J. A. Drake, H. A. Mooney, F. di Castri, R. H. Groves, F. J. Kruger, M. Rejmánek, and M. Williamson (Eds.), *Biological invasions: A global perspective.* John Wiley and Sons, New York, NY.

———. 1994. Why is Florida being invaded? Pp. 7–9 in D. C. Schmitz and T. C. Brown (Eds.), *An assessment of invasive non-indigenous species in Florida's public lands.* Florida Department of Environmental Protection, Tallahassee, FL.

———. 1996. Impacts of introduced species in the United States. *Consequences* 2(2):13–22.

———. 1997. The biology of invasions. Pp. 3–17 in D. Simberloff, D. C. Schmitz, and T. C. Brown (Eds.), *Strangers in paradise.* Island Press, Washington, DC.

Simberloff, D., D. C. Schmitz, and T. C. Brown (Eds.). 1997. *Strangers in paradise.* Island Press, Washington, DC.

Simberloff, D., and P. Stiling. 1996. How risky is biological control? *Ecology* 77:1965–1974.

Simpson, C. D., L. J. Krysl, D. B. Hampy, and G. G. Gray. 1978. The Barbary sheep: A threat to desert bighorn survival. *Transactions of the Desert Bighorn Council* 22:26–31.

Sinclair, W. A., and R. J. Campana (Eds.). 1975. Dutch elm disease: Perspectives after 60 years. *Search Agriculture* 8(5):5–52.

Sindermann, C. J. 1986. Strategies for reducing risks from introductions of aquatic organisms: A marine perspective. *Fisheries* 11(2):10–15.

Singer, F. J., W. T. Swank, and E. E. C. Clebsch. 1984. Effect of wild pig rooting in a deciduous forest. *Journal of Wildlife Management* 48:464–473.

Singer, M. C., C. D. Thomas, and C. Parmesan. 1993. Rapid human-induced evolution of insect-host associations. *Nature* 366:681–683.

Smith, B. R., and J. J. Tibbles. 1980. Sea lamprey (*Petromyzon marinus*) in Lakes Huron, Michigan, and Superior: history of invasion and control, 1936–78. *Canadian Journal of Fisheries and Aquatic Science* 37:1780–1801.

Smith, C., R. Valdez, J. L. Holechek, P. J. Swank, and M. Cardenas. 1998. Diets of native and non-native ungulates in southcentral New Mexico. *Southwestern Naturalist* 43:163–169.

Smith, C. W. 1985. Impact of alien plants on Hawaii's native biota. Pp. 180–250 in C. P. Stone and J. M. Scott (Eds.), *Hawaii's terrestrial ecosystems: Preservation and management.* Cooperative National Park Resources Studies Unit, University of Hawaii, Honolulu, HI.

———. 1989. Non-native plants. Pp. 60–69 in C. P. Stone and D. B. Stone (Eds.), *Conservation biology in Hawaii.* University of Hawaii Press, Honolulu, HI.

Smith, J. W. 1998. Boll weevil eradication: Area-wide pest management. *Annals of the Entomological Society of America* 91:239–247.

Smith, P. W. 1987. The Eurasian collared-dove arrives in the Americas. *American Birds* 41:1371–1379.

Smith S. H. 1970. Species interactions of the alewife in the Great Lakes. *Transactions of the American Fisheries Society* No. 4:754–765.

Sobhian, R., L. Fornasari, J. S. Rodier, and S. Agret. 1998. Field evaluation of natural enemies of *Tamarix* spp. in southern France. *Biological Control* 12:164–170.

Solecki, M. K. 1993. Cut-leaved and common teasel (*Dipsacus laciniatus* L. and *D. sylvestris* Huds.): Profile of two invasive aliens. Pp. 85–92 in B. N. McKnight (Ed.), *Biological pollution: The control and impact of invasive exotic species.* Indiana Academy of Science, Indianapolis, IN.

Soulé, M. E., D. T. Bolger, A. C. Alberts, R. Sauvajot, J. Wright, M. Sorice, and S. Hill. 1988. Reconstructed dynamics of rapid extinctions of chaparral-requiring birds in urban habitat islands. *Conservation Biology* 2:75–92.

Spencer, C. N., B. R. McClelland, and J. A. Stanford. 1991. Shrimp stocking, salmon collapse, and eagle displacement. *BioScience* 41:14–21.

Spencer, C. N., D. S. Potter, R. T. Bukantis, and J. A. Stanford. 1999. Impact of predation by *Mysis relicta* on zooplankton in Flathead Lake, Montana, USA. *Journal of Plankton Research* 21:51–64.

Sprugel, D. G. 1991. Disturbance, equilibrium, and environmental variability: What is natural vegetation in a changing environment? *Biological Conservation* 58:1–18.

Spurr, S. H. 1980. *Forest ecology.* 3rd edition. John Wiley and Sons, New York, NY.

St. Amant, J., F. Hoover, and G. Stewart. 1973. African clawed frog, *Xenopus laevis* (Daudin), established in California. *California Fish and Game* 55:330–331.

Stallcup, R. 1991. A reversible catastrophe. *Observer* 91(Spring/Summer):18–29.

Steadman, D. W. 1995. Prehistoric extinctions of Pacific island birds: Biodiversity meets zooarcheology. *Science* 267:1123–1131.

Stein, B. A., and S. R. Flack (Eds.). 1996. *America's least wanted: Alien species invasions of U.S. ecosystems.* The Nature Conservancy, Arlington, VA.

Stevens, W. K. 1996. Bay inhabitants are mostly aliens. *New York Times,* 20 August 1996.

Stewart, D. J., J. F. Kitchell, and L. B. Crowder. 1981. Forage fishes and their salmonid predators in Lake Michigan. *Transactions of the American Fisheries Society* 110:751–763.

Stiller, J. W., and A. L. Denton. 1995. One hundred years of *Spartina alterniflora* (Poaceae) in Willapa Bay, Washington: Random amplified polymorphic DNA analysis of an invasive population. *Molecular Ecology* 4:355–363.

Stone, C. P. 1985. Alien animals in Hawaii's native ecosystems: Toward controlling the adverse effects of introduced vertebrates. Pp. 251–297 in C. P. Stone and J. M. Scott (Eds.), *Hawaii's terrestrial ecosystems: Preservation and management.* Cooperative National Park Resources Studies Unit, University of Hawaii, Honolulu, HI.

Stone, C. P. 1989. Non-native land vertebrates. Pp. 88–95 in C. P. Stone and D. B. Stone (Eds.), *Conservation biology in Hawaii.* University of Hawaii Press, Honolulu, HI.

Stone, C. P., and L. L. Loope. 1987. Reducing negative effects of introduced animals on native biotas in Hawaii: What is being done, what needs doing, and the role of national parks. *Environmental Conservation* 14:245–258.

Stone, C. P., and J. M. Scott (Eds.). 1985. *Hawaii's terrestrial ecosystems: Preservation and management.* University of Hawaii, Honolulu.

Stone, C. P., C. W. Smith, and J. T. Tunison (Eds.). 1993. *Alien plant invasions in native ecosystems of Hawaii: Management and research.* University of Hawaii, Honolulu.

Strayer, D. L. 1991. Projected distribution of the zebra mussel, *Dreissena polymorpha,* in North America. *Canadian Journal of Fisheries and Aquatic Sciences* 48:1389–1395.

Strayer, D. L., N. F. Caraco, J. J. Cole, S. Findlay, and M. L. Pace. 1999. Transformation of freshwater ecosystems by bivalves. *Bioscience* 49:19–27.

Stromberg, J. C. 1997. Growth and survivorship of Fremont cottonwood, Goodding willow, and salt cedar seedlings after large floods in central Arizona. *Great Basin Naturalist* 57:198–208.

Stromberg, J. C., and M. K. Chew. 1997. Herbaceous exotics in Arizona's riparian ecosystems. *Desert Plants* 19:11–17.

Stromberg, J. C., L. Gengarelly, and B. F. Rogers. 1997. Exotic herbaceous species in Arizona's riparian ecosystems. Pp. 45–57 in J. H. Brock, M. Wade, P. Pysek, and D. Green (Eds.), *Plant invasions: Studies from North America and Europe.* Backhuys Publishers, Leiden, The Netherlands.

Stromberg, M. R., and J. R. Griffin. 1996. Long-term patterns in coastal California grasslands in relation to cultivation, gophers, and grazing. *Ecological Applications* 6:1189–1211.

Strong, D. R. 1979. Biogeographic dynamics of insect–host plant communities. *Annual Review of Entomology* 24:89–119.

Strong, D. R., E. D. McCoy, and J. R. Ray. 1977. Time and the number of herbivore species: The pests of sugar cane. *Ecology* 58:167–175.

Suarez, A. V., D. T. Bolger, and T. J. Case. 1998. Effects of fragmentation on native ant communities in coastal southern California. *Ecology* 79:2041–2056.

Sun, M. 1997. Population genetic structure of yellow starthistle *(Centaurea solsticialis),* a colonizing weed in the western United States. *Canadian Journal of Botany* 75:1470–1478.

Sutter, G. C., and R. M. Brigham. 1998. Avifaunal and habitat changes resulting from conversion of native prairie to crested wheatgrass: Patterns at songbird community and species levels. *Canadian Journal of Zoology* 76:869–875.

Taylor, D. J., and P. D. N. Hebert. 1993. Cryptic intercontinental hybridization in *Daphnia* (Crustacea): The ghost of introductions past. *Proceedings of the Royal Society of London* B 254:163–168.

Taylor, F. R., R. R. Miller, J. W. Pedretti, and J. E. Deacon. 1988. Rediscovery of the Shoshone pupfish, *Cyprinodon nevadensis shoshone* (Cyprinodontidae), at Shoshone Springs, Inyo County, California. *Bulletin of the Southern California Academy of Sciences* 87:67–73.

Taylor, J. N., W. R. Courtenay, Jr., and J. A. McCann. 1984. Known impacts of exotic fishes in the continental United States. Pp. 322–373 in W. R. Courtney, Jr. and J. R. Stauffer, Jr. (Eds.), *Distribution, biology, and management of exotic fishes.* Johns Hopkins University Press, Baltimore, MD.

Taylor, J. P., and K. C. McDaniel. 1998. Restoration of saltcedar (*Tamarix* sp.)–infested floodplains on the Bosque del Apache National Wildlife Refuge. *Weed Technology* 12:345–352.

Teer, J. G. 1991. Non-native large ungulates in North America. Pp. 55–66 in L. A. Renecker and R. J. Hudson (Eds.), *Wildlife production: Conservation and sustainable development.* Agricultural and Forestry Experiment Station, Fairbanks, Alaska.

Temple, S. A. 1992. Exotic birds: A growing problem with no easy solution. *Auk* 109:395–397.

Tetramura, A. H., and B. R. Strain. 1979. Localized populational differences in photosynthetic response to temperature and irradiance in *Plantago lanceolata. Canadian Journal of Botany* 57:2559–2563.

Thill, D. C., D. L. Zamora, and D. L. Kambitsch. 1985. Germination and viability of common crupina *(Crupina vulgaris)* achenes buried in the field. *Weed Science* 33:344–348.

Thomas, L. K., Jr. 1980. *The impact of three exotic plant species on a Potomac island.* USDI National Park Service Monograph Series, No. 13.

Thompson, D. Q., R. L. Stuckey, and E. B. Thompson. 1987. *Spread, impact, and control of purple loosestrife* (Lythrum salicaria) *in North American wetlands.* USDI Fish and Wildlife Service, Washington, DC.

Thompson, J. K., and L. E. Schemel. 1991. An Asian bivalve, *Potamocorbula amurensis,* invades San Francisco Bay with remarkable speed and success. *Journal of Shellfisheries Research* 10:259.

Thurber, D. K., W. R. McClain, and R. C. Whitmore. 1994. Indirect effects of gypsy moth defoliation on nest predation. *Journal of Wildlife Management* 58:493–500.

Tippie, D., J. E. Deacon, and C.-H. Ho. 1991. Effects of convict cichlids on growth and recruitment of White River springfish. *Great Basin Naturalist* 51:256–260.

Toczylowski, S. A., and R. D. Hunter. 1997. Do zebra mussels preferentially settle on unionids and/or adult conspecifics? Pp. 126–140 in F. M. D'Itri (Ed.), *Zebra mussels and aquatic nuisance species.* Ann Arbor Press, Chelsea, MI.

Trammel, M. A., and J. L. Butler. 1995. Effects of exotic plants on native ungulate use of habitat. *Journal of Wildlife Management* 59:808–816.

Tyler, T., W. J. Liss, L. M. Ganio, G. L. Larson, R. Hoffman, E. Deimling, and G. Lomnicky. 1998. Interaction between introduced trout and larval salamanders *(Amblystoma macrodactylum)* in high-elevation lakes. *Conservation Biology* 12:94–105.

Tyser, R. W., and C. H. Key. 1988. Spotted knapweed in natural area fescue grasslands: An ecological assessment. *Northwest Science* 62:151–160.

Unwin, M. J. 1997. Fry-to-adult survival of natural and hatchery-produced chinook salmon *(Oncorhynchus tshawytscha)* from a common origin. *Canadian Journal of Fisheries and Aquatic Sciences* 54:1246–1254.

Unwin, M. J., and G. J. Glova. 1997. Changes in life history parameters in a naturally spawning population of chinook salmon *(Oncorhynchus tshawytscha)* associated with releases of hatchery-reared fish. *Canadian Journal of Fisheries and Aquatic Sciences* 54:1235–1245.

Usher, M. B. 1989. Ecological effects of controlling invasive terrestrial vertebrates. Pp. 463–489 in J. A. Drake, H. A. Mooney, F. di Castri, R. H. Groves, F. J. Kruger, M. Rejmánek, and M. Williamson (Eds.), *Biological invasions: A global perspective.* John Wiley and Sons, New York, NY.

Van Bael, S., and S. Pruett-Jones. 1996. Exponential population growth of monk parakeets in the United States. *Wilson Bulletin* 108:584–588.

Vance, D. R., and R. L. Westemeier. 1979. Interactions of pheasants and prairie chickens in Illinois. *Wildlife Society Bulletin* 7:221–225.

van Riper III, C., S. G. van Riper, M. L. Goff, and M. Laird. 1986. The epizootiology and ecological significance of malaria in Hawaiian land birds. *Ecological Monographs* 56:327–344.

Van't Woudt, B. D. 1990. Roaming, stray, and feral domestic cats and dogs as wildlife problems. Pp. 291–295 in L. R. Davis and R. E. Marsh (Eds.), *Proceedings of the Fourteenth Vertebrate Pest Conference.* University of California, Davis, CA.

Van Vuren, D. 1984. Diurnal activity and habitat use by feral pigs on Santa Cruz Island, California. *California Fish and Game* 70:140–144.

Vermeij, G. 1982. Phenotypic evolution in a poorly dispersing snail after arrival of a predator. *Nature* 299:349–350.

Vermeij, G. J. 1996. An agenda for invasion biology. *Biological Conservation* 63: 3–9.

Verts, B. J., and L. N. Carraway. 1980. Natural hybridization of *Sylvilagus bachmani* and introduced *S. floridanus* in Oregon. *Murrelet* 61:95–98.

Vila, M., and C. M. D'Antonio. 1998. Fruit choice and seed dispersal of invasive vs. noninvasive *Carpobrotus* (Aizoaceae) in coastal California. *Ecology* 79:1053–1060.

Vinson, S. B., and L. Greenberg. 1986. The biology, physiology, and ecology of imported fire ants. Pp. 193–226 in S. B. Vinson (Ed.), *Economic impact and control of social insects.* Praeger, New York, NY.

Vitousek, P. M. 1986. Biological invasions and ecosystem properties: Can species make a difference? Pp. 163–176 in H. A. Mooney and J. A. Drake (Eds.), *Ecology of biological invasions of North America and Hawaii.* Springer-Verlag, New York, NY.

———. 1990. Biological invasions and ecosystem processes: Towards an integration of population biology and ecosystem studies. *Oikos* 57:7–13.

Vitousek, P. M., C. M. D'Antonio, L. L. Loope, M. Rejmánek, and R. Westbrooks. 1997. Introduced species: A significant component of human-caused global change. *New Zealand Journal of Ecology* 21:1–16.

Vitousek, P., L. Loope, and C. D'Antonio. 1994. *Biological invasion as global change.* Aspen Global Change Institute, Aspen CO.

Vitousek, P. M., and L. R. Walker. 1989. Biological invasion by *Myrica faya* in Hawai'i: Plant demography, nitrogen fixation, ecosystem effects. *Ecological Monographs* 59:247–265.

Vivrette, N. J., and C. H. Muller. 1977. Mechanism of invasion and dominance of coastal grassland by *Mesembryanthemem crystallinum. Ecological Monographs* 47:301–318.

von Broembsen, S. L. 1989. Invasions of natural ecosystems by plant pathogens. Pp. 77–83 in J. A. Drake, H. A. Mooney, F. di Castri, R. H. Groves, F. J. Kruger, M. Rejmánek, and M. Williamson (Eds.), *Biological invasions: A global perspective.* John Wiley and Sons, New York, NY.

Vredenberg, V. T., and K. R. Matthews. 1997. Introduced trout and remaining populations of yellow-legged frogs in Sequoia and Kings Canyon National

Parks: A case of co-existence or co-occurrence? Annual Meeting of the Society for Conservation Biology, Victoria, British Columbia, Canada.

Vujnovik, K., and R. W. Wein. 1997. The biology of Canadian weeds. 106. *Linaria dalmatica* (L.) Mill. *Canadian Journal of Plant Science* 77:483–491.

Wade, M. 1997. Predicting plant invasions: Making a start. Pp. 1–18 in J. H. Brock, M. Wade, P. Pysek, and D. Green (Eds.), *Plant invasions: Studies from North America and Europe.* Backhuys Publishers, Leiden, The Netherlands.

Wade, S. A. 1995. Stemming the tide: A plea for new exotic species legislation. *Journal of Land Use & Environmental Law* 10:343–370.

Wagner, W. L., D. R. Herbst, and R. S. N. Yee. 1985. Status of the native flowering plants of the Hawaiian Islands. Pp. 23–74 in C. P. Stone and J. M. Scott (Eds.), *Hawaii's terrestrial ecosystems: Preservation and management.* University of Hawaii, Honolulu.

Walker, L., and L. Hudson. 1945. *Midway in verse—or worse.* Hester Colorgraphic Studios, San Diego, CA.

Waloff, N. 1966. Scotch broom *(Sarcothamnus scoparius* (L.) Wimmer) and its insect fauna introduced into the Pacific Northwest of America. *Journal of Animal Ecology* 35:293–311.

Ward, P. S. 1987. Distribution of the introduced Argentine ant *(Iridomyrmex humilis)* in natural habitats of the lower Sacramento Valley and its effects on the indigenous ant fauna. *Hilgardia* 55(2):1–16.

Warner, R. E. 1968. The role of introduced diseases in the extinction of the endemic Hawaiian avifauna. *Condor* 70:101–120.

Warren, G. L. 1994. Ecologically disruptive non-indigenous freshwater invertebrates in Florida. Pp. 104–109 in D. C. Schmitz, and T. C. Brown (Eds.), *An assessment of invasive non-indigenous species in Florida's public lands.* Florida Department of Environmental Protection, Tallahassee, FL.

Watson, J. K. 1992. Hemlock wooly adelgid threatens eastern hemlock in Shenandoah National Park. *Park Science* 12(4):9–11.

Webb, S. L., and C. K. Kaunzinger. 1993. Biological invasion of the Drew University (New Jersey) forest preserve by Norway maple (*Acer platanoides* L.). *Bulletin of the Torrey Botanical Club* 120:343–349.

Wells, L. 1970. Effects of alewife predation on zooplankton populations in Lake Michigan. *Limnology and Oceanography* 15:556–565.

Wenner, A. M., and R. W. Thorp. 1994. Removal of feral honey bee *(Apis mellifera)* colonies from Santa Cruz Island. Pp. 513–522 in W. L. Halvorson and G. J. Maender (Eds.), *The Fourth California Islands Symposium: Update of the Status of Resources.* Santa Barbara Museum of Natural History, Santa Barbara, CA.

Werth, C. R., J. L. Riopel, and N. W. Gillespie. 1984. Genetic uniformity in an introduced population of witchweed *(Striga asiatica)* in the United States. *Weed Science* 32:645–648.

West, D. 1997. Good for karma; bad for the fish? *New York Times,* 11 January 1997, Metro Section, pp. 27, 31.

West, N. E. 1990. Structure and function of microphytic soil crusts in wildland ecosystems of arid to semi-arid regions. *Advances in Ecological Research* 20:179–223.

Wester, L. 1981. Composition of the native grasslands of the San Joaquin Valley, California. *Madroño* 28:231–241.

Westman, W. E. 1990. Park management of exotic species: Problems and issues. *Conservation Biology* 4:251–260.

White, K. L. 1967. Native bunchgrass *(Stipa pulchra)* on Hasting Reservation, California. *Ecology* 48:949–955.

Whitson, T. D., and D. W. Koch. 1998. Control of downy brome *(Bromus tectorum)* with herbicides and perennial grass competition. *Weed Technology* 12:391–396.

Whittier, T. R., D. B. Halliwell, and S. G. Paulsen. 1997. Cyprinid distributions in northeast U.S.A. lakes: Evidence of regional-scale minnow biodiversity losses. *Canadian Journal of Fisheries and Aquatic Sciences* 54:1593–1607.

Wiedemann, A. M. 1984. *The ecology of Pacific Northwest coastal sand dunes: A community profile.* U.S. Fish and Wildlife Service FWS/OBS-84/04.

Wilcove, D. S., D. Rothstein, J. Dubow, A. Phillips, and E. Losos. 1998. Quantifying threats to imperiled species in the United States. *BioScience* 48:607–615.

Wilde, G. R., and A. A. Echelle. 1997. Morphological variation in intergrade pupfish populations from the Pecos River, Texas, U.S.A. *Journal of Fish Biology* 50:523–539.

Wiley, C. J. 1997. *Aquatic nuisance species: Nature, transport, and regulation.* Canadian Coast Guard, Sarnia, Ontario, Canada.

Williams, J. D., and R. M. Nowak. 1993. Vanishing species in our own backyard: Extinct fish and wildlife of the United States and Canada. Pp. 115–148 in L. Kaufman and K. Mallory (Eds.), *The last extinction.* MIT Press, Cambridge, MA.

Williams, M. C. 1980. Purposefully introduced plants that have become noxious or poisonous weeds. *Weed Science* 28:300–305.

Williams, T. 1997. The ugly swan. *Audubon* 99(6):26–32.

Williamson, M. 1993. Invaders, weeds and the risk from genetically manipulated organisms. *Experientia* 49:219–224.

———. 1996. *Biological invasions.* Chapman and Hall, New York, NY.

Williamson, M., and A. Fitter. 1996a. The characters of successful invaders. *Biological Conservation* 78:163–170.

———. 1996b. The varying success of invaders. *Ecology* 77:1661–1666.

Wilson, E. O. 1961. The nature of the taxon cycle in the Melanesian ant fauna. *American Naturalist* 95:169–193.

Wilson, E. O. 1997. Pp. ix–x in D. Simberloff, D. C. Shultz, and T. C. Brown (Eds.), *Strangers in Paradise.* Island Press, Washington, DC.

Wilson, S. D., and J. W. Belcher. 1989. Plant and bird communities of native prairie and introduced Eurasian vegetation in Manitoba, Canada. *Conservation Biology* 3:39–44.

Witt, J. D. S., P. D. N. Hebert, and W. B. Morton. 1997. *Echinogammarus ischnus:*

Another crustacean invader of the Laurentian Lakes basin. *Canadian Journal of Fisheries and Aquatic Sciences* 54:264–268.

Wood, G. W., and R. H. Barrett. 1979. Status of wild pigs in the United States. *Wildlife Society Bulletin* 7:237–246.

Wood, G. W., and T. E. Lynn, Jr. 1977. Wild hogs in southern forests. *Southern Journal of Applied Forestry* 1:12–17.

Woods, K. D. 1993. Effects of invasion by *Lonicera tatarica* L. on herbs and tree seedlings in four New England forests. *American Midland Naturalist* 130: 62–74.

Wright, A. L., and G. D. Hayward. 1998. Barred owl range expansion into the central Idaho wilderness. *Journal of Raptor Research* 32:77–81.

Wright, A. L., and R. G. Kelsey. 1997. Effects of spotted knapweed on a cervid winter–spring range in Idaho. *Journal of Range Management* 50:487–496.

Wyckoff, P. H., and S. L. Webb. 1996. Understory influence of the invasive Norway maple *(Acer platanoides). Bulletin of the Torrey Botanical Club* 123:197–205.

Yan, N. D., and T. W. Pawson. 1997. Changes in the crustacean zooplankton community of Harp Lake, Canada, following invasion by *Bythotrephes cederstroemi. Freshwater Biology* 37:409–425.

Yong, W., and D. M. Finch. 1997a. Population trends of migratory landbirds along the middle Rio Grande. *Southwestern Naturalist* 42:137–147.

———. 1997b. Migration of the willow flycatcher along the middle Rio Grande. *Wilson Bulletin* 109:253–268.

Young, J. A. 1991. Tumbleweed. *Scientific American* 264(3):82–87.

Zaranko, D. T., D. G. Farara, and F. G. Thompson. 1997. Another exotic mollusc in the Laurentian Great Lakes: The New Zealand native *Potamopyrgus antipodium* (Gray 1843)(Gastropoda, Hydrobiidae). *Canadian Journal of Fisheries and Aquatic Sciences* 54:809–814.

Zaret, T. M., and R. T. Paine. 1973. Species introduction in a tropical lake. *Science* 182:449–455.

Zedler, P. H. 1987. *The ecology of southern California vernal pools: A community profile.* U.S. Fish and Wildlife Service Biological Report 85(7.11).

Zedler, P. H., C. R. Gautier, and G. S. McMaster. 1983. Vegetation change in response to extreme events: The effect of a short interval between fires in California chaparral and coastal scrub. *Ecology* 64:809–818.

Zimmerman, E. C. 1960. Possible evidence of rapid evolution in Hawaiian moths. *Evolution* 14:137–138.

Zuckerman, L. D., and R. J. Behnke. 1986. Introduced fishes in the San Luis Valley, Colorado. Pp. 435–453 in R. H. Stroud (Ed.), *Fish culture in fisheries management.* American Fisheries Society, Bethesda, MD.

Glossary

acid deposition The introduction of acids of sulfur or nitrogen into ecosystems via rain, snow, or dry fallout.

alleles Forms of a gene differing slightly in coding and in producing proteins of different structure and function.

alluvial Material deposited by flowing water.

amphipods An order of small, actively swimming crustaceans of freshwater and marine ecosystems.

assembly rules Criteria that define the sequence in which species can be added to a developing community.

ballast Material carried in the holds of an empty or lightly loaded cargo ship to make the vessel ride lower in the water and give it stability in heavy seas. In the past, soil and rock were commonly used. Modern ships use water as ballast.

basal area A measure of the size and abundance of one or more tree species, taken as the summed cross-sectional trunk areas of individuals per unit area.

benthic Pertaining to the floor of a body of water.

biological pesticides Pesticides consisting of microbial agents or their chemical products.

biomass A measure of the quantity of living or nonliving organic matter.

biota The plant, animal, and microbial life of a particular region.

biotype A genetic race of a species, distinct in morphology, physiology, or behavior.

bosque A local Spanish name for the riparian woodlands of rivers of New Mexico and Arizona.

byssal threads A mass of tough fibers that attach mollusks to a hard substrate.

cladocerans A class of small crustaceans, commonly known as water fleas, that are part of the zooplankton of lakes and oceans.

clone A set of genetically identical individuals resulting from vegetative or asexual reproduction of an individual.

congeners Members of the same genus.

contingent valuation A method for estimating the monetary value attached to an environmental state by surveying the willingness of the public to pay to maintain that state.

copepods A class of small crustaceans that is part of the zooplankton of lakes and oceans.

counteradaptation The ecological and evolutionary adjustments made by members of a native community of species to an invading exotic species.

cypress dome Depressions in low-lying southeastern landscapes in which cypress trees grow in nearly pure stands.

cryptogamic crust The surface layer of algae, bacteria, fungi, lichens, and mosses on soils in arid regions.

cryptogenic species A species whose status as native or exotic is uncertain because of the possibility that human-assisted dispersal occurred before scientists had studied the plant and animal life of a particular region.

detritus Dead organic matter that forms the base of food chains beginning with decomposer organisms.

diatom A class of microscopic single-celled or colonial algae.

dispersal The process by which organisms or their reproductive propagules are spread throughout available habitat or into previously unoccupied areas.

ecosystem engineer A species that can cause major restructuring of the physical features of an ecosystem.

electrophoresis A technique by which proteins or other organic molecules varying slightly in structure are caused to migrate differentially through a gel exposed to an electric field.

endemic A species or other taxonomic group that evolved its distinctive characteristics in a particular region, to which it is restricted.

epizootic A disease epidemic affecting an animal species other than humans.

estuary An aquatic ecosystem, with salinity intermediate between freshwater and ocean water, that occurs where rivers enter the ocean.

eutrophication Increase in the fertility and productivity of an ecosystem due to an increased rate of input of nutrients.

extirpation The elimination of a local population of a species or subspecies without causing the total extinction of the form.

feral population A free-living population of a domestic animal.

forb A broadleaved herbaceous plant.

fouling agents Aquatic organisms that grow on the hulls of vessels, pier pilings, or water-intake structures.

genotype The genetic constitution of an organism with respect to a certain character.

hammock A slightly elevated island of habitat in the Everglades, having soils that are well drained and capable of supporting trees. Hardwood hammocks occur in areas of freshwater marsh and in maritime hammocks in areas with saltwater influence.

heterozygosity The condition in which a gene is represented by different alleles on the two chromosomes carried by an adult animal or plant.

homozygosity The condition in which a gene is represented by the same allele on the two chromosomes carried by an adult animal or plant.

indigenous Native to a particular region. Nonindigenous, exotic, and alien are synonymous terms for species that are not native to a particular region.

invasive exotics Nonindigenous species that become abundant and influential invaders of communities of native species.

keystone species A species that can cause a major reorganization of the structure and function of an ecosystem.

kipuka In Hawaii, an isolated area of natural habitat surrounded by recent lava flows.

lampricide A pesticide chemical that is effective against the sea lamprey.

littoral zone The shallow-water region of the edge of a lake, where light reaches the bottom and rooted higher plants can grow.

locus The location of a gene on a chromosome.

mast The acorn or nut fruits of hardwood trees such as oaks, beech, and chestnut.

monomorphism A gene that shows no coding variation, that is, consists of only a single allele.

mutualism An interaction between different species that is beneficial to both.

niche The actual or potential pattern of exploitation of resources by a species.

nitrogen fixation The biological reduction of molecular nitrogen to chemical forms that can be used by organisms in the synthesis of organic molecules.

outcrossing Reproducing only by fertilization or pollination by other individuals.

phenotype The outward appearance of an organism with respect to a certain character.

phytoplankton *See* plankton.

piscivorous fish Fish that prey on other fish.

plankton Small aquatic organisms that have limited mobility and live in the water column of lakes, streams, and the ocean. Phytoplankton are planktonic plants; zooplankton planktonic animals.

polymorphism A gene exhibiting forms, known as alleles, that differ slightly in their coding pattern and produce proteins of slightly different structure and function.

recruitment The successful entry of new individuals into the breeding population.

resilience The ability of an ecosystem to recover from disruption caused by some disturbance or stress.

rhizomes Underground stems by which many vascular plants spread vegetatively.

riparian Pertaining to the banks and floodplains of a stream or river.

shade tolerance The ability of a plant to grow beneath a well-developed canopy of trees or shrubs.

sink A population that is unable to maintain itself by reproduction and that is dependent on immigration for continued existence.

successional community A community intermediate between a pioneer community and a mature community that shows relatively little change in structure and composition as long as major disturbance or habitat change does not occur.

taxon Any named systematic unit, such as subspecies, species, genus, or family.

taxon cycle An evolutionary sequence of invasion, geographical spread, evolutionary specialization and extinction of local populations shown by species that invade island archipelagos and other geographically diverse areas.

tens rule The generalization that of the species dispersing into a region, about 10 percent will appear in the wild, about 10 percent of these will establish self-reproducing populations, and about 10 percent of these will become problem invasives.

translocation The movement of individuals to a location at which the species never occurred naturally or from which it had been extirpated.

ungulate Any hoofed mammal.

veliger The planktonic larva of many mollusks.

zooplankton *See* plankton.

Appendix

Internet Sources for Exotic Species

Alien Species in Hawaii. Information Index.
http://www.hear.org/AlienSpeciesInHawaii/index.html

Alien Species Invasions of U.S. Ecosystems.
http://www.consci.tnc.org/library/pubs/dd/

America's Least Wanted: National Sea Grant. Nonindigenous Species Research and Outreach.
http://www.mdsg.umd.edu/NSGO/research/nonindigenous/index.html

APHIS. 1997. Campaign Against Nonnative Invasive Species (draft).
http://www.aphis.usda.gov?/HyperNews/get/nis.html

Biological Control of Non-Indigenous Plant Species Program Home Page.
http://www.dnr.cornell.edu/bcontrol/weeds.htm

California Exotic Pest Plant Council.
http://www.igc.apc.org/ceppc/index.html

Catalina Island Conservancy. Native and Alien Plants of the Channel Islands.
http://www.catalinas.net/seer/plants/chisle&a.htm

Center for Aquatic and Invasive Plants, University of Florida.
http://aquat1.ifas.ufl.edu/welcome.html

Exotic Plant Species of the Colorado Plateau. USGS, Biological Resources Division, Colorado Plateau Research Station.
http://www.nbs.nau.edu/FNF/Vegetation/Exotics/exotic.html

Exotic Species of the Great Lakes Region.
www.great-lakes.net/envt/exotic/exotic.html

Federal Interagency Committee for Management of Noxious and Exotic Weeds.
http://refuges.fws.gov/FICMNEWFiles/FICMNEWHomePage.html

See especially *Pulling Together: National Strategy for Invasive Plant Management.* 2nd edition.
http://bluegoose.arw.r9.fws.gov/ficmnewfiles/NatlweedStrategytoc.html

Florida Exotic Pest Plant Council.
http://www.fleppc.org/

Invaders Database System, University of Montana
http://invader.dbs.umt.edu/default.htm

Invasive Alien Plants of Virginia.
http://www.state.va.us/~der/invalien.htm

Massachusetts Institute of Technology Sea Grant College Program. Exotic Species Web Pages.
http:/massbay.mit.edu/exoticspecies/index.html

National Aquatic Nuisance Species Task Force. Nonindigenous Aquatic Species Database, USGS, Florida Caribbean Science Center, Gainesville, FL.
http://nas.nfrcg.gov/

The Nature Conservancy, Arlington, VA.
http://www.tnc.org

The Nature Conservancy Element Stewardship Abstracts. Exotic pest plants.
http:/tncweeds.ucdavis.edu/esadocs/documents/
http://www.catalinas.net/seer/er/plants/

Office of Technology Assessment. *Harmful Nonindigenous Species in the United States.* OTA-F-565 (1993).
http://www.wws.princeton.edu/~ota/disk1/1993/9325.html

Oregon Sea Grant. Exotic Invaders.
http://seagrant.orst.edu/hot/exotics.html

Problem Exotic Plants on Santa Catalina Island.
http://www.catalinas.net/seer/plants/pl_exoti.htm

Sea Grant Nonindigenous Species Site.
http://www.ansc.purdue.edu/sgnis/

Tennessee Exotic Pest Plant Council.
http://www.webriver.com/tn-eppc/

University of Minnesota. Minnesota Sea Grant Program. Exotic Species.
http://seagrant.d.umn.edu/exotic.html

USDA Natural Resources Conservation Service.
http://plants.usda.gov/plantproj/plants/index.html

Weeds of the World Project.
http://ifs.plants.ox.ac.uk/wwd/wwd.htm

Wildland Weeds Management and Research Program.
http://tncweeds.ucdavis.edu/

Wisconsin Sea Grant. Non-Indigenous Species.
http://h2o.seagrant.wisc.edu/advisory/NON_INDIGENOUS/non-indg.html

About the Author

George W. Cox was born in 1935 in Williamson, West Virginia. He attended Ohio Wesleyan University in Delaware, Ohio, receiving his B.A. in 1956. His graduate studies were at the University of Illinois, where he finished his Ph.D. in ecology in 1960.

After a year at the University of Alaska, Cox moved to San Diego, California. He joined the biology faculty at San Diego State University in 1962 as a member of the Ecology Program Area, one of the three program areas in biology. Cox is the author of *Laboratory Manual of General Ecology* (1967), which is now in its seventh edition. In 1969, he edited the book *Readings in Conservation Ecology.* He is coauthor of *Dynamic Ecology* (1973) and *Agricultural Ecology* (1979). In 1993, he authored a university textbook, *Conservation Ecology: Biosphere and Biosurvival,* and in 1997 *Conservation Biology: Concepts and Applications.* Cox's teaching and research interests include conservation ecology, the ecology of grassland and savanna ecosystems, the evolution of migratory birds, and the ecology and evolution of island birds. His publication record includes more than fifty research articles.

During leaves of absence, Cox has conducted post-doctoral research at the University of Pennsylvania, taught at the Universidad Católica in Santiago, Chile, investigated grassland and savanna ecology in Africa and South America, and directed a grant program in ecology at the National Science Foundation in Washington, DC.

Cox is currently Emeritus Professor of Biology at San Diego State University and carries on research and writing at his home in Santa Fe, New Mexico.

Index